Identification of
Time-varying
Processes

Identification of Time-varying Processes

Maciej Niedźwiecki
Technical University of Gdańsk, Poland

JOHN WILEY & SONS, LTD
Chichester · New York · Weinheim · Brisbane · Singapore · Toronto

Copyright © 2000 by John Wiley & Sons, Ltd
      Baffins Lane, Chichester,
      West Sussex, PO19 1UD, England

      National 01243 779777
      International (+44) 1243 779777

e-mail (for orders and customer service enquiries): cs-books@wiley.co.uk

Visit our Home Page on http://www.wiley.co.uk or http://www.wiley.com

Other Wiley Editorial Offices

John Wiley & Sons, Inc., 605 Third Avenue,
New York, NY 10158-0012, USA

WILEY-VCH Verlag GmbH
Pappelallee 3, D-69469 Weinheim, Germany

Jacaranda Wiley Ltd, 33 Park Road, Milton,
Queensland 4064, Australia

John Wiley & Sons (Asia) Pte Ltd, 2 Clementi Loop #02-01,
Jin Xing Distripark, Singapore 129809

John Wiley & Sons (Canada) Ltd, 22 Worcester Road
Rexdale, Ontario, M9W 1L1, Canada

British Library Cataloguing in Publication Data

A catalogue record for this book is available from the British Library

ISBN 0 471 98629 1

Produced from PostScript files supplied by the author.
Printed and bound in Great Britain by Antony Rowe Ltd, Chippenham, Wiltshire.
This book is printed on acid-free paper responsibly manufactured from sustainable forestry, in which at least two trees are planted for each one used for paper production.

To my Mom and Dad
my best teachers

And to Barbara
my love

Nothing can ever happen twice.
In consequence, the sorry fact is
that we arrive here improvised
and leave without the chance to practice.
Even if there is no one dumber,
if you're the planet's biggest dunce,
you can't repeat the class in summer:
this course is only offered once.
No day copies yesterday,
no two nights will teach what bliss is
in precisely the same way,
with exactly the same kisses.

Reproduced from the poem 'Nothing Twice'
by Wisława Szymborska [†]

† translated by Stanisław Barańczak and Clare Cavanagh in 'View with a Grain of Sand', 1996, by permission of Faber and Faber Ltd.

Contents

Preface

When building a mathematical model of a physical process (signal, system, phenomenon) one can proceed in two different ways. Taking the analytic approach one uses the basic laws of physics to describe the dynamic process behavior. The other approach is based on process identification: the form of the model is to some extent arbitrary and its coefficients are determined experimentally using statistical procedures similar to curve fitting. Akaike calls such models 'instrumental'. Instrumental models are not phenomenologically justified and therefore their coefficients have no physical significance. However, they have some obvious practical advantages: they are easy to build and update without any physical insights and, more importantly, due to their relative simplicity they allow one to formulate and solve problems in a mathematically tractable way, including practically important problems such as adaptive prediction, predictive coding of signals, adaptive filtering, spectrum estimation, change detection and adaptive control. Solutions like these are usually known as model-based solutions.

Over short time intervals most of the processes can be satisfactorily approximated by linear dynamic time-invariant models, but over longer time intervals they reveal nonstationary features or characteristics, hence they call for models with time-dependent coefficients. Physical phenomena exhibit nonstationary behavior for a number of reasons, mainly due to the variation of internal (aging, fatigue) and external (setpoint changes, time-dependent disturbances) operating conditions. The associated time constants range from tens of milliseconds (audio signals) to hours (telecommunication channels) and days (technological processes).

The past three decades have brought a large number of really challenging applications of system identification in different areas such as telecommunications (channel equalization, predictive coding, noise canceling), signal processing (prediction of time series, signal reconstruction, outlier elimination, spectrum estimation) and automatic control (adaptive control, failure detection). Most of the proposed solutions have been obtained using the so-called certainty equivalence principle. First, the problem is solved analytically for a known model of process dynamics. The model coefficients appearing in the resulting solution are then replaced with the corresponding estimates obtained via system identification. Quite obviously, in a time-varying environment the performance of all adaptive schemes mentioned above depends critically on the parameter tracking capabilities of the identification routines employed.

Some special classes of nonstationary processes, such as processes with periodically time-varying coefficients, can be described by models with unknown but constant parameters. Identification of these reducible processes is in some ways analogous to identification of stationary processes – the larger the amount of data, the more accurate the parameter estimates (at least theoretically). The book is devoted solely to identification of irreducible nonstationary processes, i.e. processes with time-varying coefficients which don't change in a totally predictable way. No matter how large the data size and identification technique, the parameter estimates derived for an irreducible nonstationary process never actually converge to their true values; instead they follow parameter changes with some finite accuracy. Estimators with this property are usually called finite memory estimators, since they gradually 'forget' information coming from the remote past as the new data becomes available. The common feature of all finite memory parameter tracking algorithms is their ability to compromise between estimation accuracy (variance) and awareness to parameter changes (bias). One of the key issues in identification of time-varying systems, the trade-off between variance and bias should be made in accordance with the degree of nonstationarity of the analyzed process.

Even though several good monographs and textbooks have recently been published on adaptive filtering, none of them gives a comprehensive treatment of the key issue in adaptive system design – identification of time-varying characteristics for dynamic processes. This book attempts to fill the gap. To the best of my knowledge, it is the first systematic presentation and comparison of all three major approaches to time-varying identification: the local estimation approach, the basis function approach and the approach based on Kalman filtering or smoothing. All important aspects of the problem are tackled, such as assessment of the estimation memory and estimation bandwidth of different identification algorithms, comparison of their tracking capabilities, the estimation bias/variance trade-offs and optimization of adaptive filters, computational complexity and numerical stability of recursive identification algorithms, and selected practical applications of time-varying process identification.

The two most important concepts throughout the book are the estimation memory of an adaptive filter and its associated time and frequency responses. Quantification of the estimation memory allows one to objectively compare different tracking algorithms. Comparing estimators with different memory spans barely makes any sense as it resembles comparing runners that specialize in different distances.

The concept of the associated time and frequency characteristics is equally important. Analyzing the tracking capabilities of different estimation algorithms reveals intriguing dualities between time-varying identification and signal processing. We show that the averaged trajectory of parameter estimates yielded by parameter tracking algorithms can be approximately viewed as an output of a linear filter (associated filter) excited by the signal made up of the true parameter changes. Hence, in the mean, the estimated parameter trajectory can be regarded as a filtered version of the true parameter trajectory. This signal processing perspective allows one to build a unified framework for all major approaches to time-varying identification. Analysis of the time and frequency characteristics of linear filters associated with different parameter tracking algorithms provides many insights into the tracking capabilities (as well as tracking limitations) of adaptive filters, allowing one to solve a number of practically important problems such as quantification of the estimation variance/bias trade-off, determination of the estimation bandwidth, i.e. the frequency range in which the time-varying parameters can be followed successfully, and optimization of design parameters for tracking algorithms. Even though most of the analytical results are derived for a special class of nonstationary processes – finite impulse response systems subject to a stationary excitation – they seem to provide useful qualitative guidelines to understanding the behavior of parameter tracking algorithms working under less constrained conditions.

What this book is not about

- *Analysis of time-varying dynamic systems* is barely treated. Basically, I explain how to track and use slowly varying systems, i.e. systems which can be regarded as locally stationary. The dynamics of such systems can be analyzed, with sufficient practical accuracy, using the tools and frameworks developed for time-invariant systems, including frequency-domain concepts such as frequency responses and spectral density functions. Most practical applications of time-varying identification fall into this category.
- *Identification of reducible nonstationary processes* (such as cyclostationary processes or processes described by the so-called integrated linear models) is not treated at all. Identification of these processes is in many ways analogous to identification of time-invariant systems (one looks for a set of constant coefficients characterizing the time-varying system dynamics), so it does not really fit into this book.
- *Identification of multivariable and/or complex systems.* Discussion of different identification approaches was deliberately restricted to single-input single-output (SISO) systems excited by real signals (signals which can be represented by real numbers). I intended to keep the results as simple as possible. Most of them can be easily extended to multivariate systems with complex input/output by replacing some of the vectors with the appropriately structured matrices and by adding complex conjugate signs in the right places.
- *Continuous-time and/or nonlinear models.* The book intentionally focuses on identification of discrete-time linear instrumental models. Estimation of time-varying parameters for continuous-time and/or nonlinear models is performed mainly during physical modeling (with simulation in mind), hence it lies outside the scope of this book.

Acknowledgments

Many people have had all kind of influences on my research in time-varying identification and I owe them my deep gratitude: colleagues from the Department of Systems Engineering, Australian National University, where I spent three very inspiring and enjoyable years of my life; colleagues from the Department of Automatic Control, Technical University of Gdańsk, where I have been working for more than 20 years in a friendly and stimulating environment; many students who over the past decade have listened to my lectures on system identification and always challenged me to talk simply about complicated matters; and many, many researchers in the field of system identification who made me understand what this is really all about.

But rather than giving a long list of academics I feel indebted to, I would like to take this opportunity to express my gratitude to someone who does not belong to the scientific community. These special thanks go to Mr Zbyszek Karwowski. In 1993 Zbyszek, a cofounder of a small high-tech US company, asked me if I could think of using my knowledge on system identification to solve a real problem of some kind. He didn't have any particular application in mind at that time. I answered, after a while, that I had some ideas which could be useful in restoration of archived audio recordings but that I was not sure they would really work. He risked. We started from scratch and ended up with DART^{TM} (Digital Audio Restoration Technology) – a professional sound restoration system. Working on the DART project has been a very interesting experience for me. I learned how inspiring a good theory can be when solving practical problems. But I also learned that to make it really work in practice one has to turn many little knobs in the right directions, and do things one will never find in the books because, being so case dependent, they have no stand-alone scientific value. Finally, I realized how rewarding a successful application can be, especially an application on the borderline between science and art. Thanks Zbyszek!

I am very indebted to my PhD student Tomasz Kłaput for conducting computer simulations (in MATLAB) and to Mrs Iwona Sosińska for preparing all the remaining figures. Last but not least, I would like to express my gratitude to the Wiley staff for their encouragement, help and patience.

Gdańsk, February 2000   Maciej Niedźwiecki

1

Modeling Essentials

1.1 Physical and instrumental approaches to modeling

When building a mathematical model of a physical process (a signal or a system) one can proceed in two different ways. Taking the analytic approach, one uses the basic rules of physics – Newton's laws, the energy conservation law, moments conservation law, Kirchoff's law, etc. – to describe the process behavior [47]. The coefficients of such 'physical models' usually have a clear physical meaning as they are functions of universal constants (characterizing the laws), system parameters (masses, densities, resistances, etc.) and environmental parameters (temperatures, friction coefficients, etc.). When properly calibrated, physical models are important tools for process simulation. Simulation may serve many different purposes.

When designing a controller for a technological process, rarely is an automatic control engineer allowed to experiment with a real plant. High costs of such experiments combined with safety and reliability concerns suggest a more rational solution: a physical (in the sense described above) model of a plant is built and regarded as its 'substitute' at the preliminary stage of controller design, i.e. for the purpose of tuning, testing and initial evaluation of the control loop. Only when computer simulations involving a realistic model of the plant confirm good properties of the proposed control scheme is one allowed to test it on a real system. Many industrial companies have developed very accurate simulation models of their plants, e.g. planes, ships, missiles, engines, mills, furnaces. So accurate are some of these models that the new designs can be tested by means of computer simulation only. The Boeing 777 is the first civil aircraft belonging to this new generation of computer-verified constructions; the entire design was conceived 'behind the desk' without conducting costly experiments in aerodynamic tunnels (the same models are used in flight simulators). For commercial and/or national security reasons many details of the existing simulation packages are proprietary or classified.

The second aim of physical models, however strange it may sound, is replacement or substitution of real systems. The sound generation mechanism of many existing acoustic instruments can be very accurately modeled using the sets of appropriately structured wave propagation equations [174], [175], [176]. Since the recent advances in computer technology allow one to solve these equations in real time, a new family of virtual instruments (such as a trumpet [41] or a church organ [39]) should soon be available, surpassing the existing synthesizers in terms of the quality and naturalness of the produced sound. The unique feature of virtual instruments is their ability to simulate real instruments in all regimes of their operation, including snaps and misuse (e.g. trumpet overblowing).

Even though physical models can produce good approximations to the behavior of real systems, three reasons limit their direct use for solving practical problems:

- They provide a continuous-time description of the modeled processes (sets of ordinary or partial differential equations).
- They are usually fairly complex (the number of parameters/variables can range from tens to hundreds).
- Most of them are nonlinear.

An entirely different approach to model building is based on process identification: the form of the model is to some extent arbitrary and its coefficients are determined experimentally using statistical procedures similar to curve fitting. Akaike calls such models 'instrumental' [1]. Instrumental models are not phenomenologically justified and therefore their coefficients have no physical significance. However, they have some obvious practical advantages: they are easy to build and update without any physical insights and, more importantly, due to their relative simplicity they allow mathematically tractable formulations and then solutions for many practically important problems such as adaptive prediction, predictive coding of signals, adaptive filtering, spectrum estimation, change detection and adaptive control. These solutions are usually known as model-based solutions. The model-based approach is the pragmatic one. In this framework, building a mathematical model of a dynamic process is not a goal in itself; models are only the means of solving practical problems in the sense that the corresponding prediction, filtering and control formulas depend explicitly on model coefficients.

Example 1.1 (dynamic weighing of vehicles)

When static weighing is performed, a vehicle has to stop on the scales platform and remain there until indications of the strain gauges (placed under the platform) settle down to their steady-state values (Figure 1.1). In many practical situations it is more convenient and economically better justified to weight vehicles 'in motion', i.e. without stopping them on the platform. This approach is usually known as *dynamic weighing*. When a vehicle crosses the scales platform at a constant speed, the observed system response may be highly oscillatory, not even reaching its steady-state value corresponding to the static weight of consecutive axles (Figure 1.2). The problem of dynamic weighing can be formulated as follows: determine the steady-state value of the system step response based on its transient (pulse) response (Figure 1.3). A very simple model-based solution to this problem was proposed in [147]. The weight estimation is performed in three steps (Figure 1.4).

Step 1

Assuming that the input signal $u(t)$ is constant and equal to u_0 in the time interval $T = [t_{on}, t_{off}]$, the three coefficients a_1, a_2 and b_1 of a second-order linear model of the dynamic weighing system (vehicle + platform)

$$H(z) = \frac{b_1}{1 + a_1 z^{-1} + a_2 z^{-2}} \, , \tag{1.1}$$

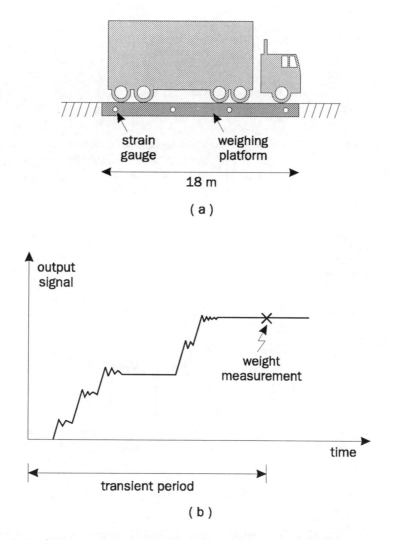

(a)

(b)

Figure 1.1 A typical static weighing system and the corresponding measurements.

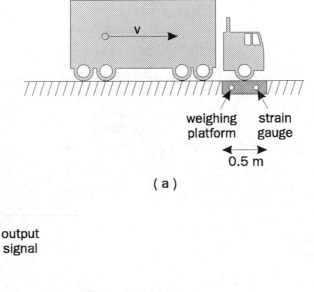

(a)

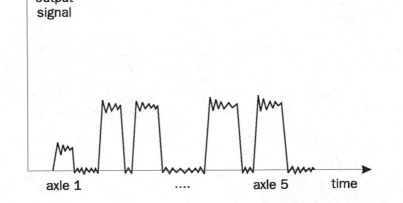

(b)

Figure 1.2   A typical dynamic weighing system and the corresponding measurements.

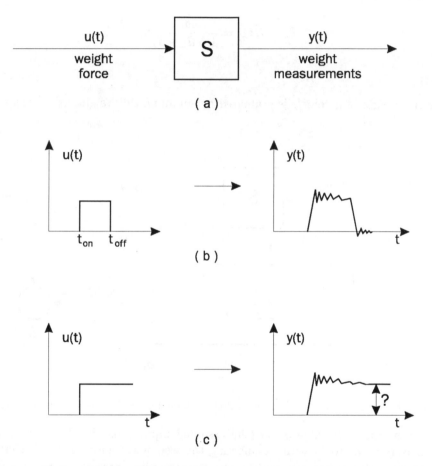

Figure 1.3 (a) Formulation of the problem for the dynamic weighing system S. Given the pulse response of the system (b), estimate the steady-state value of its step response (c); the moments t_{on} and t_{off} are determined by a set of optic sensors.

are estimated from the available data $\{y(t), t \in T\}$ (for every axle of the car).

Step 2

Since the steady-state value of the step response of a system governed by (1.1) can be obtained from

$$y(\infty) = H(1)u(\infty),$$

the weight w_a of a particular axle is estimated using the formula

$$\widehat{w}_a = \widehat{H}(1) = \frac{\widehat{b}_1}{1 + \widehat{a}_1 + \widehat{a}_2} u_0. \tag{1.2}$$

Step 3

The static weight of a vehicle is evaluated by summing the weights of all axles.

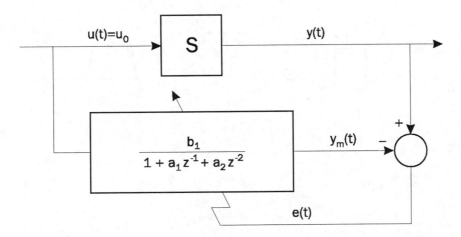

Figure 1.4   Identification of the second-order model of the dynamic weighing system.

Note that that the value of u_0 (the assumed input pulse height) can be chosen arbitrarily (e.g. set to 1) without affecting the weight estimate (1.2); all changes in u_0 are automatically compensated by the corresponding changes in $\widehat{b}_1$ obtained via system identification.

The technique described above, yielding very good results in practice, is a good illustration of the model-based (or parametric) approach. At first glance, estimation of the static weight of a vehicle from dynamic measurements seems to be a very difficult task. The measured signal – the response of the weighing platform – depends on many factors such as properties of the weighing system, characteristics of the vehicle's suspension system, the vehicle's speed, the road and weather conditions. Physical insights suggest that a pretty complex, nonlinear mathematical model is needed to adequately represent the overall system (vehicle + platform). It may thus seem surprising that the weight estimates based on a simplistic instrumental model

(1.1) yield satisfactory results. The explanation is simple: despite its simplicity the second-order model seems to successfully extract the weight-related information from the available data. Remember that our problem is to evaluate the steady-state response of a dynamic system based on its transient response. The mathematical model of the system provides a natural and simple means for performing such a mapping. From a practical viewpoint, the degree to which the model (1.1) can approximate the transient response of a real system is immaterial as long as (1.2) yields satisfactory weight estimates.

1.2 The Titius–Bode law and the method of least sqares

The history of the empirical formula known as the Titius–Bode law of planetary distances [151], [207] is an excellent example of the fact that instrumental models can play an important and stimulating role in scientific research. In his translation of the Charles Bonnet's book *Contemplation de la Nature* (1764), the German astronomer Johann Daniel Titius von Wittenberg included his own observations on some intriguing regularities of planetary geometry in our solar system. Assuming that the average distance of Saturn (the most distant of all planets known at that time) from the Sun is equal to 100 units, Titius remarked that the average planetary radii d_n could be described by a very simple mathematical formula:

$$d_n = 2 + 3 \cdot 2^n, \tag{1.3}$$

where n is a planet's 'serial number' assigned in the following way.

n	planet
$-\infty$	Mercury
0	Venus
1	Earth
2	Mars
3	?
4	Jupiter
5	Saturn

Table 1.1 Average distance of different planets from the Sun (as known in the middle of the eighteenth century)

Table 1.1 shows that when Titius was alive, there was no planet corresponding to $n = 3$. 'Can it be that the Grand Constructor left that place empty?' asked Titius. 'Never!' was his emphatic answer.

When in 1781 William Herschel discovered Uranus, encircling the Sun at an average distance of 192 units ($d_6 = 196$), confidence in the Titius formula, popularized in the meantime by his fellow astronomer Johann Elert Bode, increased even more. But the orbit corresponding to $n = 3$ still remained empty. In 1801, soon after the international search had begun, the Italian amateur astronomer Giovanni Piazzi spotted an unknown heavenly body later named Ceres. Piazzi continued to trace the trajectory of 'his' planet for almost six weeks before a bad cold forced him to stop observations. When he recovered from his illness Ceres was lost! Nobody could confirm his findings (closeness to Sun and cloudy skies were partially to blame).

Then a German mathematician entered the scene; it was the young but already well-known Carl Friedrich Gauss. Using a new method he called least squares, Gauss was able to take the scarce and inaccurate measurements provided by Piazzi and estimate eight parameters characterizing the orbit of Ceres. The lost planet was soon tracked down. Its average distance from the Sun was equal to 27.67 units, in very good agreement with the law of Titius–Bode ($d_3 = 28$). Shortly after spotting Ceres, other small planets with similarly shaped orbits were discovered: Juno (1804), Vesta (1807), Astraea (1845), etc. The modern solar catalog contains several thousands of these objects and they are known as planetoids, a much better description. According to one of the hypotheses, all planetoids are remnants of a small planet torn into pieces under circumstances that are still something of a mystery.

The history of the Titius–Bode law is in fact a history of a very simple instrumental model that has had a major influence on two scientific disciplines: astronomy and statistics. When first proposed by Titius, the formula (1.3) had no solid physical foundations and almost no scientific significance; it was nothing but a simple rule showing some sort of order (whether true or fictitious) behind the empirical data. The Titius–Bode law supported and in a concise way summarized the existing empirical evidence. However, its main advantage was that it suggested the existence of some unknown elements of the solar puzzle. Like every good mathematical model, it had the *generalization* property, i.e. the ability to predict unknown observations from things that are already known.

1.3   The principle of parsimony

There are numerous examples showing that complex physical processes can be successfully dealt with using simple instrumental models. A nonlinear model of the distillation column, based on mass and energy balance, results in 30 differential equations containing 140 variables. Despite this, a very simple adaptive predictive regulator, based on a second-order linear model, can be used for effective control of column dynamics [122].

Similarly, even though human voice production is a pretty complex phenomenon, most speech analysis/synthesis applications rely on linear models of order 10 to 15 [161]. This brings us to one of the most important modeling principles – the *principle of parsimony*.

In simple words the principle of parsimony says that 'when describing a dynamic process one should not use extra parameters if not necessary' [31], [179]. Parsimony is clearly reminiscent of a philosophical principle known as an Occam's razor, after William of Occam: a person should not increase, beyond what is necessary, the number of entities required to explain anything.

It might seem that the parsimony principle contradicts the common sense observation that 'extra parameters (degrees of freedom) allow one to describe physical phenomena more accurately', which suggests that more complex models can't be inferior to simple ones. Even though agreeable for physical models, i.e. models obtained as a result of physical inference, the above statement is false for instrumental models obtained via system identification, i.e. as a result of statistical inference. The point is that parameter estimates obtained from a limited number of experimental data (finite observation history) are of finite accuracy; their estimation variance is limited from below by the celebrated Cramer–Rao bound [179]. Loosely speaking, when the standard deviation of a coefficient estimate exceeds the true coefficient value, the benefits stemming from its inclusion in the process model become problematic from the statistical viewpoint. It can be shown, under fairly general assumptions [179], that estimation of superfluous parameters, i.e. parameters which are de facto zero, decreases the average predictive capability of the corresponding models. Note that for such fictitious parameters the standard deviation of the estimation error always exceeds the true (zero) value irrespective of the length of the observation history. However, the principle of parsimony says more than that. Pending experimental conditions and the size of the available data set, one may obtain a better model (in terms of its predictive ability) by *not* estimating certain process parameters even if they are known to be nonzero. In other words, the simplified (approximate) process model built from a finite data sample may turn out to be better than a full-size model with an adequate structure. The size of the available data set is one of the key issues here. One of the implications of the parsimony principle is that the number of estimated process coefficients should be much smaller than the number of data points. The rule of thumb is to build models for which

$$\text{number of estimated coefficients} \leq 0.2 \times \text{number of observations} \qquad (1.4)$$

According to (1.4), at least five 'independent' observations should correspond to each degree of freedom of the identified model. Even though, as with all heuristic rules, inequality (1.4) should be regarded as no more than a general guideline, it should still prove quite useful to practically oriented readers.

When coping with nonstationary processes one often has access to very large data sets. Consider the adaptive filtering of audio signals. Each second of the audio signal sampled at the CD rate of 44.1 kHz brings over 40 000 new samples! Should one care about the principle of parsimony under such circumstances? Surprisingly, the answer turns out to be yes. For reasons that will soon become clear, all parameter tracking algorithms have the so-called finite memory property. The parameter estimates at any time instant effectively depend on only a limited number of past data points. Therefore, if the number of samples in (1.4) is replaced with the 'equivalent memory length' of the tracking algorithm, the model complexity guidelines summarized above remain valid for nonstationary identification.

1.4   Mathematical models of stationary processes

According to the fundamental theorem due to Wold [183], any zero-mean second-order wide-sense stationary discrete-time stochastic process can be written down as a sum of two mutually uncorrelated components. The first component is a deterministic or linearly predictable process, so called as it can be predicted from its own past with zero mean square error, using a suitable linear filter. The second component, a purely nondeterministic (or regular) process, can be written down in the form

$$y(t) = \sum_{i=0}^{\infty} h_i v(t - i), \qquad t = \ldots, -1, 0, 1, \ldots \qquad (1.5)$$

where

$$h_0 = 1, \qquad \sum_{i=0}^{\infty} h_i^2 < \infty,$$

and $\{v(t)\}$ denotes a white noise sequence, i.e. a sequence of zero-mean uncorrelated random variables of constant variance

$$E[v(t)] = 0,$$

$$E[v(t)v(s)] = \begin{cases} \sigma_v^2 & \text{if} \quad t = s \\ 0 & \text{if} \quad t \neq s \end{cases} \qquad (1.6)$$

According to (1.6) any wide-sense stationary process can be regarded as a result of passing white noise through a causal (one-sided) shaping filter with square summable impulse response (Figure 1.5)

$$H(q^{-1}) = \sum_{i=0}^{\infty} h_i q^{-i},$$

where the symbol q^{-1} denotes the backward shift operator

$$q^{-1}y(t) = y(t-1).$$

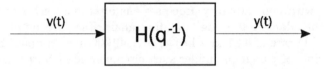

Figure 1.5 Linear filter described by Wold's theorem.

The filter postulated by Wold's theorem is characterized using an infinite number of coefficients, clearly impractical when it comes to system identification. When a general noise shaping filter is replaced with a finite-order rational filter, then depending on the assumed structure of $H(q^{-1})$, one obtains three parsimonious representations of stationary processes known as the autoregressive model, the moving average model and the mixed autoregressive-moving average model.

1.4.1   Autoregressive model

When the noise shaping filter $H(q^{-1})$ has the form

$$H(q^{-1}) = \frac{1}{A(q^{-1})} = \frac{1}{1 - \sum_{i=1}^{n_A} a_i q^{-i}}, \qquad (1.7)$$

i.e. it is an all-pole filter of order n_A, the process $y(t)$ observed at its output under the white noise excitation is called *autoregressive* (AR). The term 'autoregressive' stems from the time-domain representation of an AR process.

$$y(t) = \sum_{i=1}^{n_A} a_i y(t-i) + v(t). \tag{1.8}$$

According to (1.8) the value of the AR signal observed at instant t can be partially explained in terms of its own history; it can be 'regressed' on the past n_A values $y(t-1), \ldots, y(t-n_A)$, i.e., self-regressed. The second 'unpredictable' term on the right-hand side of (1.8) is due to a random perturbation $v(t)$.

Yule, who introduced the autoregressive model in 1927 [202], used a very simple example to rationalize it – a clock pendulum disturbed with random noise pulses (some children pelted it with peas). He was therefore the first one to realize that autoregressive models can be used for describing pseudoperiodic stochastic processes.

Autocorrelation function of an autoregressive process

The second-order characteristics of an autoregressive process depend solely on the values of autoregressive coefficients $a_1, \ldots, a_{n_A}$ and the driving noise variance σ_v^2. Denote by $q_1, \ldots, q_{n_A}$ the poles of the noise shaping filter (1.7), i.e. the roots of the *characteristic polynomial* $A(q^{-1})$:

$$A(q_k^{-1}) = 1 - \sum_{i=1}^{n_A} a_i q_k^{-i} = 0, \qquad k = 1, \ldots, n_A.$$

The necessary and sufficient condition for the asymptotic weak stationarity of an $AR(n_A)$ process takes the following form.

A1 All zeros of $A(q^{-1})$, i.e. all poles of $H(q^{-1})$, lie inside the unit circle in the complex plane:

$$|q_k| < 1, \qquad k = 1, \ldots, n_A.$$

When the stability condition is fulfilled and the noise shaping filter (1.7) is excited in the 'infinite past', autocorrelation function of an $AR(n_A)$ process is, irrespective of the initial conditions, translation invariant:

$$E[y(t)y(t-\tau)] = r_y(\tau).$$

The recursive expression for computation of $r_y(\tau)$ can be easily obtained by multiplying both sides of (1.8) by $y(t-\tau)$ and taking expectations:

$$E[y(t)y(t-\tau)] = \sum_{i=1}^{n_A} a_i E[y(t-i)y(t-\tau)] + E[v(t)y(t-\tau)]. \tag{1.9}$$

For $\tau > 0$ the second term on the right-hand side of (1.9) is zero (since the random variable $v(t)$ is uncorrelated with past signal values $y(t-i), i > 0$) and this gives

$$r_y(\tau) = \sum_{i=1}^{n_A} a_i r_y(\tau - i), \tag{1.10}$$

or alternatively

$$A(q^{-1})r_y(\tau) = 0. \tag{1.11}$$

For $\tau = 0$ it holds that

$$E[\,v(t)y(t)\,] = \sigma_v^2,$$

and therefore, since $r_y(\tau) = r_y(-\tau)$,

$$r_y(0) = \sum_{i=1}^{n_A} a_i r_y(-i) + \sigma_v^2 = \sum_{i=1}^{n_A} a_i r_y(i) + \sigma_v^2. \tag{1.12}$$

For all $\tau > n_A$ equation (1.10) constitutes a linear difference equation in $r_y(\tau)$ with initial conditions $r_y(0), \ldots, r_y(n_A)$. The initial conditions can be determined as functions of autoregressive coefficients $a_1, \ldots, a_{n_A}$ and the driving noise variance σ_v^2 can be obtained by solving for $r_y(0), \ldots, r_y(n_A)$ the set of autocorrelation equations corresponding to $\tau = 0, 1, \ldots, n_A$:

$$
\begin{aligned}
r_y(0) &= a_1 r_y(1) + a_2 r_y(2) + \ldots + a_{n_A} r_y(n_A) + \sigma_v^2, \\
r_y(1) &= a_1 r_y(0) + a_2 r_y(1) + \ldots + a_{n_A} r_y(n_A - 1), \\
&\ \vdots \\
r_y(n_A) &= a_1 r_y(n_A - 1) + a_2 r_y(n_A - 2) + \ldots + a_{n_A} r_y(0).
\end{aligned}
\tag{1.13}
$$

Since equations (1.13) are linear with respect to $r_y(0), \ldots, r_y(n_A)$ the solution is straightforward. The same set of linear equations, known in a slightly rearranged form as Yule–Walker equations, can also be used to solve an inverse problem: determining the parameters of an autoregressive process from the known autocorrelation coefficients. According to (1.13) there is a unique relationship between the two sets of coefficients.

Since the autocorrelation function of an autoregressive process obeys, under (A1), the stable, homogeneous, linear difference equation (1.10), the explicit expression for $r_y(\tau)$ takes the form

$$r_y(\tau) = \sum_{i=1}^{s} \sum_{j=1}^{m_i} A_{ij} \tau^{j-1} q_i^{\tau}, \qquad \tau \geq 0, \tag{1.14}$$

where m_i denote multiplicities of the poles q_i, A_{ij} are complex amplitudes which can be uniquely determined from $a_1, \ldots, a_{n_A}$ and

$$\sum_{i=1}^{s} m_i = n_A.$$

When all poles of the noise shaping filter are single, which is a typical situation in practice, expression (1.14) can be rewritten in the simpler form

$$r_y(\tau) = \sum_{i=1}^{n_A} A_i q_i^{\tau}, \qquad \tau \geq 0, \tag{1.15}$$

where $A_1, \ldots, A_{n_A}$ is the set of parameter-dependent coefficients.

The contribution of real system poles takes the form of damped exponentials, i.e. terms that decay exponentially to zero. All complex poles appear in conjugate pairs (if $\text{Im}(q_i) \neq 0$ both q_i and $q_i^\star$ must be the roots of $A(q^{-1})$), yielding terms of a damped sinusoidal nature:

$$A_i q_i^\tau + A_i^\star (q_i^\star)^\tau = \alpha_i \rho_i^\tau \cos(\omega_i \tau + \beta_i),$$

where $\alpha_i = 2|A_i|$, $\beta_i = \arg(A_i)$, $\rho_i = |q_i|$ and $\omega_i = \arg(q_i)$.

Example 1.2

Consider a second-order autoregressive process governed by

$$y(t) = a_1 y(t-1) + a_2 y(t-2) + v(t).$$

Here the asymptotic stationarity condition (A1) takes the following form (Figure 1.6):

$$a_1 + a_2 \leq 1,$$
$$a_2 - a_1 \leq 1,$$
$$|a_2| \leq 1.$$

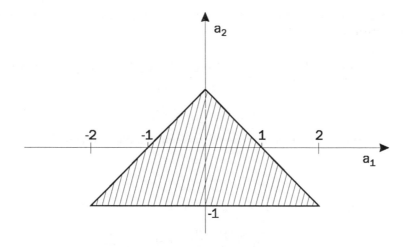

Figure 1.6 Stability region (shaded) for the second-order AR process in terms of autoregressive coefficients.

After solving the set of Yule–Walker equations

$$
\begin{aligned}
r_y(0) &= a_1 r_y(1) + a_2 r_y(2) + \sigma_v^2, \\
r_y(1) &= a_1 r_y(0) + a_2 r_y(1), \\
r_y(2) &= a_1 r_y(1) + a_2 r_y(0),
\end{aligned}
$$

one obtains

$$r_y(0) = \frac{(1 - a_2)\sigma_v^2}{(1 + a_2)\left[(1 - a_2)^2 - a_1^2\right]},$$

$$r_y(1) = \frac{a_1}{1 - a_2} r_y(0) = \frac{a_1 \sigma_v^2}{(1 + a_2)\left[(1 - a_2)^2 - a_1^2\right]},$$

and using equation (1.10),

$$r_y(\tau) = a_1 r_y(\tau - 1) + a_2 r_y(\tau - 2), \qquad \tau \geq 2.$$

Denote by

$$q_1 = \frac{a_1 + \sqrt{a_1^2 + 4a_2}}{2},$$

$$q_2 = \frac{a_1 - \sqrt{a_1^2 + 4a_2}}{2}$$

the two roots of the characteristic equation

$$1 - a_1 q^{-1} - a_2 q^{-2} = 0.$$

The shape of the autocorrelation function $r_y(\tau)$ depends on whether these roots are real or complex.

Case (i)

If $a_1^2 + 4a_2 > 0$ the characteristic equation has two real roots and the solution for $r_y(\tau)$ can be expressed in the form

$$r_y(\tau) = \frac{r_y(0)}{(q_1 - q_2)(1 + q_1 q_2)} \left[q_1(1 - q_2^2)q_1^\tau - q_2(1 - q_1^2)q_2^\tau\right], \qquad \tau > 0.$$

Case (ii)

If $a_1^2 + 4a_2 = 0$ the characteristic equation has one double real root

$$q_0 = \frac{a_1}{2},$$

and the corresponding expression for $r_y(\tau)$ becomes

$$r_y(\tau) = r_y(0)\frac{q_0(1 - q_0^2)}{1 + q_0^2}\tau q_0^{\tau - 1}, \qquad \tau > 0.$$

Case (iii)

If $a_1^2 + 4a_2 < 0$ the characteristic equation has a pair of complex conjugate roots, leading to

$$r_y(\tau) = r_y(0)\alpha\rho^\tau \cos(\omega\tau + \beta), \qquad \tau > 0,$$

where

$$\alpha = \frac{2}{1 - a_2}\sqrt{\frac{a_2\left[(1 - a_2)^2 - a_1^2\right]}{a_1^2 + 4a_2}},$$

$$\rho = \sqrt{-a_2},$$

$$\omega = \arctan\frac{\sqrt{-a_1^2 - 4a_2}}{a_1},$$

$$\beta = \arctan\frac{-a_1(1 + a_2)}{(1 - a_2)\sqrt{-a_1^2 - 4a_2}}.$$

Spectrum of an autoregressive process

When a noise shaping filter has complex poles, the autocorrelation function of the corresponding AR process contains damped sine components, which is an indication of signal periodicity, even though no explicit sinusoidal terms are present in the system equation. Yule was the first one to notice that autoregressive models can be used to describe processes with pseudoperiodic components. As a matter of fact, his historic paper [202] analyzing sunspot activity numbers was also the first attempt to perform what is now called the parametric estimation of a spectrum of a random process.

The power spectral density of a discrete-time wide-sense stationary stochastic process $y(t)$ can be defined as a Fourier transform of its autocorrelation function

$$S_y(\omega) = \sum_{\tau=-\infty}^{\infty} r_y(\tau)e^{-j\omega\tau}, \tag{1.16}$$

$$-\pi < \omega \le \pi,$$

where ω denotes the normalized (dimensionless) angular frequency. When $\{y(t)\}$ is obtained as a result of sampling a band-limited continuous-time signal, then

$$\omega = 2\pi\frac{f}{f_s},$$

where f denotes frequency (in hertz) and f_s is the sampling rate, which has to be greater than twice the highest frequency component of the sampled signal.

The autocorrelation function and the power spectral density function form a Fourier transform pair; the relationship dual to (1.16) is

$$r_y(\tau) = \frac{1}{2\pi}\int_{-\pi}^{\pi} S_y(\omega)e^{j\omega\tau}\,d\omega \tag{1.17}$$

Even though the power spectrum can be computed directly from (1.16) there is a much simpler way of doing this, based on the following well-known result from the theory of linear stochastic systems [179].

Theorem 1.1

Consider a discrete-time, linear, time-invariant and stable system characterized by the transfer function $H(q^{-1})$ excited by a wide-sense stationary signal $\{x(t)\}$ with power spectrum $S_x(\omega)$. Then the output signal $\{y(t)\}$ is also asymptotically wide-sense stationary and its power spectrum can be expressed in the form

$$S_y(\omega) = |H(e^{-j\omega})|^2 S_x(\omega). \tag{1.18}$$

Since the spectral density of white noise is

$$S_v(\omega) = \sigma_v^2, \qquad \omega \in (-\pi, \pi],$$

the power spectrum of an autoregressive process can be expressed in the form

$$S_y(\omega) = \frac{\sigma_v^2}{|A(e^{-j\omega})|^2} = \frac{\sigma_v^2}{|1 - \sum_{i=1}^{n_A} a_i e^{-j\omega i}|^2} . \tag{1.19}$$

Example 1.3

Consider a first-order autoregressive process governed by

$$y(t) = ay(t-1) + v(t), \qquad |a| < 1,$$

and with an autocorrelation function

$$r_y(\tau) = \frac{\sigma_v^2}{1 - a^2} a^{|\tau|}.$$

Straightforward calculations yield

$$S_y(\omega) = \sum_{\tau=-\infty}^{\infty} r_y(\tau) e^{-j\omega\tau} = \frac{\sigma_v^2}{1 - a^2} \sum_{\tau=-\infty}^{\infty} a^{|\tau|} e^{-j\omega\tau}$$

$$= \frac{\sigma_v^2}{1 - a^2} \left[\sum_{\tau=0}^{\infty} (ae^{-j\omega})^\tau + \sum_{\tau=1}^{\infty} (ae^{j\omega})^\tau \right] = \frac{\sigma_v^2}{1 - a^2} \left[\frac{1}{1 - ae^{-j\omega}} + \frac{ae^{j\omega}}{1 - ae^{j\omega}} \right]$$

$$= \frac{\sigma_v^2}{|1 - ae^{-j\omega}|^2} ,$$

which is of course identical with (1.19) since

$$A(e^{-j\omega}) = 1 - ae^{-j\omega}.$$

∎

The resonant structure of $S_y(\omega)$ – the number and location of spectral peaks characterizing different frequency components of the analyzed signal – depends on the number and localization of complex poles of the noise shaping filter $H(q^{-1})$.

For well-separated complex poles that are sufficiently close to the unit circle, the following rules of thumb can be used for *qualitative* shape evaluation of the spectral density function (Figure 1.7):

- Each pair of complex conjugate poles can introduce one spectral peak (resonance).
- The angular frequency coordinate of a spectral peak is approximately equal to the phase angle of the corresponding pole.
- The height of a spectral peak is inversely proportional to the distance from the corresponding pole to the unit circle.

These 'rules' should be used with caution: several complex poles clustered together may create a single spectral peak, and poles with small radii may not create any peak at all. However, irrespective of the pole distribution, the number of spectral resonances may not exceed $n_A/2$, i.e. half the order of autoregression.

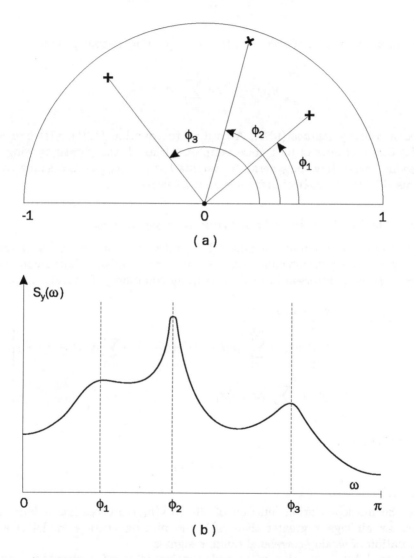

Figure 1.7 Approximate relationship between the location of noise shaping filter poles in the complex plane (a) and the shape of the spectral density function (b) of the corresponding AR process.

1.4.2 Moving average model

If the noise shaping filter $H(q^{-1})$ is adopted in the form

$$H(q^{-1}) = 1 + \sum_{i=1}^{n_C} c_i q^{-i} = C(q^{-1}), \tag{1.20}$$

i.e. it is an all-zero filter of order n_C, the corresponding output process

$$y(t) = v(t) + \sum_{i=1}^{n_C} c_i v(t-i) \tag{1.21}$$

is called a *moving average* (MA). According to equation (1.21), $y(t)$ is a weighted sum of a finite number of past noise samples. Although the phrase 'moving average' is somewhat misleading (in general the weights $1, c_1, \ldots, c_{n_C}$ do not sum to one), it is widely used in the statistical literature on time series.

Autocorrelation function of a moving average process

Since (1.20) is a finite impulse response (FIR) filter, the wide-sense stationarity of an MA(n_C) process is guaranteed for every $t > n_C$, irrespective of initial conditions.

Let $c_0 = 1$. Straightforward calculations using equation (1.6) yield

$$\begin{aligned}
r_y(\tau) &= \mathrm{E}[\,y(t)y(t-\tau)\,] \\
&= \mathrm{E}\left[\left(v(t) + \sum_{i=1}^{n_C} c_i v(t-i)\right)\left(v(t-\tau) + \sum_{i=1}^{n_C} c_i v(t-\tau-i)\right)\right] \\
&= \left(c_\tau + \sum_{i=1}^{n_C-\tau} c_i c_{\tau+i}\right)\sigma_v^2, \tag{1.22}
\end{aligned}$$

for $0 \le \tau \le n_C$ and

$$r_y(\tau) = 0,$$

for $\tau > n_C$.

Since the autocorrelation function of the MA(n_C) process has a finite span (it vanishes for all lags τ greater than n_C), the moving average model is a natural representation of weakly correlated random signals.

Unlike the AR case, the autocorrelation equations for the MA process are nonrecursive. They allow explicit computation of autocorrelation coefficients $r_y(\tau)$ given the set $c_1, \ldots, c_{n_C}$ and σ_v^2. However, solution of the inverse (identification) problem – determination of the MA coefficients based on the set of known autocorrelation coefficients – is much more difficult due to the fact that (1.22) is not linear in the process parameters. Additionally, in order to obtain practically meaningful identification results, some constraints have to be imposed on the admissible parameter space; see invertibility condition (A2).

Spectrum of a moving average process

Since the MA process is a result of passing white noise through a linear all-zero filter $C(q^{-1})$, its power spectral density function can be expressed in this form (cf. (1.18)):

$$S_y(\omega) = |C(e^{-j\omega})|^2 \sigma_v^2 = |1 + \sum_{i=1}^{n_C} c_i e^{-j\omega i}|^2 \sigma_v^2. \tag{1.23}$$

The power spectrum of an AR process can be characterized in terms of spectral peaks (resonances). The spectrum of an MA process is in some sense composed of spectral valleys (antiresonances). Based on the distribution of zeros of $C(q^{-1})$ one can give qualitative assesment for the shape of the MA spectrum (Figure 1.8):

- Each pair of complex-conjugate zeros can introduce one spectral valley (antiresonance).
- The angular frequency coordinate of a spectral valley is approximately equal to the phase angle of the corresponding zero.
- The depth of a spectral valley is inversely proportional to the distance of the corresponding zero from the unit circle.

Like the rules for the AR case, they should be used with caution.

Invertibility of a moving average process

Consider the problem of one-step-ahead prediction of the MA(1) process governed by

$$y(t) = v(t) + cv(t-1), \tag{1.24}$$

that is, the problem of estimating $y(t)$ based on its observation history $\mathcal{Y}(t-1) = \{y(t-1), y(t-2), \ldots\}$.

First af all, observe that $v(t) \perp \mathcal{Y}(t-1)$ and hence the first term on the right-hand side of (1.24) is 'unpredictable' in the sense that the prediction error variance can't be smaller than the variance of the variable $v(t)$ itself. As to the second term, the situation is different: since the random variable $v(t-1)$ is correlated with $y(t-1)$ there is certainly a way of inferring about $v(t-1)$ from $\mathcal{Y}(t-1)$. A straightforward solution to this problem is based on the concept of *inverse filtering*. Since

$$y(t) = (1 + cq^{-1})v(t), \tag{1.25}$$

one can attempt to recover the driving (input) sequence using the inverse filter $H^{-1}(q^{-1})$:

$$v(t, c) = \frac{1}{1 + cq^{-1}} y(t),$$

or equivalently

$$v(t, c) = -cv(t-1, c) + y(t). \tag{1.26}$$

Observe that equation (1.25) can be rewritten in an analogous form

$$v(t) = -cv(t-1) + y(t), \tag{1.27}$$

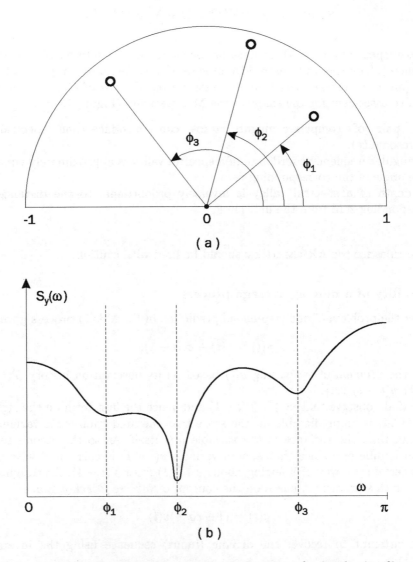

Figure 1.8 Approximate relationship between the location of noise shaping filter zeros in the complex plane (a) and the shape of the spectral density function (b) of the corresponding MA process.

and that the quantities $v(t,c)$ and $v(t)$ coincide if the assumed initial condition $v(0,c)$ matches the true one. Subtracting (1.27) from (1.26) and denoting

$$\eta(t) = v(t,c) - v(t),$$

one can easily study how mismatch in the initial conditions affects the evolution of the estimation error

$$\eta(t) = -c\eta(t-1) = (-c)^t\eta(0). \tag{1.28}$$

Quite clearly if $|c| < 1$, i.e. if the inverse filter is stable, then

$$v(t,c) \xrightarrow[t\to\infty]{} v(t),$$

and the initial estimation error decays to zero at an exponential rate. On the other hand, if $|c| > 1$ the inverse filtering scheme will diverge; the estimation error $\eta(t)$ will grow exponentially fast, irrespective of the value adopted for $v(0,c)$ (unless the initial conditions are known exactly).

All observations made so far can be easily extended to MA processes of arbitrary order. The error equation associated with the inverse filter

$$v(t,\theta) = -\sum_{i=1}^{n_C} c_i v(t-i,\theta) + y(t), \tag{1.29}$$

where $\theta = [c_1, \ldots, c_{n_C}]^T$, takes the form

$$\eta(t) = -\sum_{i=1}^{n_C} c_i \eta(t-i), \tag{1.30}$$

or equivalently

$$C(q^{-1})\eta(t) = 0,$$

leading to the following *invertibility condition*.

A2 All zeros of $C(q^{-1})$, i.e. all poles of the inverse filter $H^{-1}(q^{-1})$, lie inside the unit circle in the complex plane.

If the invertibility condition holds, the one-step-ahead prediction of an $MA(n_C)$ process $y(t)$, given its observation history $\mathcal{Y}(t-1)$, can be obtained from

$$\widehat{y}(t|t-1;\theta) = \sum_{i=1}^{n_C} c_i v(t-i,\theta), \tag{1.31}$$

where $v(t,\theta)$ is the estimate yielded by (1.29). But what should one do if the invertibility condition is not fulfilled? The answer is somewhat surprising: every moving average process with no spectral zeros on the unit circle has an invertible representation.

Theorem 1.2

Consider a noninvertible MA(n_C) process

$$C(q^{-1})v(t) = v(t) + \sum_{i=1}^{n_C} c_i v(t-i) \qquad (1.32)$$

such that $C(q^{-1})$ has no zeros on the unit circle and at least one zero outside the unit circle in the complex plane. Let $\{v(t)\}$ denote a white noise sequence of variance σ_v^2. There exists a unique invertible second-order-equivalent representation of (1.32),

$$\tilde{C}(q^{-1})\tilde{v}(t) = \tilde{v}(t) + \sum_{i=1}^{n_C} c_i \tilde{v}(t-i), \qquad (1.33)$$

such that $\tilde{C}(q^{-1})$ has all zeros inside the unit circle and $\{\tilde{v}(t)\}$ is another (different from $\{v(t)\}$) white noise sequence of variance $\sigma_{\tilde{v}}^2 > \sigma_v^2$.

Proof

Two zero-mean processes are second-order-equivalent if they have identical autocorrelation functions (or, equivalently, the same power spectral density functions). Therefore, to prove Theorem 1.2 it is sufficient to show how to construct an invertible polynomial $\tilde{C}(q^{-1})$ such that (cf. (1.23))

$$|\tilde{C}(q^{-1})|^2 \sigma_{\tilde{v}}^2 = |C(q^{-1})|^2 \sigma_v^2.$$

Rewrite $C(q^{-1})$ in a factorized form

$$C(q^{-1}) = (1 - q_1 q^{-1}) \cdots (1 - q_{n_C} q^{-1}),$$

where $q_1, \ldots, q_{n_C}$ are the zeros of $C(q^{-1})$. Note that

$$|C(e^{-j\omega})|^2 = \prod_{i=1}^{n_C} |1 - q_i e^{-j\omega}|^2.$$

Suppose that the ith zero of $C(q^{-1})$ lies outside the unit circle, i.e. $|q_i| > 1$. Since

$$|1 - q_i e^{-j\omega}|^2 = |q_i|^2 |1 - \frac{1}{q_i^*} e^{-j\omega}|^2,$$

one can rewrite $|C(e^{-j\omega})|^2$ in the form

$$|C(e^{-j\omega})|^2 = \delta |\tilde{C} e^{-j\omega}|^2,$$

where

$$\tilde{C}(q^{-1}) = \prod_{i=1}^{n_C} (1 - \tilde{q}_i q^{-1}),$$

$$\tilde{q}_i = \begin{cases} 1/q_i^\star & \text{if} \quad |q_i| > 1 \\ q_i & \text{if} \quad |q_i| < 1 \end{cases}$$

and

$$\delta = \prod_{i=1}^{n_C} \delta_i$$

$$\delta_i = \begin{cases} |q_i|^2 & \text{if} \quad |q_i| > 1 \\ 1 & \text{if} \quad |q_i| < 1 \end{cases}$$

Observe that all zeros of $\tilde{C}(q^{-1})$ lie inside the unit circle and that $\delta > 1$, since at least one zero of $C(q^{-1})$ was assumed to lie outside the unit circle. Finally, note that putting

$$\sigma_{\tilde{v}}^2 = \delta \sigma_v^2 > \sigma_v^2$$

one can rewrite (1.32) as (1.33), which completes our construction of an invertible MA representation.

■

If $C(q^{-1})$ has n_1 real zeros and n_2 pairs of complex conjugate zeros, all of multiplicity one $(n_1 + 2n_2 = n_C)$, the corresponding MA process has $2^{n_1+n_2}$ different representations. All these representations are equivalent in the sense that they characterize stochastic processes with an identical covariance structure. Even though invertible and noninvertible models can be used to generate an MA process, only the invertible model can be effectively used to predict its future. Consequently, the invertible representation is the only one that has a practical significance. Note that the driving noise variance, and hence also the limiting value of the mean square one-step-ahead prediction error, takes the largest value for the invertible process representation.

According to Theorem 1.2, the question of whether or not a given MA process is invertible is an ill-posed question. Invertibility is a property that has to be *enforced* on an MA process in order to obtain its 'operational' model.

Example 1.4

Consider the first-order moving average process governed by

$$y(t) = v(t) + 2v(t-1). \tag{1.34}$$

Since

$$S_y(\omega) = |1 + 2e^{-j\omega}|^2 \sigma_v^2 = 4|1 + 0.5e^{-j\omega}|^2 \sigma_v^2,$$

the invertible representation of $\{y(t)\}$ has the form

$$y(t) = \tilde{v}(t) + 0.5\tilde{v}(t-1), \tag{1.35}$$

where

$$\sigma_{\tilde{v}}^2 = 4\sigma_v^2.$$

Note how there is a slight abuse of notation when passing from (1.34) to (1.35): the corresponding stochastic processes are only second-order-equivalent rather than strictly equivalent, hence formally, they should be described by different symbols.

1.4.3   Equivalence of autoregressive and moving average models

Based on the material in the previous two sections, one might conclude there are some qualitative differences between the AR and MA processes. The following example will show this is not the case.

Example 1.5

Consider the AR(1) process governed by

$$y(t) = ay(t-1) + v(t), \qquad |a| < 1, \tag{1.36}$$

or equivalently

$$y(t) = \frac{1}{1 - aq^{-1}} v(t).$$

Since for $|q| > |a|$ it holds that

$$\frac{1}{1 - aq^{-1}} = 1 + \sum_{i=1}^{\infty} a^i q^{-i},$$

one can rewrite (1.36) as an infinite-dimensional moving average process

$$y(t) = v(t) + \sum_{i=1}^{\infty} a^i v(t-i). \tag{1.37}$$

Similarly, the invertible MA(1) process

$$y(t) = v(t) + cv(t-1), \qquad |c| < 1, \tag{1.38}$$

has an equivalent autoregressive representation of infinite order:

$$y(t) = \sum_{i=1}^{\infty} (-c)^i y(t-i) + v(t). \tag{1.39}$$

■

These results can be easily generalized to processes of arbitrary order [31], [89], [121].

Theorem 1.3

Every finite-order wide-sense stationary autoregressive process can be described by an invertible infinite-order moving average model and vice versa.

■

In view of the asymptotic equivalence of AR and MA models, should one care which modeling option to choose? According to the principle of parsimony certainly yes. Processes that can be satisfactorily approximated by low-order autoregressive models may require tens of MA parameters to reach a comparable degree of approximation and vice versa (Figures 1.9 and 1.10). A good trade-off between the approximation quality and the model complexity can usually be achieved by combining in one description both autoregressive and moving average terms; this is the so-called ARMA model.

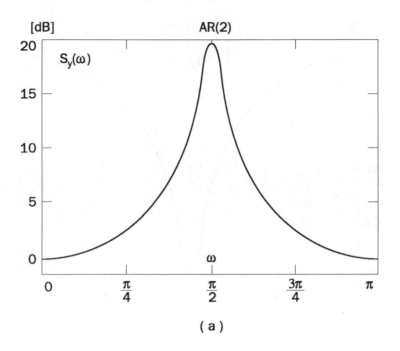

(a)

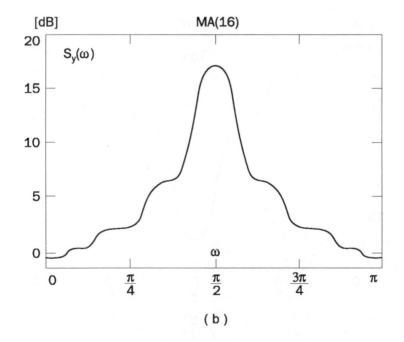

(b)

Figure 1.9 Spectrum of a second-order AR process (a) and its 16th-order MA approximation (b).

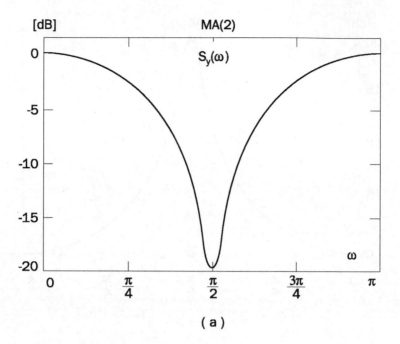

(a)

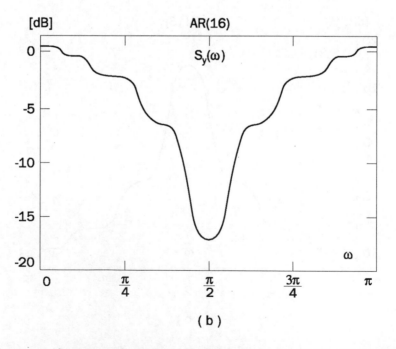

(b)

Figure 1.10 Spectrum of a second-order MA process (a) and its 16th-order AR approximation (b).

1.4.4 Mixed autoregressive moving average model

The mixed autoregressive moving average (ARMA) model combines both AR and MA terms:

$$y(t) = \sum_{i=1}^{n_A} a_i y(t-i) + v(t) + \sum_{i=1}^{n_C} c_i v(t-i), \tag{1.40}$$

which is equivalent to adopting the following noise shaping filter with both poles and zeros:

$$H(q^{-1}) = \frac{C(q^{-1})}{A(q^{-1})} = \frac{1 + \sum_{i=1}^{n_C} c_i q^{-i}}{1 - \sum_{i=1}^{n_A} a_i q^{-i}}. \tag{1.41}$$

The ARMA process is asymptotically stationary if all zeros of $A(q^{-1})$ lie inside the unit circle in the complex plane (condition A1); it is invertible if its moving average part is invertible, i.e. if all zeros of $C(q^{-1})$ lie inside the unit circle (condition A2).

Autocorrelation function of a mixed process

Consider an ARMA(n_A, n_C) process governed by (1.40). An equation for $r_y(\tau)$ can be obtained by multiplying both sides of (1.40) by $y(t-\tau)$ and taking the expectation:

$$r_y(\tau) = \sum_{i=1}^{n_A} a_i r_y(\tau-i) + r_{vy}(\tau) + \sum_{i=1}^{n_C} c_i r_{vy}(\tau-i), \tag{1.42}$$

where

$$r_{vy}(\tau) = \mathrm{E}[\,v(t)y(t-\tau)\,].$$

Denote by $\{h_i\}$ the impulse response of the noise shaping filter $H(q^{-1})$:

$$H(q^{-1}) = \sum_{i=0}^{\infty} h_i q^{-i}.$$

Then

$$r_{vy}(\tau) = \begin{cases} 0 & \text{if} \quad \tau > 0 \\ h_{-\tau} \sigma_v^2 & \text{if} \quad \tau \leq 0 \end{cases}$$

and consequently

$$r_y(\tau) = \sum_{i=1}^{n_A} a_i r_y(\tau-i) + \sigma_v^2 \sum_{i=\tau}^{n_C} c_i h_{i-\tau}, \tag{1.43}$$

for $0 \leq \tau \leq n_C$ and

$$r_y(\tau) = \sum_{i=1}^{n_A} a_i r_y(\tau-i), \tag{1.44}$$

for $\tau > n_C$.

Note that equation (1.44) is identical with the equation derived for an autoregressive process (1.10). Hence, if all zeros $q_1, \ldots, q_{n_A}$ of $H(q^{-1})$ have multiplicity one, the explicit expression for $r_y(\tau)$ is

$$r_y(\tau) = \sum_{i=1}^{n_A} B_i q_i^\tau, \qquad \tau \geq \max(0, n_A - n_C),$$

where $B_1, \ldots, B_{n_A}$ is the set of coefficients dependent on $a_1, \ldots, a_{n_A}$ and $c_1, \ldots, c_{n_C}$.

Example 1.6

Consider the ARMA(1,1) process governed by

$$y(t) = ay(t-1) + v(t) + cv(t-1), \qquad |a| < 1, |c| < 1,$$

Since

$$\frac{1 + cq^{-1}}{1 - aq^{-1}} = 1 + (a+c)q^{-1} + \cdots,$$

we have $h_0 = 1$ and $h_1 = a + c$, leading to the following set of equations:

$$
\begin{aligned}
r_y(0) &= ar_y(1) + \sigma_v^2 + \sigma_v^2 c(a+c), \\
r_y(1) &= ar_y(0) + \sigma_v^2 c,
\end{aligned}
\tag{1.45}
$$

which can be solved for $r_y(0)$ and $r_y(1)$:

$$r_y(0) = \frac{1 + c^2 + 2ac}{1 - a^2}\sigma_v^2, \quad r_y(1) = \frac{(1+ac)(a+c)}{1-a^2}\sigma_v^2.$$

According to (1.44)

$$r_y(\tau) = ar_y(\tau - 1)$$

for $\tau > 1$.

Note that equations (1.45) are not linear in ARMA parameters a and c, which is a consequence of incorporating the MA term. For this reason, identification of ARMA processes is considerably more difficult than identification of AR processes.

Spectrum of a mixed process

The spectral density function of an ARMA signal can be expressed in the form

$$S_y(\omega) = \frac{|C(e^{-j\omega})|^2}{|A(e^{-j\omega})|^2}\sigma_v^2 = \frac{|1 + \sum_{i=1}^{n_C} c_i e^{-j\omega i}|^2}{|1 - \sum_{i=1}^{n_A} a_i e^{-j\omega i}|^2}\sigma_v^2. \tag{1.46}$$

Since the noise shaping filter $H(q^{-1})$ has both poles and zeros, the spectrum of an ARMA process can easily match the spectral peaks and spectral valleys of the modeled signal (Figure 1.11). It is interesting to note that in certain applications, such as speech processing, capturing spectral antiresonances may be equally important as modeling resonances; for example, our perception of nasal sounds depends not only on what we hear but also on what we do not hear [42].

1.4.5 A bridge to continuous-time processes

Suppose that $\{y(t)\}$ is a continuous-time stochastic process governed by the following linear state-space equation:

$$
\begin{aligned}
dx(t) &= Ax(t) + b\,dw(t), \\
y(t) &= c^T x(t),
\end{aligned}
\tag{1.47}
$$

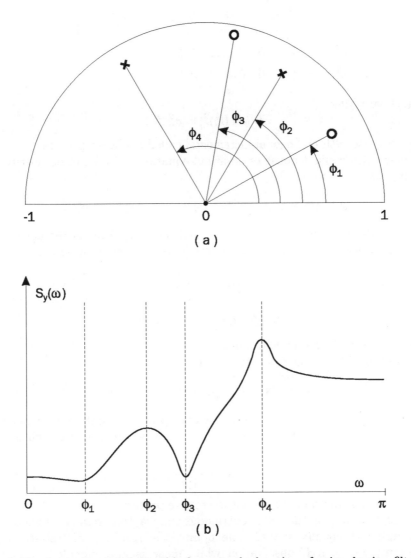

Figure 1.11 Approximate relationship between the location of noise shaping filter poles and zeros in the complex plane (a) and the shape of the spectral density function (b) of the corresponding ARMA process.

where $x(t)$ is the n_A-dimensional state vector,

$$
A = \begin{bmatrix}
0 & 1 & 0 & \cdots & 0 \\
0 & 0 & 1 & \cdots & 0 \\
\vdots & & & \ddots & \\
0 & 0 & 0 & \cdots & 1 \\
-\alpha_0 & -\alpha_1 & -\alpha_2 & \cdots & -\alpha_{n_A-1}
\end{bmatrix},
$$

$$
b^T = [\,0 \ 0 \ 0 \ \cdots \ 1\,],
$$

$$
c^T = [\,1 \ 0 \ 0 \ \cdots \ 0\,],
$$

and $\{w(t)\}$ satisfying

$$
\mathrm{E}[(dw(t))^2] = \sigma_v^2\, dt,
$$

denotes a process with orthogonal increments, called a Wiener process.

Note that, with a slight abuse of notation, equations (1.47) can be rewritten in an explicit input/output form:

$$
y^{(n_A)}(t) + \alpha_{n_A-1} y^{(n_A-1)}(t) + \cdots + \alpha_0 y(t) = v(t),
$$

where $y^{(i)}(t)$ denotes the ith derivative of $y(t)$ and $v(t)$ denotes the derivative of a Wiener process $w(t)$ (often known as a continuous-time white noise):

$$
v(t) = \frac{dw(t)}{dt},
$$

$$
\mathrm{E}[\,v(t)v(s)\,] = \sigma_v^2 \delta(t-s),
$$

where $\delta(t)$ denotes the Dirac delta function.

When all roots of the characteristic polynomial

$$
A(s) = s^{n_A} + \alpha_{n_A-1} s^{n_A-1} + \cdots + \alpha_0
$$

have negative real parts, the process $y(t)$ is asymptotically stationary. Since $y(t)$ can be viewed as a result of passing a continuous-time white noise through an all-pole analog filter

$$
H(s) = \frac{1}{A(s)},
$$

it is often called a continuous-time autoregressive process.

It is well known [14] that if a continuous-time autoregressive process of order n_A is sampled at equally spaced time points, the resulting discrete-time process is ARMA($n_A, n_A - 1$). If observation white noise is also included in (1.47), the adequate model becomes ARMA(n_A, n_A). Hence a discrete-time autoregression *cannot* be regarded as a sampled version of a continuous-time autoregression.

Since identical results can be derived in the case where

$$
H(s) = \frac{C(s)}{A(s)},
$$

$$
C(s) = s^{n_C} + \gamma_{n_C-1} s^{n_C-1} + \cdots + \gamma_0, \qquad n_C < n_A,
$$

i.e. for continuous-time mixed processes, it is clear that the discrete-time ARMA model is a *generic* representation of processes obtained by sampling wide-sense stationary finite-dimensional continuous-time stochastic signals. According to the results summarized above, there seems to be no 'physical' reason for fitting $\text{ARMA}(n_A, n_C)$ models with $n_A < n_C$. Note that in the continuous-time framework $n_A > n_C$ is a condition on the physical realizability of a noise shaping filter $H(s)$.

1.4.6 Models with exogenous inputs

When the signal $y(t)$ is observed at the output of a plant driven by a measurable input signal $u(t)$, some input-dependent terms have to be added to the models described above to account for the presence of such exogenous excitation. The model governed by

$$y(t) = \sum_{i=1}^{n_A} a_i y(t-i) + \sum_{i=1}^{n_B} b_i u(t-i) + v(t) \qquad (1.48)$$

is usually termed ARX (autoregressive with exogenous input) and the model

$$y(t) = \sum_{i=1}^{n_A} a_i y(t-i) + \sum_{i=1}^{n_B} b_i u(t-i) + v(t) + \sum_{i=1}^{n_C} c_i v(t-i) \qquad (1.49)$$

is called ARMAX.

Note that the ARX and ARMAX models can also be expressed in a polynomial form as

$$A(q^{-1})y(t) = B(q^{-1})u(t) + v(t)$$

and

$$A(q^{-1})y(t) = B(q^{-1})u(t) + C(q^{-1})v(t),$$

respectively, where

$$B(q^{-1}) = \sum_{i=1}^{n_B} b_i q^{-i}.$$

Both ARX and ARMAX models have an infinite impulse response (IIR). There are several applications where a simpler, finite impulse response (FIR) model can be used instead of (1.48) or (1.49):

$$y(t) = \sum_{i=1}^{n_B} b_i u(t-i) + v(t), \qquad (1.50)$$

1.4.7 The shorthand notation

All models introduced in Section 1.4 can be written down in the following shorthand form:

$$y(t) = \varphi^T(t)\theta + v(t)$$

where $\varphi(t)$ denotes the regression vector (Table 1.2) and θ is the vector of model coefficients (Table 1.3)

Model	Regression vector $\varphi(t)$
AR	$[y(t-1),\ldots,y(t-n_A)]^T$
FIR	$[u(t-1),\ldots,u(t-n_B)]^T$
MA	$[v(t-1),\ldots,v(t-n_C)]^T$
ARX	$[y(t-1),\ldots,y(t-n_A),u(t-1),\ldots,u(t-n_B)]^T$
ARMA	$[y(t-1),\ldots,y(t-n_A),v(t-1),\ldots,v(t-n_C)]^T$
ARMAX	$[y(t-1),\ldots,y(t-n_A),u(t-1),\ldots,u(t-n_B),$ $v(t-1),\ldots,v(t-n_C)]^T$

Table 1.2 Structure of the regression vector for different models

Model	Parameter vector θ
AR	$[a_1,\ldots,a_{n_A}]^T$
FIR	$[b_1,\ldots,b_{n_B}]^T$
MA	$[c_1,\ldots,c_{n_C}]^T$
ARX	$[a_1,\ldots,a_{n_A},b_1,\ldots,b_{n_B}]^T$
ARMA	$[a_1,\ldots,a_{n_A},c_1,\ldots,c_{n_C}]^T$
ARMAX	$[a_1,\ldots,a_{n_A},b_1,\ldots,b_{n_B},c_1,\ldots,c_{n_C}]^T$

Table 1.3 Structure of the parameter vector for different models

Note that, for models containing the moving average part, some of the components of the regression vector are not directly observable. However, if the MA part of the description is invertible then the corresponding variables – past noise samples $v(t-1),\ldots,v(t-n_C)$ – can be recovered from the semi-infinite observation history.

1.5   The model-based approach to adaptive signal processing and control

The system is called adaptive if it can adapt its behavior to unknown and/or time-varying characteristics of the analyzed or controlled process. The past three decades have brought a large number of really challenging applications of adaptive systems in different areas such as telecommunications (channel equalization, predictive coding, noise canceling), signal processing (prediction of time series, signal reconstruction, outlier elimination, spectral analysis) and automatic control (adaptive control, failure detection).

A majority of known adaptive systems are model-based, i.e. they rely on the mathematical model of the underlying physical process. Most of these systems are designed using the so-called certainty equivalence principle. First the problem is solved analytically for a known model of process dynamics. Then the model coefficients in the resulting solution are replaced with the corresponding estimates obtained via system identification. Dynamic weighing of vehicles (Section 1.1) is a simple example.

Adopting the model-based approach, one attempts to parameterize solution of the problem at hand (prediction, estimation, control, etc.) in terms of the process model coefficients. Physical models are usually not suitable for this purpose. First, being complex and nonlinear, they often lead to problem formulations that are mathematically not tractable. Second, physical models are often not parsimonious, i.e. they contain many 'nuisance' parameters that are of secondary importance when solving a particular problem. The popularity of instrumental models such as AR(X) or ARMA(X) stems from the fact that, freed from these drawbacks, they allow for relatively simple solutions to many practical problems, some of which are summarized below.

1.5.1 Prediction

An ability to predict its future behavior is fundamental to understanding any system. Most of the time forecasting is control-oriented as decisions depend on results. Weather forecasts, despite their notoriously poor quality, affect the lives of many professionals (farmers, aircrew, seafarers) as well as millions of ordinary people. Prediction of time series arising in econometrics (sales figures), biology (animal populations), biomedicine (physiological data), power management (power demand), water management (rainfall, river flows), etc., allows early detection of abnormal or emergency situations requiring some sort of compensation or human intervention.

If the input signal $u(t)$ is independent of $v(s)$ for all $t < s$ then the optimal, in the mean square sense, one-step-ahead prediction of the output of an ARX system governed by (1.48) can be expressed in the form

$$\widehat{y}(t|t-1;\theta) = \mathrm{E}[y(t)|\Xi(t-1),\theta] = \sum_{i=1}^{n_A} a_i y(t-i) + \sum_{i=1}^{n_B} b_i u(t-i), \qquad (1.51)$$

where $\Xi(t-1) = \{\mathcal{U}(t-1), \mathcal{Y}(t-1)\}$ denotes the observation history available at instant $t-1$: $\mathcal{U}(t-1) = \{u(1), \ldots, u(t-1)\}$, $\mathcal{Y}(t-1) = \{y(1), \ldots, y(t-1)\}$.

Note that equation (1.51) is intuitively pretty obvious. The right-hand side of (1.51) contains all terms of the ARX equation (1.48) that are known at instant $t-1$; the noise term $v(t)$ is missing as under the whiteness assumption it is 'unpredictable', namely

$$\mathrm{E}[v(t)|\Xi(t-1),\theta] = \mathrm{E}[v(t)] = 0.$$

Note also that the orthogonality condition imposed on the input does not preclude feedback from $y(\cdot)$ to $u(\cdot)$.

Using the certainty equivalence principle it is possible to derive the following adaptive version of the ARX predictor (1.51):

$$\widehat{y}(t|t-1;\widehat{\theta}(t-1)) = \varphi^T(t)\widehat{\theta}(t-1), \qquad (1.52)$$

where $\varphi(t) = [y(t-1), \ldots, y(t-n_A), u(t-1), \ldots, u(t-n_B)]^T$ and $\widehat{\theta}(t-1) = f[\Xi(t-1)]$ is an estimate of the unknown parameter vector $\theta = [a_1, \ldots, a_{n_A}, b_1, \ldots, b_{n_B}]^T$.

For an invertible ARMAX process governed by (1.49) the (asymptotically) optimal one-step-ahead predictor is

$$\widehat{y}(t|t-1;\theta) = \sum_{i=1}^{n_A} a_i y(t-i) + \sum_{i=1}^{n_B} b_i u(t-i) + \sum_{i=1}^{n_C} c_i v(t-i,\theta), \qquad (1.53)$$

where $\{v(t,\theta)\}$ is a sequence recovered from the measurable quantities $\{y(t)\}$ and $\{u(t)\}$ by means of inverse filtering:

$$v(t,\theta) = y(t) - \sum_{i=1}^{n_A} a_i y(t-i) - \sum_{i=1}^{n_B} b_i u(t-i) - \sum_{i=1}^{n_C} c_i v(t-i,\theta) \qquad (1.54)$$

with arbitrary, usually zero, initial conditions (for an invertible system $v(t,\theta)$ tends to $v(t)$ irrespective of initial conditions).

A simpler formula, asymptotically equivalent to (1.53), can be obtained if equations (1.53) and (1.54) are expressed in a polynomial form

$$\hat{y}(t|t-1;\theta) = [1 - A(q^{-1})]y(t) + B(q^{-1})u(t) + [C(q^{-1}) - 1]v(t,\theta),$$
$$v(t,\theta) = \frac{A(q^{-1})}{C(q^{-1})}y(t) - \frac{B(q^{-1})}{C(q^{-1})}u(t)$$

and combined together. Substituting $v(t,\theta)$ from the second equation into the first equation gives

$$\hat{y}(t|t-1;\theta) = \frac{C(q^{-1}) - A(q^{-1})}{C(q^{-1})}y(t) + \frac{B(q^{-1})}{C(q^{-1})}u(t),$$

or equivalently

$$\hat{y}(t|t-1;\theta) = \sum_{i=1}^{n_A} a_i y(t-i) + \sum_{i=1}^{n_B} b_i u(t-i) + \sum_{i=1}^{n_C} c_i(y(t-i) - \hat{y}(t-i|t-i-1)). \quad (1.55)$$

The adaptive version of the one-step-ahead predictor for an ARMAX system (an adaptive counterpart of (1.53)) can be obtained from

$$\hat{y}(t|t-1;\hat{\theta}(t-1)) = \hat{\varphi}^T(t)\hat{\theta}(t-1), \quad (1.56)$$

where $\hat{\varphi}(t) = [y(t-1),\ldots,y(t-n_A), u(t-1),\ldots,u(t-n_B), v(t-1,\hat{\theta}(t-1)),\ldots, v(t-n_C,\hat{\theta}(t-n_C))]^T$,

$$v(t,\hat{\theta}(t)) = y(t) - \hat{\varphi}^T(t)\hat{\theta}(t), \quad (1.57)$$

and $\hat{\theta}(t)$ denotes an estimate of the vector of ARMAX coefficients $\theta = [a_1,\ldots,a_{n_A}, b_1,\ldots,b_{n_B}, c_1,\ldots,c_{n_C}]^T$.

Remark

Invertibility of the process model is a prerequisite for the inverse filter (1.57) to work properly. To enforce this condition the roots of the polynomial

$$\hat{C}(q^{-1},\hat{\theta}(t)) = 1 + \sum_{i=1}^{n_C} \hat{c}_i(t)q^{-i}$$

must be checked and, if necessary, projected inside the unit circle before using (1.57).

1.5.2   *Predictive coding of signals*

Predictive coding is a technique which exploits redundancy of transmitted signals. It is widely used, for example, in voice communication systems. Instead of sending the receiver encoded samples of the original speech signal, one can use an indirect 'differential coding' scheme. According to this scheme, the transmitted signal is divided into consecutive blocks or frames. A simple mathematical model is fitted to each frame, yielding a sequence of residuals (modeling errors). The model parameters and a suitably quantized sequence of residuals are sent to the receiver, where the original signal frame is recovered.

The higher coding efficiency of the ADPCM (adaptive predictive pulse code modulation) scheme described above, compared to a standard PCM scheme, stems from the fact that for highly correlated sources, such as speech, the residuals are usually much smaller than the original signal (Figure 1.12) and hence they can be encoded using a smaller number of bits per sample. A typical ADPCM coder operates at 32kbit/s (8 kHz sampling, 20 ms frames) and yields a similar performance as a 64kbit/s PCM device. Taking into account the 'anatomy' of speech signals, much

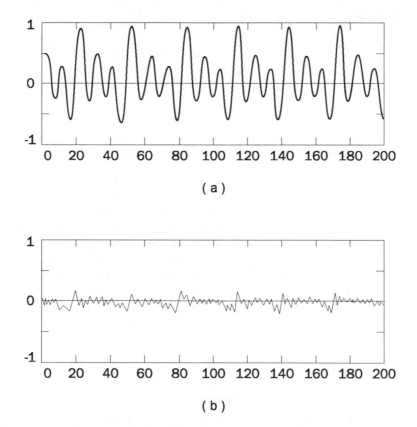

(a)

(b)

Figure 1.12 A fragment of a speech signal (a) and the corresponding sequence of residuals obtained by inverse filtering (b).

higher compression rates can be achieved. Voiced speech is produced by exciting the vocal tract (consisting of pharynx and mouth cavity) with quasi-periodic glottal air pulses generated by the vibrating vocal chords. Unvoiced speech is produced by forcing air through a constriction in the vocal tract. Different sounds are formed by varying the shape of the vocal tract through changing the position of speech articulators: velum, tongue, teeth and lips. Physical models of speech production are quite complex and involve a set of partial differential equations based on acoustic theory and low-viscosity (air) fluid mechanics [42]. Despite this, a relatively simple instrumental model of speech generation (Figure 1.13) is successfully used in practice – a linear (slowly time-varying) all-pole filter, representing the vocal tract, excited by a periodic train of pulses (for voiced sounds) or white noise (for unvoiced sounds).

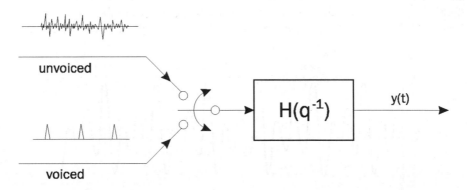

Figure 1.13   A simple speech production model used for adaptive speech coding.

Speech can therefore be regarded as a sort of generalized autoregressive process with switched noise/pulse excitation. The model described above has interesting practical implications. If, in addition to filter parameters (typically 8–12 autoregressive coefficients), the transmitted data contains information about the type of excitation (voiced/unvoiced) and its characteristics (gain, pitch frequency) then the corresponding speech frame can be *synthesized* at the receiver (Figure 1.14). Even though the synthesized speech generally differs from the original speech, then depending on the compression rate and the technical details, its perceptual quality is good or very good. In addition to high coding efficacy, reaching the rate of 2kbit/s, the LPC (linear predictive coding) vocoders offer interesting encryption opportunities exploited in secure (e.g. military) communications. For more details on speech coding, see the excellent survey by Spanias [181].

1.5.3   Detection and elimination of outliers

The term 'outlier' refers to a sample which does not fit the other, usually neighboring, samples. Most of the outliers can be attributed to some kind of impulsive noise corrupting the measuring equipment, communication channel and/or the storing medium.

Restoration of old audio files, such as archived gramophone records, is an interesting example of impulsive noise filtering. Clicks, pops and scratches are caused by local groove damage due to aging or mishandling of the record material. The width of noise pulses can be anything between several samples and several hundreds of samples (for old 78 rpm records). Isolation of large clicks and scratches is relatively easy since

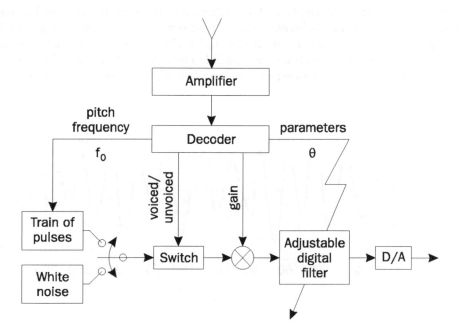

Figure 1.14 Schematic of a receiver in the LPC system.

they correspond to samples with unusually large values. Some sort of signal bound-checking can be used for their detection. Localization of small noise pulses is less straightforward. Since the corresponding samples don't stand out from the others, a more subtle detection technique has to be used.

Most of the modern declicking algorithms are based on adaptive prediction. Denote by

$$\epsilon(t) = y(t) - \widehat{y}(t|t-1)$$

the one-step-ahead prediction error (called innovation) and denote by

$$\widehat{\sigma}_\epsilon^2(t) = \frac{1}{M} \sum_{i=1}^{M} \epsilon^2(t-i)$$

the local estimate of the prediction error variance (M can be set to 20, for example). Finally, let $d(t)$ be the binary output of the outlier detector

$$d(t) = \begin{cases} 0 & \text{noise input present} \\ 1 & \text{noise input absent} \end{cases}$$

The following simple decision rule can be used:

$$\widehat{d}(t) = \begin{cases} 0 & \text{if } |\epsilon(t)| \leq \mu \widehat{\sigma}_\epsilon(t) \\ 1 & \text{if } |\epsilon(t)| > \mu \widehat{\sigma}_\epsilon(t) \end{cases} \tag{1.58}$$

According to (1.58) the sample is classified as an outlier and scheduled for reconstruction if the magnitude of the corresponding prediction error exceeds μ times

($\mu > 0$) its standard deviation. In audio applications the detection threshold μ has to be determined empirically; in most cases the best results are obtained for $\mu \in [3,5]$.

Figure 1.15 shows a fragment of an archived audio signal and the corresponding sequence of prediction errors. Small but audible clicks that are hardly visible on the signal plot can be easily localized on the innovation plot. The same mathematical

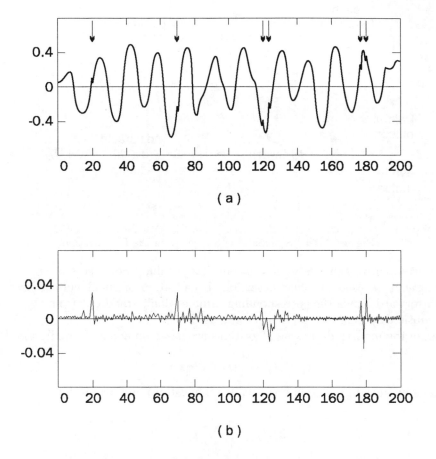

(a)

(b)

Figure 1.15   A fragment of an archived audio signal (a) and the corresponding sequence of innovations (b). Arrows indicate positions of audible clicks.

model that is the basis for adaptive prediction can be used for reconstruction (interpolation) of samples called into question by the outlier detector. Suppose one has to reconstruct a block of k samples missing from the stationary autoregressive time series governed by (1.8). The optimal linear estimate, in the mean square sense, for the vector of missing samples

$$y_m = [y(t+1), \ldots, y(t+k)]^T$$

can be obtained [145] as follows:

$$\hat{y}_m = [A_m^T(\theta) A_m(\theta)]^{-1} A_m^T(\theta) A_o(\theta) y_o, \tag{1.59}$$

where $A_m(\theta)$ is the $(n_A + k) \times k$ matrix which can be obtained by removing the first n_A and last n_A columns from the $(n_A + k) \times (2n_A + k)$ matrix

$$A(\theta) = \begin{bmatrix} a_{n_A} & a_{n_A-1} & \cdots & -1 & 0 & \cdots & 0 & 0 \\ 0 & a_{n_A} & \cdots & a_1 & -1 & \cdots & 0 & 0 \\ \vdots & & & & & & \vdots & \vdots \\ 0 & 0 & \cdots & & & & a_1 & -1 \end{bmatrix},$$

$A_o(\theta)$ is the $(n_A + k) \times 2n_A$ matrix composed of the first n_A and last n_A columns of $A(\theta)$ and

$$y_o = [\, y(t - n_A + 1), \ldots, y(t), y(t + k + 1), \ldots, y(t + k + n_A)\,]^T$$

denotes the vector of known (uncorrupted) samples preceding and succeeding the gap.

We note that (1.59) can be interpreted as an orthogonal projection of the vector of unknown samples on the subspace spanned by known samples

$$\{\ldots, y(t - 1), y(t), y(t + k + 1), y(t + k + 2), \ldots\}.$$

Example 1.7

If a single (isolated) sample $y(t)$ is missing, the following estimate can be obtained from (1.59):

$$\widehat{y}(t) = \sum_{i=1}^{n_A} \gamma_i [\, y(t - i) + y(t + i)\,],$$

where

$$\gamma_1 = \frac{-a_{n_A}}{1 + \sum_{i=1}^{n_A} a_i^2}$$

and

$$\gamma_i = \frac{-a_{n_A-i+1} + \sum_{j=1}^{i-1} a_{n_A-j+1} a_{i-j}}{1 + \sum_{i=1}^{n_A} a_i^2},$$

$$1 < i \leq n_A.$$

∎

Replacing the autoregressive coefficients in (1.59) with their estimates, the adaptive version of the reconstruction formula can be obtained.

Remark

The adaptive detection/reconstruction scheme sketched above will work satisfactorily provided that the identification algorithm yielding the estimates $\widehat{a}_1(t), \ldots, \widehat{a}_{n_A}(t)$ is robust to outliers. Robustness is also a desired feature of the local estimator of the prediction error variance $\widehat{\sigma}_\epsilon^2(t)$.

Quite obviously, the problems of detection/reconstruction and identification are mutually coupled and therefore require a joint treatment. An attempt to give such a joint solution, based on the extended Kalman filter (EKF), was made in [148]. The

EKF algorithm proposed in [148] can be viewed as a combination of two Kalman filters coupled in a nonlinear fashion. The first filter is designed to perform the online process identification and the second one is used to reconstruct those variables in the regression vector (made up of past signal measurements) which were called into question by the outlier detection routine. The outlier detector is based on monitoring prediction errors and its decisions are carried out by the proper covariance assignment. At each time instant the decision threshold of the detector is determined in accordance with some accuracy measures updated by the EKF algorithm.

1.5.4 Equalization of communication channels

In an ideal communication system the signal $y(t)$ measured at the receiver is identical with the transmitted signal $u(t)$. In reality, due to the dispersive nature of the propagation media, communication channels introduce into the transmitted signals a certain degree of distortion and a certain amount of noise. In modern wireless systems using the atmospheric or ionospheric high-frequency (HF) radio channels, distortion is caused by the so-called multipath effect.

Consider the HF mobile radio system depicted in Figure 1.16. In a typical mobile radio situation one station is fixed in position and the other station is moving. In urban environments the direct line between transmitter and receiver is usually obstructed by buildings, so propagation of the electromagnetic energy to and from the mobile unit is largely by scattering, either by reflection from the flat sides of buildings or by diffraction around obstacles (natural or artificial).

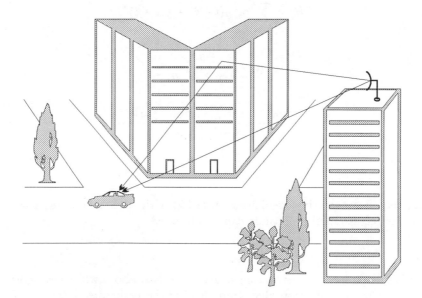

Figure 1.16   Communication through an HF radio channel.

In the line-of-sight Hertzian transmission between fixed ground stations (Figure 1.17) the multipath effect is caused by reflections of electromagnetic waves from different layers of the atmosphere and ionosphere.

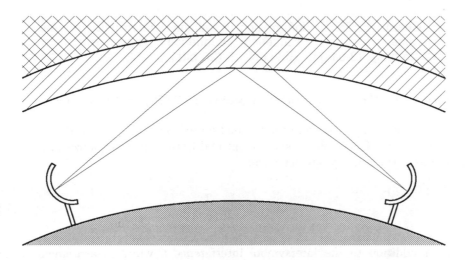

Figure 1.17 Communication through a line-of-sight Hertzian link.

In both cases mentioned above, interference of electromagnetic waves reaching the receiver by different paths results in the so-called fading effect – due to wave cancellation the total received signal, in certain time intervals, may be null. For this reason multipath channels are often known as fading channels or, since fading affects specific frequencies, as frequency-selective channels.

The linear FIR model

$$y(t) = \sum_{i=0}^{n_B} b_i u(t-i) + v(t),$$ (1.60)

a slight modification of (1.50), is successfully used to describe the relationship between $y(t)$ and $u(t)$ in a wireless communication system over sufficiently short time intervals. The coefficients $b_0, \ldots, b_{n_B}$ characterize the impulse response of the channel along with the transmitter and receiver filters. In digital communication systems, where signals are encoded and transmitted over the channel as a sequence of pulses, one can assume that

$$u(t) = \pm 1,$$

or, if quadrature amplitude modulation is used,

$$u(t) = \pm 1 \pm j.$$

The first term $b_0 u(t)$ of the sum on the right hand side of (1.60) can be identified as the desired symbol (up to the scale coefficient b_0), while the remaining terms $\sum_{i=1}^{n_B} b_i u(t-i)$ constitute the so-called intersymbol interference (ISI), a distortion which causes spreading and overlapping of adjacent pulses.

Since in the simplest case the decision device at the output of the data transmission system compares the received signal $y(t)$ with some preassigned threshold, the ISI component, when left unchecked, can seriously degrade the overall system performance. To overcome this problem the receiver is equipped with a special compensating filter, called a channel equalizer, which can remove most of the intersymbol interference.

Guard time (8.25 bits)	Tail (3 bits)	Information sequence (58 bits)	Training sequence (26 bits)	Information sequence (58 bits)	Tail (3 bits)

Figure 1.18   The user data packet adopted in the ETSC/GSM mobile radio system.

If the channel is invertible and the channel noise is negligible ($\sigma_v^2 \cong 0$), which is the case in many practical applications, the optimal linear output equalizer, in the mean square sense, has a pretty obvious form:

$$E(q^{-1}) = \frac{1}{B(q^{-1})} \, . \tag{1.61}$$

When, in addition to the intersymbol interference a white measurement noise is present, the best results can be obtained by applying the maximum likelihood (ML) sequence estimation technique. The ML estimation can be carried out by means of the Viterbi algorithm [190], [49]. If noise is additive, white and Gaussian, the Viterbi algorithm minimizes, in a computationally efficient manner, the Euclidean distance between the received sequence and all possible transmitted sequences.

Naturally, the quality of such demodulation relies heavily on the knowledge of channel coefficients. To enable receiver estimation of the channel response, a special known sequence of data symbols – the so-called training sequence – is transmitted along with the information sequence.

If channel characteristics remain stable during transmission, the training sequence takes the form of a preamble sent at the beginning of the communication session. In cases like that, channel identification is performed only once during the entire session.

To ensure reliability of communication over fast-fading radio channels, data is transmitted in fixed-length blocks or bursts, each of which contains its own training sequence. For example, in the high-capacity pan-European ETSC/GSM system the user data packets consist of two groups of 58 information bits interleaved with a 26-bit training sequence (Figure 1.18).

In more advanced communication systems the channel identification is extended beyond the training periods [162], [96], [35]. Since the true (transmitted) symbols are not known under these circumstances, to estimate the channel coefficients the system output is replaced with the actual output of the decision device (whether based on simple thresholding or employing the Viterbi algorithm). Adaptive equalizers based on this principle are called *decision-oriented* or *blind*.

1.5.5   *Spectrum estimation*

Suppose spectral analysis of an observed process $\{y(t)\}$ has to be performed based on a finite data segment $\mathcal{Y}(N)$ of length N. The problem of spectrum estimation can be solved in two different ways.

The first approach is known as nonparametric and refers directly to the Wiener–Khinchin relationship (1.16). Replacing unknown autocorrelation coefficients $r_y(\tau)$ in

(1.16) with their estimates

$$\hat{r}_y(\tau) = \hat{r}_y(-\tau) = \frac{1}{N-\tau} \sum_{t=1}^{N-\tau} y(t)y(t+\tau), \qquad 0 \le \tau \le M, \qquad (1.62)$$

one arrives at the following expression for the 'empirical' spectral density function:

$$\hat{S}_y(\omega) = \sum_{\tau=-M}^{M} \hat{r}_y(\tau)e^{-j\omega\tau}. \qquad (1.63)$$

In accordance with the principle of parsimony M, the maximum autocorrelation lag in (1.63) should be kept much smaller than the number of available data items N ($M = 0.1N$ is a typical recommendation).

Truncation of the autocorrelation function, a consequence of adopting a finite summation index in (1.63), results in distortions known as spectral leakage. In nonparametric analysis such distortions are practically unavoidable as only a finite set of autocorrelation lag estimates can be obtained from a finite observation set. It is possible, however, to significantly reduce spectral leakage by replacing the implicit rectangular window

$$w(\tau) = \begin{cases} 1 & \text{if } |\tau| \le M \\ 0 & \text{if } |\tau| > M \end{cases}$$

implied by (1.63) with a suitably designed tapered (bell-shaped) window, such as the Hann raised cosine window, for example

$$w(\tau) = \begin{cases} 0.5[1 + \cos \frac{\pi\tau}{M}] & \text{if } |\tau| \le M \\ 0 & \text{if } |\tau| > M \end{cases}$$

See [73] for an up-to-date overview of windows used in spectral analysis. The Fourier transform of the windowed estimate of the autocorrelation function

$$\hat{S}_y(\omega) = \sum_{\tau=-M}^{M} w(\tau)\hat{r}_y(\tau)e^{-j\omega\tau} \qquad (1.64)$$

is known as the Blackman–Tukey estimate of a signal power spectrum. It is straightforward to show that the variance of autocorrelation estimates $\hat{r}(\tau)$ given by (1.62) increases with τ. Hence windowing in (1.64) has the effect of downweighting the high-variance autocorrelation lag estimates around the endpoints.

The alternative, model-based approach to spectrum estimation is called parametric. The concept goes back to the original paper of Yule [202], who used parameters of the second-order AR model of sunspot series to determine its 'dominant' period of approximately 10.6 years. However, it was not until the work of Burg [34], Parzen [155] and Hannan [72] that the significance of this new approach was fully recognized [90], [165]. The idea (like all good ideas) is very simple. Instead of parameterizing the spectral density function using an *infinite* number of autocorrelation coefficients as in the Wiener–Khinchin formula (1.16), one can attempt to express it in terms of a *finite* number of AR or ARMA coefficients as in (1.19) or (1.46), respectively. Spectrum estimates can then be calculated by replacing the true model coefficients in (1.19)

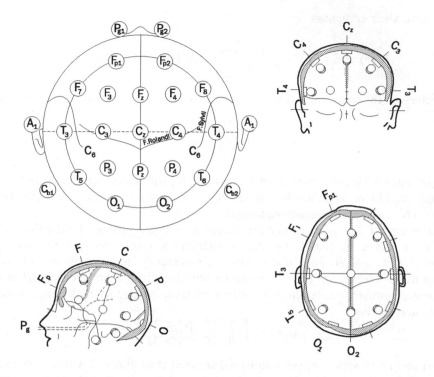

Figure 1.19 Localization of electrodes on the patient's skull (reproduced from Electroencephalographic atlas 2nd edition, by Jerzy Majkowski, 1991, with permission from PZWL, Warsaw).

or (1.46) with their estimates obtained via process identification. For example, if the autoregressive model of time series is adopted, the parametric spectrum estimate is

$$\widehat{S}_y(\omega) = \frac{\widehat{\sigma}_v^2(N)}{|1 - \sum_{i=1}^{n_A} \widehat{a}_i(N) e^{-j\omega i}|^2} \cdot \tag{1.65}$$

For appropriately chosen orders of the rational process model, parametric estimates guarantee good accuracy and high spectral resolution even for very short data sequences. Moreover, the parametric approach yields smooth and easily interpretable spectral plots, which is important in many applications [89], [121].

Analysis of electroencephalographic (EEG) signals is an interesting example of modern spectral estimation techniques. The human brain produces electrical activity which can be recorded using electrodes positioned upon the skull (Figures 1.19 - 1.21). EEG signals carry information about the state of the brain and its reaction to internal and external stimuli. Despite advances made recently in the area of computer-aided tomography (CAT), EEG remains a practically useful tool for studying the functional states of the brain (e.g. sleep-stage analysis) and for diagnosing functional brain disturbances (e.g. epilepsy). The frequency content of electroencephalographic signals is crucial for their assessment. According to the traditional clinical nomenclature introduced in a pioneering work of Berger [22], EEG signals are described and classified in terms of several rhythmic activities called alpha (8–13 Hz), beta (14–30 Hz), delta

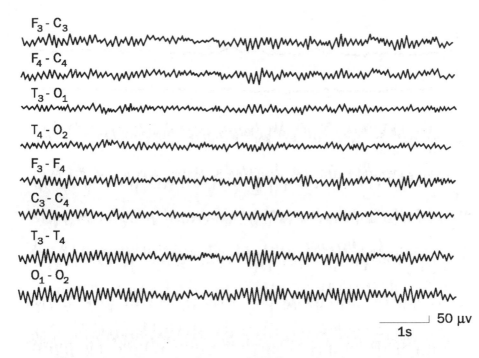

Figure 1.20   Example of a normal EEG recording (reproduced from Electroencephalographic atlas 2nd edition, by Jerzy Majkowski, 1991, with permission from PZWL, Warsaw).

(1–3 Hz) and theta (4–7 Hz). The normal EEG of an adult in a state called relaxed wakefulness is usually dominated by rhythmic alpha activity and a less pronounced beta activity. The contribution of delta and theta activities is arrhythmic and much smaller (Figure 1.22).

The presence or absence of different frequency components and their prominence are important factors considered by EEG-based diagnosis. For example, the presence of strong low-frequency components (delta, theta) is usually a sign of abnormal brain operation, due to disease, injury or functional disturbances (thromboses, inflammations, tumors, etc). Another group of anomalous EEG features, appearing in patients suffering from some kind of epileptic seizures, consists of specific transient signals classified as spikes, sharp waves and spike-and-wave activity. Spikes and sharp waves are usually superimposed on a more regular (background) EEG activity. Detection of such abnormal elements may be crucial for diagnosis and early treatment of epilepsy.

Several physical models describing the neurophysiological generation of EEG were proposed in the literature. Some of them, such as the local neuron population model described by Zetterberg et al. [203] (including three types of neurons involved in both temporal and spatial interactions), are capable of simulating the electrical brain activity and they are virtually indistinguishable from the true EEG. Not surprisingly, such models are fairly complex and nonlinear (since individual neurons are nonlinear processing elements).

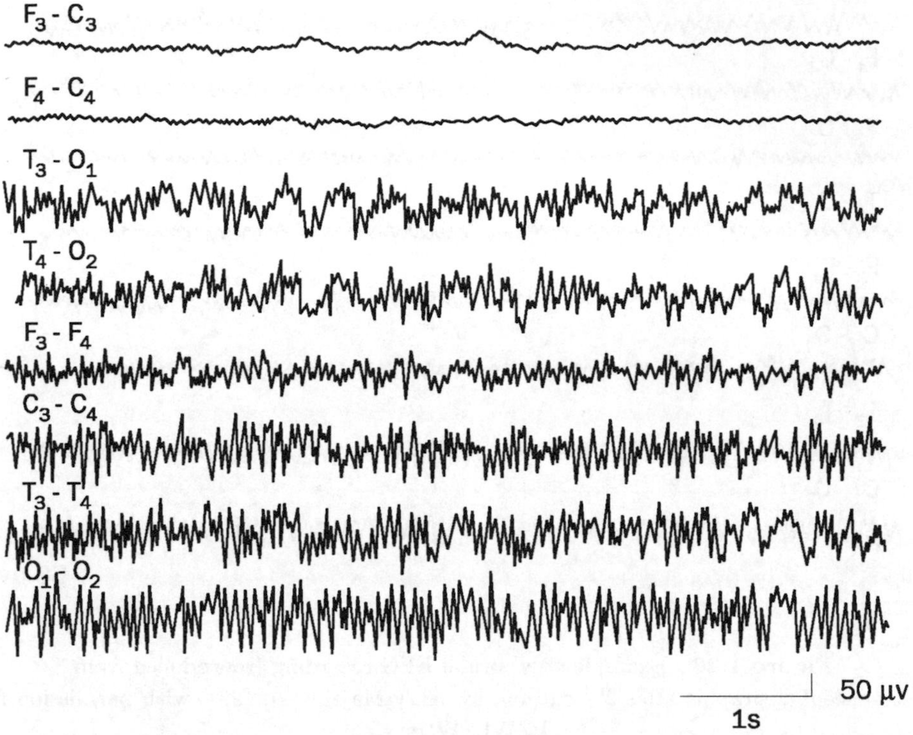

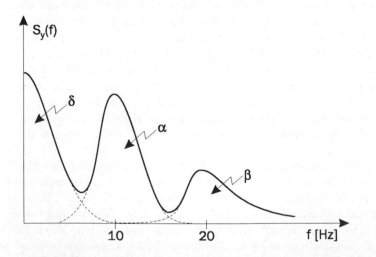

Figure 1.21   Example of a pathological EEG recording (reproduced from
Electroencephalographic atlas 2nd edition, by Jerzy Majkowski, 1991, with permission from
PZWL, Warsaw).

Figure 1.22   Spectrum of a typical 'normal' EEG recording.

When it comes to EEG analysis, complex phenomenological models like the one described above are of little help; simple, purely descriptive models such as AR or ARMA are used instead [204], [55], [79]. In addition to yielding easily interpretable spectral plots, the parametric analysis allows one to determine, directly from the model coefficients, important features such as the bandwidth and power of different spectral components, and the coherence between different EEG channels [56], [78]. Parametric models have also been successfully applied to detection of short transients such as spikes and sharp waves (using the time-domain techniques already described in Section 1.5.3).

1.5.6 Adaptive control

Many technological processes require stabilization of certain internal and/or output variables at prescribed values.

Consider a single-input single-output plant where the output should be kept at a desired (nominal) level y_0. In the absence of disturbances such control objective can be easily met by applying to the plant's input a suitably chosen constant control signal u_0.

When noise is present, more sophisticated dynamic stabilization techniques have to be used to achieve good results. In order to make the analysis mathematically tractable, the plant dynamics is linearized around the set point (u_0, y_0). Denoting by $u(t)$ a difference between the actual control signal and its nominal value u_0 and defining the output signal $y(t)$ likewise, one can express the linearized model in a familiar ARX or ARMAX form. The control task can then be formulated as that of minimizing the variance of the system output

$$u(t) \; : \; E[y^2(t)] \longmapsto \min .$$ (1.66)

For an ARX plant, it is pretty straightforward to derive the minimum variance control rule. The control signal is chosen so as to compensate all known terms on the right-hand side of (1.48) (i.e. everything except the noise term $v(t)$), leading to the following control rule:

$$u(t) = -\frac{1}{b_1} \left[\sum_{i=1}^{n_A} a_i y(t-i+1) + \sum_{i=2}^{n_B} b_i u(t-i+1) \right] .$$ (1.67)

For ARMAX plants the corresponding rule becomes

$$
\begin{aligned}
u(t) \;=\; & -\frac{1}{b_1} \left[\sum_{i=1}^{n_A} a_i y(t-i+1) + \sum_{i=2}^{n_B} b_i u(t-i+1) \right. \\
& \left. + \; \sum_{i=1}^{n_C} c_i v(t-i+1, \theta) \right],
\end{aligned}
$$ (1.68)

which can be also rewritten in an explicit (asymptotically equivalent) form

$$u(t) = -\frac{1}{b_1} \left[\sum_{i=1}^{n_A} a_i y(t-i+1) + \sum_{i=1}^{n_C} c_i y(t-i+1) + \sum_{i=2}^{n_B} b_i u(t-i+1) \right] .$$ (1.69)

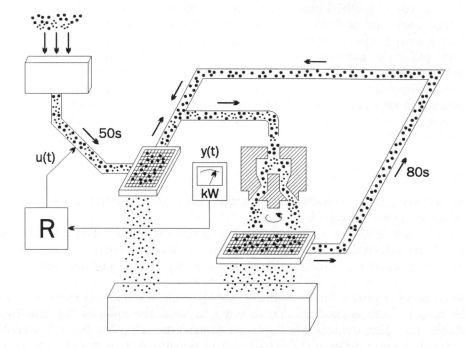

Figure 1.23 Schematic diagram of an ore crushing plant (Reprinted from Automatica Vol 12.1, Borisson U. and Syding R., *Schematic diagram of ore crushing plant* tpp 1-7, January 1976, with permission from Elsevier Science).

Adaptive minimum variance regulators, often called self-tuning, can be obtained by replacing known system parameters a_i, b_i and c_i with their estimates $\widehat{a}_i(t)$, $\widehat{b}_i(t)$ and $\widehat{c}_i(t)$.

One of the early successful applications of self-tuning minimum variance regulators, reported here after Borisson and Syding [30], was for control of the ore crushing plant depicted in Figure 1.23. The aim of this plant, consisting of two screens and a crusher driven by an electric motor, is to crush the raw ore material (discharged from the refillable input bin through a system of conveyor belts) to lumps with a maximum dimension of 25 mm. Lumps which already have the desired dimension are separated by the first screen. The rest of the ore is fed into the crusher. The second screen, placed at the output of the crusher, selects the material which needs to be recycled because its grains are the wrong size. The control objective is to change the flow of the raw material supplied by the electromechanical feeder so as to maximize the power of the crusher without overheating its driving motor. The overload protection mechanism (a special slip coupling) automatically stops the crushing line if the load becomes too high. Since it may take about an hour before the full production rate is restored after a standstill, the probability of such forced stops must be kept sufficiently small if production is to be maintained at a high level.

By minimizing the fluctuations in the instantaneous power of the driving motor around its nominal value, it is possible to bring the set point power closer to the maximum capacity of the crusher for a given probability of an overload (Figure 1.24).

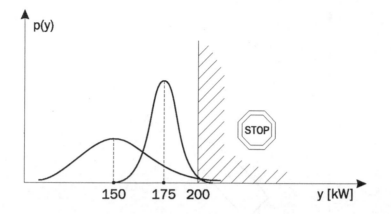

Figure 1.24 How the set point of the ore crushing plant depends on the variance of fluctuations in the instantaneous power: smaller variance allows one to increase the nominal power from 150 kW to 175 kW.

Hence the optimal control of the ore crusher falls pretty naturally into the minimum variance category.

The self-tuning minimum variance controller of the crusher described in [30] was based on a simple ARX model of the plant:

$$y(t) = \sum_{i=1}^{3} a_i y(t - i - 3) + \sum_{i=1}^{3} b_i u(t - i - 3) + v(t).$$

To account for long transportation delays in the plant (40–50 s in the main loop and 70–80 s in the recycle loop) the dead time of 3 sampling intervals was incorporated. The sampling rate of $T = 20$ s was commeasurate with the time constants of the feeder (12 s) and the crusher (20 s).

Experimental results reported in [30] were very promising. Adaptive control made it possible to raise the set point power from 170 kW (PI control) to 200 kW yielding a production increase of approximately 10%. Even though these preliminary tests never developed into a full-scale application (the ore-mining industry started to face severe economic problems in the mid 1970s), the work of Borisson and Syding remains an excellent reference for practically oriented control engineers.

Control of the ore crusher belongs to the class of so-called *cheap control* problems; note that neither the energy of the input signal nor its variability is incorporated in the control objective (1.66). Many other applications exist where some control over the input signal must be established to account for the steering cost, input signal limitations or other input-related factors. Consider, for example, the problem of keeping a ship on a steady course. Denoting by $u(t)$ the rudder command and by $y(t)$ the yaw rate, i.e. the difference between the reference course and the actual heading (Figure 1.25), the average increase in drag due to yawing and rudder motions can be approximately described by

$$\alpha \left(E[u^2(t)] + \beta E[y^2(t)] \right),$$

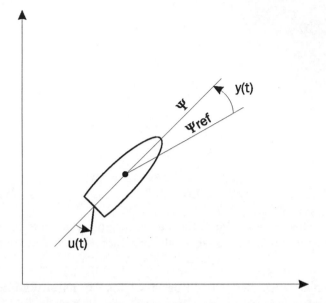

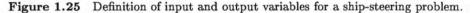

Figure 1.25 Definition of input and output variables for a ship-steering problem.

where α and β are positive coefficients depending on the ship and its operating conditions [10].

Hence, if the minimum fuel consumption is attempted (for large vessels fuel consumption governs the operating costs), the control task can be formulated in the following way:

$$u(t) \;=\; \mathrm{E}[u^2(t)] + \beta \mathrm{E}[y^2(t)] \longmapsto \min. \tag{1.70}$$

If the linear (ARX or ARMAX) model of the ship's dynamics is available, or established by process identification, the problem can be solved using linear quadratic (LQ) control theory. Adaptive autopilots based on this principle (some of them using a simpler minimum variance strategy or gain scheduling methods) were proposed in the late 1970s [10], [153] and gradually became an almost standard piece of equipment onboard ship.

2

Models of Nonstationary Processes

2.1 The origins of time dependence

Even though it was not stated explicitly, nonstationarity of the underlying physical processes is the common feature of all applications discussed in the previous chapter. Speech (Section 1.5.2) can be viewed as a sequence of different sounds obtained by varying the shape of the vocal tract and the type of excitation. Consequently, speech signals can be regarded as stationary only in relatively short time intervals of 10–20 ms. The same remark concerns other audio signals, such as music, where nonstationarity is a consequence of the time-varying nature of sound formation and/or propagation.

The time variation of communication channels (Section 1.5.4) is caused by several 'internal' and 'external' factors. Nonstationarity of high-frequency ionospheric channels results from slow variation of magnetoionic components (reflecting layers). Most band-limited channels are nearly stationary over periods of 10–15 minutes, although faster variation can also be observed under certain circumstances, e.g. during sunrise or sunset, when propagation conditions undergo rapid changes.

More significant changes of channel characteristics can be observed in mobile radio systems. Since the multipath effect is caused by scattering to the mobile antenna from many objects in its vicinity, channel characteristics may change dramatically even for small changes in position of the mobile unit. Quite obviously, the rate of time variation of a fading radio channel depends on the speed of the vehicle (the Doppler effect).

Most EEG recordings (Section 1.5.5) lasting longer than a few minutes show changes in essential characteristics such as the power level and frequency content. There are many factors which are responsible for this: changes in the functional state of the brain (degree of wakefulness or vigilance, transitions between different sleep stages), influences on the brain (diseases, drugs, stimuli) or disturbances produced by the examined subject (eye movements, arterial pulsations, scalp-muscle activity). The changes are usually slow but sometimes they appear as sudden jumps.

Nonstationarity of the ore crushing plant (Section 1.5.6) originates from variations in crushability of the raw material (depending mainly on the waste rock content), the average lump size and the wear of the jackets (the gap width of the crusher is adjusted every second day and the jackets are replaced every third month). Some of these factors can change significantly over a short period of time. For example, each time the input bin is filled, small lumps tend to gather at the bottom of the container, which may cause a temporary drop in the amount of raw material entering the crusher (as all small lumps are separated by the first screen).

Finally, the ship-steering dynamics (Section 1.5.6) depends on external disturbances such as wind, waves and current. Additionally it may change with the operating conditions such as speed, trim and loading of the vessel as well as the depth of the water. Note that set point changes, when affecting plants with nonlinear dynamics, usually result in a nonstationary-like system behavior: coefficients of the linearized (local) system model are set point dependent, hence they are also time dependent – even if the true (nonlinear) system description is time invariant!

Summing up, physical phenomena exhibit nonstationary behavior for a number of reasons: due to the time-varying nature of signal formation and propagation (audio signals, communication channels), disturbances (audio signals, EEG signals, ore crusher, ship), changes of the functional state (EEG signals), wear, aging and fatigue (ore crusher) or set point changes (ship). The associated time constants range from milliseconds (speech, HF radio channels) to hours (ship) and even days (many technological processes). In addition to slow persistent variation, many processes are subject to occasional jump changes of their characteristics (EEG signals, HF radio channels, ore crusher).

2.2 Characteristics of nonstationary processes

Denote by

$$F(y_1,\ldots,y_n;t_1,\ldots,t_n) = P(y(t_1) \le y_1,\ldots,y(t_n) \le y_n) \tag{2.1}$$

the nth order distribution of a stochastic process $\{y(t)\}$. The process is called strictly stationary if its distributions are time shift invariant, i.e. the following equality holds

$$F(y_1,\ldots,y_n;t_1,\ldots,t_n) = F(y_1,\ldots,y_n;t_1 - \tau,\ldots,t_n - \tau) \tag{2.2}$$

for every selection of n, $t_1,\ldots,t_n$ and τ.

As a direct consequence of (2.2) for $n = 1$ and $n = 2$, one obtains

$$E[y(t)] = m_y(t) = \int_{-\infty}^{\infty} y \, dF(y;t)$$

$$= \int_{-\infty}^{\infty} y \, dF(y;t - \tau) = m_y(t - \tau) = m_y, \tag{2.3}$$

and

$$E[\, y(t)y(t - \tau)\,] = r_y(t,\tau) = \int_{-\infty}^{\infty} y_1 y_2 \, dF(y_1,y_2;t,t - \tau)$$

$$= \int_{-\infty}^{\infty} y_1 y_2 \, dF(y_1,y_2;s,s - \tau) = r_y(s,\tau) = r_y(\tau). \tag{2.4}$$

Processes obeying (2.3) and (2.4) are called second-order stationary or wide-sense stationary.

For nonstationary processes the time invariance condition (2.1) fails at least for some n, $t_1,\ldots,t_n$ and τ. From a practical viewpoint the most important class of nonstationary processes comprises signals which show nonstationarity in their mean

$$E[y(t)] = m_y(t) \ne \text{const}$$

and/or covariance structure

$$E\left[(y(t) - m_y(t))(y(t - \tau) - m_y(t - \tau))\right] = c_y(t, \tau) \neq c_y(\tau).$$

Processes whose mean value is not constant are known as nonstationary mean time series. If the covariance structure of the process also changes with time (i.e. its autocovariance is a function of both correlation lag *and* time), it is called the nonstationary covariance time series.

Nonstationary mean values, or trends, occur in many meteorological and econometric time series. In a majority of cases they can be successfully handled by augmenting the AR and ARMA models (Section 1.4) with the deterministic trend component

$$\sum_{i=1}^{n_D} d_i f_i(t),$$

(added to the right-hand side of (1.8) or (1.40)), where $f_i(t)$, $i = 1, \ldots, n_D$, denote *known* functions of time. The systematic growth or decay of the mean is usually described by the polynomial model ($f_1(t) = 1$, $f_2(t) = t$, $f_3(t) = t^2$, etc.) while its seasonal (oscillatory) components are captured by using periodic functions of time ($f_1(t) = \sin \omega_1 t$, $f_2(t) = \cos \omega_1 t$, $f_3(t) = \sin \omega_2 t$, $f_4(t) = \cos \omega_2 t$, etc.). The problem of modeling deterministic trends 'superimposed' on stochastic models is beyond the scope of this monograph (see e.g. Kitagawa and Gersch [95], and the references therein). We will further assume that all trends are removed from the data prior to modeling a time series or a dynamic system, i.e. that the corresponding processes are zero-mean nonstationary covariance. In many cases detrending can be achieved by subtracting from the signal its local average.

Even though there is no general theory of nonstationary processes, some interesting results were derived for zero-mean nonstationary covariance time series. Cramer [38] has shown that a purely nondeterministic second-order, zero-mean nonstationary covariance process $\{y(t)\}$ possesses a one-sided linear representation similar to the Wold's decomposition of a stationary process:

$$y(t) = \sum_{i=0}^{\infty} h_i(t)v(t - i), \tag{2.5}$$

where $\{v(t)\}$ denotes white noise and

$$\sum_{i=0}^{\infty} h_i^2(t) < \infty, \qquad \forall t.$$

According to (2.5) the nonstationary process $y(t)$ can be viewed as a result of passing white noise through a causal time-varying filter of the form

$$H(t, q^{-1}) = \sum_{i=0}^{\infty} h_i(t)q^{-i}.$$

Based on the Cramer–Wold decomposition Tjøstheim [184] proposed the following definition of the *evolutionary spectrum* of a nonstationary process:

$$S_y(\omega, t) = |H(t, e^{-j\omega})|^2 \sigma_v^2 = \left| \sum_{i=0}^{\infty} h_i(t) e^{-j\omega i} \right|^2 \sigma_v^2, \tag{2.6}$$

which can be interpreted as the spectrum of a stationary process tangent, at time t, to the nonstationary process under study. The quantity $H(t, e^{-j\omega})$ is sometimes called a frozen frequency characteristic of a time-varying filter. The Tjøstheim evolutionary spectrum has most of the desirable properties of a time-varying spectral density function, postulated by Loynes [116].

Since the noise shaping filter in (2.5) is characterized in terms of an infinite number of impulse response coefficients $h_i(t)$, the Cramer–Wold decomposition cannot be directly used for identification purposes. By truncating the impulse response of the Cramer–Wold filter

$$H(t, q^{-1}) = 1 + \sum_{i=1}^{n_C} c_i(t) q^{-i} = C(t, q^{-1}), \tag{2.7}$$

one obtains the following finite-order time-varying MA representation of a nonstationary process:

$$y(t) = v(t) + \sum_{i=1}^{n_C} c_i(t) v(t - i). \tag{2.8}$$

If the shaping filter has the form

$$H(t, q^{-1}) = \frac{1}{1 - \sum_{i=1}^{n_A} a_i(t) q^{-i}} = \frac{1}{A(t, q^{-1})}, \tag{2.9}$$

one arrives at a time-varying AR representation

$$y(t) = \sum_{i=1}^{n_A} a_i(t) y(t - i) + v(t). \tag{2.10}$$

Finally, if a general finite-order rational (pole and zero) model is adopted

$$H(t, q^{-1}) = \frac{1 + \sum_{i=1}^{n_C} c_i(t) q^{-i}}{1 - \sum_{i=1}^{n_A} a_i(t) q^{-i}} = \frac{C(t, q^{-1})}{A(t, q^{-1})}, \tag{2.11}$$

the process is described in terms of a time-varying ARMA model

$$y(t) = \sum_{i=1}^{n_A} a_i(t) y(t - i) + v(t) + \sum_{i=1}^{n_C} c_i(t) v(t - i). \tag{2.12}$$

The rational evolutionary spectrum (an 'instantaneous' spectrum of a time-varying MA, AR or ARMA process) can be defined analogously to (2.6); see Grenier [63]. Extensions of (2.10) and (2.12) to nonstationary ARX and ARMAX models are straightforward.

Remark

Sometimes we will refer to (2.7), (2.9) and (2.11) as linear filters with time-varying 'zeros' and/or 'poles'. By time-varying zeros and poles we mean the roots of $C(t, q^{-1})$ and $A(t, q^{-1})$, respectively. Strictly speaking, it is only correct to talk about zeros and poles for time-invariant systems. However, if a system varies slowly with time, it may be useful conceptually to think of it as possessing time-varying zeros and poles.

2.3   Irreducible nonstationary processes and parameter tracking

We will call a nonstationary process 'reducible' if it can be described, in the entire time domain, by a model with constant coefficients.

Consider, for example, the AR(1) process with a periodically varying coefficient

$$y(t) = a(t)y(t-1) + v(t), \tag{2.13}$$

$$a(t) = a_0 \sin \omega t.$$

If the angular frequency ω of parameter oscillations is known, one can rewrite (2.13) in the form

$$y(t) = a_0 z(t) + v(t), \tag{2.14}$$

where

$$z(t) = y(t-1) \sin \omega t,$$

i.e. one can describe $y(t)$ by a time-invariant regression-like equation. Quite clearly, the same remark concerns a wider class of processes with deterministically (i.e. predictably) varying coefficients.

The so-called integrated time series models (IMA, IAR, ARIMA) describe another practically useful class of nonstationary stochastic signals represented by time-invariant models. The integrated models, popularized by Box and Jenkins [31], describe the evolution of the first-order or higher-order differences of an analyzed process, i.e. they constitute a special subclass of a class of stochastic processes with stationary increments discussed by Yaglom [198]. For example, the first-order IAR process is governed by

$$\nabla y(t) = a \nabla y(t-1) + v(t), \qquad |a| < 1 \tag{2.15}$$

where

$$\nabla y(t) = y(t) - y(t-1).$$

Note that (2.15) is nothing but an AR model applied to the first-order difference of $y(t)$.

Since reducible nonstationary processes can be parameterized in terms of time-invariant coefficients, their identification is analogous to identification of stationary processes; the more data is available, the more accurate the parameter estimates. Consider a periodically varying AR process (2.13). Using the method of least squares one can easily derive a consistent estimator of the coefficient a_0 in (2.14), yielding estimates that converge to a true parameter value as the number of data points increases:

$$\widehat{a}_0(t) \underset{a.s.}{\overset{t \mapsto \infty}{\longmapsto}} a_0.$$

This in turn means that one can consistently estimate the time-varying coefficient $a(t)$ in (2.13), namely

$$\widehat{a}(t) - a(t) \quad \underset{a.s.}{\overset{t \to \infty}{\longmapsto}} \quad 0,$$

where

$$\widehat{a}(t) = \widehat{a}_0(t) \sin \omega t. \tag{2.16}$$

According to (2.16), after sufficiently long observation, one can determine parameters of the process model with arbitrarily small errors.

The book is devoted to identifying irreducible nonstationary processes only; that is, processes whose coefficients are neither constant nor varying in a totally predictable way. No matter how large the data size and no matter the choice of identification technique, the parameter estimates derived for an irreducible nonstationary process never actually converge to their true values; instead they follow parameter changes with some finite accuracy. Estimators with this property are usually called *finite memory* estimators since they gradually 'forget' information coming from the remote past as the new data becomes available. For many types of estimators the notion of 'estimation memory' can be further quantified in terms of the adaptation gain, the effective size of the analysis window, etc. All finite memory parameter tracking algorithms have to compromise between the estimation accuracy (variance) and awareness to parameter changes (bias). The first component decreases while the second component increases with the growing filter memory. The trade-off between variance and bias, which should be made in accordance with the degree of nonstationarity of the analyzed process, is clearly one of the key issues in identifying processes with time-varying characteristics.

Remark

An interesting class of reducible nonstationary processes, frequently encountered in practical applications (especially in telecommunications), are the *cyclostationary* processes. The (wide-sense) cyclostationary processes exhibit periodicity in their mean, correlation or spectral descriptors. The well-established theory of cyclostationary processes can be found in the literature - see e.g. Gardner [54] for continuous-time processes and Giannakis [59] for discrete-time processes.

2.4   Measures of tracking ability

Quite obviously, the performance of all adaptive systems described in Section 1.5.6 depends on the accuracy of the process parameter estimates. When system coefficients are both unknown and time-varying, some objective measures are needed to quantify and compare the parameter tracking capabilities of different identification algorithms.

First of all, it is important to realize that obtaining a good mathematical model of an analyzed process is not a goal in itself. It is the effectiveness of the model-based predictive coding, channel equalization, spectrum estimation, control, etc. that governs the usefulness of the parametric approach to adaptive signal processing. Depending on the application at hand, the performance of an adaptive scheme can be measured in terms of *direct* quality indicators such as the coding efficiency (kilobytes per second), numerical complexity (number of MIPS) and subjective quality (MOS, DAM and

DRT scores) of speech coders [181], the probability of a binary symbol error during transmission over an equalized channel [160], spectral distortion measures, rate/cost efficiency, etc. Even though such direct application-oriented quality measures are the ultimate performance indicators (as they characterize true benefits that stem from application of adaptive procedures), they usually depend on parameter tracking errors in a complex and nonlinear fashion, hence they may be difficult to analyze.

Mean square prediction and tracking errors

The most frequently used *indirect* quality measure is the mean square one-step-ahead prediction error

$$E[\epsilon^2(t)] = E\left[\left(y(t) - \widehat{y}(t|t-1; \widehat{\theta}(t-1))\right)^2\right], \tag{2.17}$$

characterizing the 'predictive capabilities' of an adaptive scheme. There are at least two good reasons supporting this choice:

- It seems to be a meaningful quality measure in most of the prediction-oriented applications. The coding efficiency of an adaptive predictive coder depends on the average rate of the signal variance versus the prediction error variance; under Gaussian assumptions the probability of symbol error can be linked to the prediction error variance, etc.
- It is an *unprejudiced* quality measure in the sense that it allows one to confront a real (physical) process with its mathematical model without making any simplifying assumptions about the nature of the analyzed data.

In order to analyze the relation between the mean square prediction error and the parameter tracking error, consider a time-varying process governed by

$$y(t) = \varphi^T(t)\theta(t) + v(t).$$

Since the one-step-ahead certainty equivalence predictor for this system is

$$\widehat{y}(t|t-1; \widehat{\theta}(t-1)) = \varphi^T(t)\widehat{\theta}(t-1),$$

one gets

$$\epsilon(t) = \varphi^T(t)(\theta(t) - \widehat{\theta}(t-1)) + v(t). \tag{2.18}$$

The first term on the right-hand side of (2.18),

$$\epsilon_{ex}(t) = \varphi^T(t)(\theta(t) - \widehat{\theta}(t-1)),$$

is often called the excess prediction error and describes the component of the prediction error which can be solely attributed to nonideal parameter tracking capabilities of an adaptive system. Since the process noise $v(t)$ was assumed to be independent of past data $\Xi(t-1)$ (hence independent of $\varphi(t)$ and $\widehat{\theta}(t-1)$), the mean square prediction error can be expressed in the form

$$E[\epsilon^2(t)] = E[\epsilon_{ex}^2(t)] + \sigma_v^2. \tag{2.19}$$

Note that if the time-varying process coefficients were known exactly, one could use the following predictor:

$$\widehat{y}(t|t-1) = \varphi^T(t)\theta(t),$$

forcing the excess prediction error to zero and hence reducing the mean square prediction error to its minimum value of σ_v^2.

To obtain more insight into the structure of the excess mean square prediction error, we shall use some known results of the so-called averaging theory – an asymptotic theory of adaptive systems developed under the assumption that the process of adaptation is sufficiently slow [7], [12]. If the process parameters vary 'sufficiently slowly' and the parameter tracking algorithm has a 'sufficiently long' estimation memory (the notion of an estimation memory will be clarified in Chapter 4), the corresponding adaptive system can usually be regarded as a two-timescale process with fast and slow dynamics corresponding to the analyzed process loop and the adaptive loop, respectively. When the conditions of such an asymptotic analysis are fulfilled, variations of process parameters $\theta(t)$ and their estimates $\widehat{\theta}(t)$ are much slower than variations of the components of the regression vector $\varphi(t)$, leading to

$$\theta(t) - \widehat{\theta}(t-1) \cong \theta(t) - \widehat{\theta}(t) \tag{2.20}$$

and

$$\overline{\epsilon_{ex}^2(t)} \cong (\theta(t) - \widehat{\theta}(t))^T \,\overline{\varphi(t)\varphi^T(t)}\,(\theta(t) - \widehat{\theta}(t)), \tag{2.21}$$

where $\overline{(\cdot)}$ denotes local averaging.

If regression vectors $\varphi(t)$ form a (locally) wide-sense stationary process with a covariance matrix $\Phi_o = \mathrm{E}[\varphi(t)\varphi^T(t)]$, one obtains

$$\mathrm{E}[\epsilon_{ex}^2(t)] \cong \mathcal{P}[\widehat{\theta}(t)]$$

where

$$\mathcal{P}[\widehat{\theta}(t)] = \mathrm{E}\left[(\theta(t) - \widehat{\theta}(t))^T \Phi_o(\theta(t) - \widehat{\theta}(t))\right] = \mathrm{E}[\|\theta(t) - \widehat{\theta}(t)\|_{\Phi_o}^2]. \tag{2.22}$$

Note that the weighting matrix in (2.22) characterizes the covariance structure of the regression vector.

When Φ_o is similar to the identity matrix,

$$\Phi_o = \sigma_\varphi^2 I_n, \tag{2.23}$$

which means that different components of the regression vector are mutually uncorrelated, the right-hand side of (2.22) is proportional to

$$\mathcal{T}[\widehat{\theta}(t)] = \mathrm{E}[\|\theta(t) - \widehat{\theta}(t)\|^2], \tag{2.24}$$

i.e. it can be expressed in terms of the mean square value of the parameter tracking error yielded by the identification routine.

Remark

There are situations where condition (2.23) is fulfilled in a pretty natural way. For example, in adaptive equalization applications the regression vector is comprised of past measurements of the transmitted signal – the sequence of binary symbols obtained by encoding the data. The lack of intersymbol correlation is guaranteed by the choice of a particular training sequence (in the training part of the data packet) and by the decorrelation properties of binary encoders, often enhanced by data scrambling (in the information part of the data packet).

∎

The excess mean square prediction error $\mathcal{P}[\widehat{\theta}(t)]$ and the mean square parameter tracking error $\mathcal{T}[\widehat{\theta}(t)]$ are the most popular indirect measures of the tracking ability of adaptive systems. Both measures should be used with caution. In particular, one should remember three things

- Under typical operating conditions, e.g. for slowly varying processes, the excess mean square prediction error constitutes only a fraction of the total mean square prediction error. One should therefore realize that even if the contribution due to the excess prediction error is significantly reduced, the overall performance improvement will be much smaller.
- According to (2.22), convergence of parameter estimates in certain directions of the parameter space is more important than convergence in some other directions. Directions which are important from the adaptive prediction viewpoint are determined by the large-eigenvalue eigenvectors of the matrix Φ_o. When the eigenvalues of Φ_o are widely spread, the mean square parameter tracking error may be a very poor measure of the predictive ability of an identification algorithm.
- The approximation (2.22) is valid for a restricted class of dynamic processes, e.g. for time-varying FIR systems subject to a stationary excitation or for time-varying MA processes. For nonstationary AR(X) processes, the covariance matrix of the regression vector is parameter dependent and hence also time dependent, which significantly complicates analysis of the mean square prediction error.

Mean square interpolation and matching errors

One of the general conclusions from the qualitative analysis of different tracking algorithms presented in the next three chapters is that the expected trajectory of parameter estimates can be regarded, to a certain extent, as a delayed version of the true parameter trajectory. This means that $\widehat{\theta}(t)$ can be viewed as an estimate of $\theta(t - \tau_e)$ rather than of $\theta(t)$, where τ_e is an *estimation delay*. Whether or not the estimation delay is an undesirable phenomenon depends on the particular application at hand. Clearly, in adaptive prediction or control applications, delay (phase) distortions are as harmful as the scale (amplitude) distortions; after all we need a good estimate of $\theta(t)$ and *not* of $\theta(t - \tau_e)$ in order to predict or control the system's behavior at instant t. There are some other applications, however, where the constant delay of the estimated parameter trajectory with respect to the true parameter trajectory can be mitigated by incorporating into the adaptive loop a *decision delay* of τ_e samples.

An interesting application of this technique can be found in [162], [96], [35], where the so-called continuously adaptive receivers for fast time-varying mobile communication channels are proposed. Continuously adaptive receivers estimate channel characteristics in training and tracking modes. The idea is to use the preamble-based channel response estimates as a starting point for a continuous decision-oriented channel parameter tracking throughout the entire burst. The aim of such continuous adaptation is to provide the best possible estimates of channel coefficients (changing due to fading and Doppler effects) to the Viterbi processor. In principle, the Viterbi processor operates in concurrence with the channel identification processor. It turns out, however, that it is worthwhile to feed the parameter tracking algorithm with an appropriately *delayed* sequence of symbols obtained via a maximum likelihood analysis of the Viterbi trellis [35].

In cases like this the mean square interpolation error

$$E[(y(t - \tau_e) - \varphi^T(t - \tau_e)\widehat{\theta}(t))^2] \tag{2.25}$$

is a more adequate performance measure than the mean square prediction error. The excess mean square interpolation error can be similarly defined and, at least in certain cases, linked to the weighted mean square parameter fitting error

$$E[(\varphi^T(t - \tau_e)(\theta(t - \tau_e) - \widehat{\theta}(t)))^2] \cong \mathcal{I}[\widehat{\theta}(t)], \tag{2.26}$$

$$\mathcal{I}[\widehat{\theta}(t)] = E[\|\theta(t - \tau_e) - \widehat{\theta}(t)\|_{\Phi_o}^2]. \tag{2.27}$$

Finally, the mean square parameter matching error

$$\mathcal{M}[\widehat{\theta}(t)] = E[\|\theta(t - \tau_e) - \widehat{\theta}(t)\|^2] \tag{2.28}$$

can be used as an analog of the mean square parameter tracking error (2.24).

2.5   Prior knowledge in identification of nonstationary processes

The prior knowledge about the way system parameters vary with time depends on the application considered and may range from fairly specific information which can be incorporated into the process of parameter tracking to almost complete ignorance.

Parameter changes observed for physical processes can be broadly classified into four categories:

- Slow persistent changes (sometimes known as a parameter drift)
- Infrequent abrupt changes (occasional parameter jumps)
- Mixed-mode variations (slow drift plus occasional jumps)
- All other changes – call them *fast* parameter variations

Almost all successful applications of adaptive systems fall into the first three categories. When the process parameter changes become too fast, most adaptive systems fail in the sense that they cannot provide an acceptable behavior, unless one has a fairly specific knowledge of process nonstationarities and/or one can measure some auxiliary variables, or detect certain events, which either reflect or influence the way process parameters change with time. Quite obviously, there is no sharp division between 'slow' and 'fast' parameter variations in this classification. Section 2.6 sheds some light on this problem.

2.5.1   Events and auxiliary measurements

In many applications there are some events which cause or accompany abrupt changes of process characteristics: external stimuli or the patient's muscle activity may strongly influence the measured EEG signal [79]; the free/working day status is an important factor which should be taken into consideration when forecasting the power demand from consumers [29] or predicting the sales figures (the trading day effect) [95]; soon after filling the input bin, the flow of the raw material into the ore crusher is likely to drop temporarily [30], and so on.

In some cases the influence of certain measurable quantities on process coefficients is well understood and can be used to reduce variability of the model by means of proper parameter scaling. Consider the ship model, for example. When a ship makes small deviations from a straight-line course, it admits the following simplified linear model (characterizing the rudder-heading dynamics):

$$H(s) = \frac{b}{s(s+a)} ,$$

known as the Nomoto model. Both coefficients in the Nomoto model depend on the speed of the ship:

$$a \sim v^2, \qquad b \sim v.$$

By scaling the parameters of the autopilot according to speed, it is therefore possible to obtain an adaptive system which is almost insensitive to speed variations [10].

The regulation technique based on incorporating measurements of operating conditions of the plant is known as *gain scheduling* (since it was originally used to compensate for the process gain changes only). To apply gain scheduling, one should find suitable scheduling variables and this usually requires a good knowledge of the physics behind the controlled process.

The event-oriented analysis and gain scheduling are knowledge-based techniques providing a feedforward compensation of directly measurable process nonstationarities. Since both methods are inherently open-loop they can only partially reduce the 'degree of system nonstationarity'; it is usually impossible to completely eliminate the sources of process time variation.

2.5.2   Probabilistic models

Parameter drift

Adaptive equalization of fading telecommunication channels is a good example of an application where a pretty detailed probabilistic description of an underlying time-varying process is available.

Under stationary uncorrelated scattering and Rayleigh fading, the time-varying channel coefficients can be modeled as mutually uncorrelated stochastic processes with identical autocorrelation functions specified as

$$r_\theta(\tau) = \sigma_\theta^2 J_o(2\pi f_n \tau), \tag{2.29}$$

where σ_θ^2 denotes variance of a single channel coefficient, $J_o(\cdot)$ is the zeroth-order

Bessel function

$$J_o(x) = \frac{1}{2\pi} \int_{-\pi}^{\pi} e^{jx \sin \phi} d\phi,$$

and f_n denotes the normalized Doppler frequency – the ratio of the Doppler frequency f_d to the symbol rate f_s. Finally, the Doppler frequency f_d can be obtained from

$$f_d = \frac{v}{c} f_c,$$

where v is the vehicle's speed, c is the speed of light and f_c is the carrier frequency. The spectral density function corresponding to (2.29) – the Doppler power spectrum – can be expressed as

$$S_\theta(\omega) = \left\{ \begin{array}{ll} \frac{2\sigma_\theta^2}{\omega_n} \left[1 - \left(\frac{\omega}{\omega_n}\right)^2\right]^{-1/2} & \text{for} \quad |\omega| < \omega_n \\ 0 & \text{for} \quad |\omega| \geq \omega_n \end{array} \right. \tag{2.30}$$

where $\omega_n = 2\pi f_n$.

Note that the maximum Doppler shift f_d depends on the vehicle's speed, For example, when a vehicle moves at 10 m/s ($\sim$ 36 km/h) and the carrier frequency is $f_c = 900$ MHz (GSM), the maximum Doppler shift becomes $f_d = 30$ Hz. If a typical symbol rate of 24 kbd is applied, the maximum normalized Doppler shift is

$$f_n = \frac{f_d}{f_s} = \frac{30}{24 \times 10^3} = 1.25 \times 10^{-3}.$$

The nonrational model (2.29) and (2.30), known as the Jakes model [81], is often approximated by simple rational models such as the second-order autoregression [185]

$$\theta_i(t) = 2\alpha\theta_i(t-1) + \alpha^2\theta_i(t-2) + w_i(t), \quad i = 1, \ldots, n, \tag{2.31}$$

or the first-order autoregression [44]

$$\theta_i(t) = \alpha\theta_i(t-1) + w_i(t), \quad i = 1, \ldots, n, \tag{2.32}$$

where α, $0 < \alpha < 1$, is the scalar coefficient determining the position of the pole of the rational shaping filter and $\{w_i(t)\}$ denotes white (complex) Gaussian noise.

In order to use the simplified models of parameter variation, such as (2.31) or (2.32), one has to specify their design coefficients such as the scalar gain α (deciding upon the bandwidth of parameter changes) and the variance of driving noise σ_w^2. When α in (2.32) is set to one, the model of parameter changes becomes

$$\theta_i(t) = \theta_i(t-1) + w_i(t), \quad i = 1, \ldots, n, \tag{2.33}$$

which is known as the *random walk* model. According to (2.33) the evolution of each system coefficient is modeled as a random walk process with a mean square rate of change equal to σ_w^2. The random walk model was successfully used in several different applications such as channel equalization (Goddard [60], Falconer and Ljung [45]), estimation of a time-varying spectrum for EEG signals (Bohlin [29], Isaksson, Wennberg and Zetterberg [79]) or forecasting power consumption (Bohlin [29]).

Parameter jumps

There are several ways of describing jump parameter changes. If each system coefficient can take a finite number of prescribed values, evolution of $\theta(t)$ can be modeled as a Markov chain over a finite number of 'states' $\theta_1, \ldots, \theta_N$. The first-order homogeneous Markov chain can be characterized in terms of the $N \times N$ state transition probability matrix

$$S = [s(k|l)],$$

where

$$s(k|l) = P\left(\theta(t) = \theta_k | \theta(t-1) = \theta_l\right).$$

Estimation based on the Markov chain assumption is usually known as the hidden Markov model (HMM) approach - see e.g. [42] and the references therein. The HMM technique is effective only in cases where the number of states of the Markov chain is small (some gain scheduling applications fulfill this requirement).

If parameter jumps are not restricted to several known values, a more adequate model of parameter variation is

$$\theta_i(t) = \theta_i(t-1) + \delta_i(t)z_i(t), \quad i = 1, \ldots, n, \tag{2.34}$$

where

$$\delta_i(t) = \begin{cases} 1 & \text{with probability} \quad p_i \\ 0 & \text{with probability} \quad 1 - p_i \end{cases}$$

p_i, $p_i \ll 1$ denotes the (small) jump probability and $z_i(t)$ is a random variable with a prescribed probability density function.

Finally, when system parameters are subject to mixed-mode variations (slow drift plus occasional jumps) one can combine (2.33) and (2.34) into one equation [46]:

$$\theta_i(t) = \theta_i(t-1) + w_i(t) + \delta_i(t)z_i(t), \quad i = 1, \ldots, n, \tag{2.35}$$

describing a process with orthogonal increments (it is assumed that the processes $\{w_i(t)\}$ and $\{z_i(t)\}$ are mutually uncorrelated).

2.5.3 Deterministic models

Stochastic modeling of multipath communication channels is well motivated when time-varying path delays arise due to a large number of scatters. When the multipath effects are caused by a few strong reflectors and when the path delays change linearly with time due to the receiver/transmitter motion, the communication channel (such as a mobile radio channel or underwater acoustic channel) can be regarded as almost periodically time-varying. In cases like this, the time-varying impulse response of the channel $\{\theta_i(t), i = 1, \ldots, n_B\}$ can be approximated by a weighted sum of complex exponentials [108], [58], [186]:

$$\theta_i(t) \cong \sum_{j=1}^{k} c_{ij} e^{j\omega_i t}, \tag{2.36}$$

where k is the number of dominant reflectors and $\omega_i, i = 1, \ldots, k$ denote the corresponding Doppler shifts. Even though the angular frequencies ω_i are usually not known a priori, they can be easily estimated in the startup phase of equalization [186]. Consequently, in the case considered, the channel parameters can be approximated as linear combinations of *known* functions of time. The prior knowledge encapsulated in (2.36) allows one to design adaptive equalizers for rapidly varying communication channels.

2.6   Slowly varying systems and the concept of local stationarity

In a majority of cases our knowledge of the true mechanisms for process time variation is very limited and the available prior information can be summarized in statements like these: speech signals can be regarded as stationary in time intervals up to 10–20 ms long; the ore crushing plant can be treated as time invariant over periods not exceeding 1–2 hours. Nonstationary processes with such characteristics are usually referred to as locally stationary or quasi-stationary - see e.g. Silverman [171] and Snyder [177]. The time constants, such as those mentioned above, carry important information about the bandwidth of process time variation. We will demonstrate very soon how to use such knowledge for the purpose of tuning parameter tracking algorithms.

   All applications discussed in Section 1.5 were derived under the assumption that the underlying physical systems are stationary. The solutions obtained were then rewritten in terms of time-varying parameters to account for system nonstationarity. Even though not optimal in any sense, this approach usually yields satisfactory results provided that the analyzed processes change slowly with time. To put the word 'slowly' in the right perspective, consider a causal linear time-varying system characterized by the time-dependent impulse response $h(t,\tau)$ and obeying the uniform stability condition

$$\sum_{\tau=0}^{\infty} |h(t,\tau)| \leq c < \infty, \qquad \forall t. \tag{2.37}$$

Denote by τ_o a certain measure of the 'effective length' of $h(t,\tau)$ regarded as a function of τ and by $\omega_o = 2\pi f_o$ the 'effective spectral bandwidth' of $h(t,\tau)$ regarded as a function of t.

   Definition of the effective length τ_o can be based on the dominant mass criterion

$$\tau_o = \inf \overline{\tau} \ : \ \sum_{\tau=0}^{\overline{\tau}} |h(t,\tau)| \geq (1-\epsilon) \sum_{\tau=0}^{\infty} |h(t,\tau)|, \qquad \forall t, \tag{2.38}$$

where ϵ, $0 < \epsilon \ll 1$ is a small positive constant (e.g. 0.05 or 0.01). We note that the stability condition (2.37) alone does not guarantee there exists $\tau_o < \infty$ obeying (2.38).

   The 'effective bandwidth' of the process nonstationarity can be defined analogously. If, for example, $h(\cdot,\tau)$ can be regarded as a stationary stochastic process with a spectral density function $S_h(\tau,\omega)$ one can set

$$\omega_o = \inf \overline{\omega} \ : \ \int_0^{\overline{\omega}} S_h(\tau,\omega)\, d\omega \geq (1-\epsilon) \int_0^{\pi} S_h(\tau,\omega)\, d\omega, \qquad \forall \tau, \tag{2.39}$$

which insures that the dominant portion of the power of $r_h(t,\tau)$ for any given τ is contained within the spectral bandwidth $[-\omega_o, \omega_o]$.

Note that τ_o can be interpreted as the maximum normalized time constant of the system and $t_o = 1/f_o$ as the minimum normalized time constant of its nonstationarity. Both time constants defined above are *dimensionless*; to obtain quantities expressed in units of time one should multiply τ_o and t_o by the sampling period T_s. A system can be called locally stationary (or quasi-stationary) if one can find T such that

$$t_o \gg T \gg \tau_o. \tag{2.40}$$

The first inequality $(T \ll t_o)$ insures that the system can be regarded as (approximately) time invariant in time intervals of length T or shorter. The second inequality $(T \gg \tau_o)$ insures that, when subject to a stationary excitation, the system will reach (again approximately) its steady state in time intervals of length T, or longer. If the term 'much smaller' is interpreted as 'at least ten times smaller', condition (2.40) is equivalent to

$$\tau_o f_o \leq 0.01. \tag{2.41}$$

When the local stationarity conditions are met, the behavior of the nonstationary system (2.5) can be approximately analyzed, in sufficiently short time intervals, using the tools and concepts developed for stationary systems, including frequency domain concepts such as the instantaneous (frozen) frequency response

$$H(t, e^{-j\omega}) = \sum_{\tau=0}^{\infty} h(t, \tau) e^{-j\omega\tau}, \tag{2.42}$$

or instantaneous spectrum.

If the effective length of the system impulse response is replaced with the effective length of the absolutely summable autocorrelation function $r_y(t, \tau)$ of a nonstationary process $\{y(t)\}$,

$$\tau_o = \inf \bar{\tau} \; : \; \sum_{\tau=0}^{\bar{\tau}} |r_y(t, \tau)| \geq (1 - \epsilon) \sum_{\tau=0}^{\infty} |r_y(t, \tau)|, \qquad \forall t, \tag{2.43}$$

and if $\omega_o = 2\pi f_o$ is defined as an effective bandwidth of time variations of $r_y(t, \tau)$ (regarded as a function of t), inequality (2.41) becomes a condition of local stationarity for a nonstationary stochastic process.

Remark 1

Note that under (2.40) one can estimate the autocorrelation function of $y(t)$ using the local averaging technique

$$\hat{r}_y(t, \tau) = \frac{1}{T} \sum_{s=t+1-T/2}^{t+T/2} y(s) y(s - \tau). \tag{2.44}$$

The inequality $T \gg \tau_o \geq 1$ insures a small variance of the estimate and the inequality $T \ll t_o$ insures a small bias due to the 'time smearing' effects – see Gardner [53].

■

Remark 2

The concepts of 'locally stationary systems' and 'locally stationary stochastic processes' are closely related. Consider a Cramer–Wold representation of a nonstationary process (2.5). Vaguely speaking, to yield a locally stationary process at its output, the Cramer–Wold filter should vary sufficiently slowly with time. Unfortunately, in the general case (for an IIR filter) it is pretty difficult to derive quantitative results along these lines, i.e. to find conditions which when imposed on $h(t, \tau)$ guarantee local stationarity of $y(t)$. It is not difficult to show that condition (2.40) is not sufficient in this respect.

The analysis is much easier to handle for FIR filters. Note, for example, that if the impulse response of the noise forming filter obeys $h(t, \tau) = 0, \forall \tau \geq n_H, \forall t$, one obtains

$$r_y(t, \tau) = 0, \qquad \forall \tau \geq n_H, \forall t,$$

i.e. the finite length of the impulse response $h(t, \tau)$ imposes a trivial upper bound on the effective length of the autocorrelation of the resulting time-varying MA process.

Remark 3

Note that the qualitative statement 'the system can be regarded as time invariant in analysis intervals shorter than T_o seconds' allows one to formulate the following rough estimate of the normalized bandwidth of system nonstationarity:

$$f_o = \frac{1}{\tau_o} \ll \frac{T_s}{T_o},$$

i.e. $\tau_o \gg T_o/T_s$ where T_s denotes the length of the sampling interval.

2.7   Rate of process time variation

Quite obviously, the speed of system time variations puts limits on the performance of adaptive systems. If changes are too fast, all adaptive algorithms will fail to follow them satisfactorily, unless a very specific prior knowledge of system time variation is available.

2.7.1   Speed of variation and sampling frequency

When quantifying the speed of system variation one should remember that, as long as discrete-time models are considered, the corresponding measures should be normalized with respect to the length of the sampling interval T_s. The fact that the ore crushing plant remains 'nearly stationary' in 1 hour intervals does not mean its dynamics is easier to track than the dynamics of a vocal tract which exhibits stationary behavior in 20 ms periods. Since the sampling interval adopted in the first case was equal to 20 s [30], a 1 hour period of crusher operation is equivalent to 180 samples. A very similar estimate can be obtained for the second case: under 10 kHz sampling (adequate for speech signals) every 20 ms speech frame consists of 200 samples. When measured in numbers of samples rather than units of time, the (non)stationarity constants are approximately the same in both cases. One should realize that it is the bandwidth of a *continuous-time* system that really matters, so faster sampling does not really slow down the observed time variations. If the sampling frequency is increased twice, the effective length of the impulse response of the process or the effective length of its autocorrelation function, measured in samples, increase likewise. Moreover, to yield a comparable performance the adaptive filter designed to track changes in process dynamics should also be scaled up appropriately, namely the order of the model and the estimation memory of the parameter tracking algorithm should be increased by a factor of 2.

Even though oversampling does not really help one to cope with processes that vary too fast, changing the *timescale* of the analyzed phenomenon may be beneficial in certain applications. Consider sending data through a fading telecommunication channel. Increasing the symbol rate is an obvious way of decreasing the rate of channel variation observed during transmission of each fixed-length data frame; a higher rate means that symbols are encoded by shorter pulses, hence the time span of each frame becomes shorter. This in turn allows one to guarantee a pre-specified symbol error rate for higher Doppler frequencies, i.e. for greater speeds of mobile units – see [206] for more details.

2.7.2 Nonstationarity degree

Several authors have attempted to construct quantitative measures of the degree of system time variation.

The analytic approach

Consider a time-varying plant governed by

$$y(t) = \varphi^T(t)\theta(t) + v(t).$$

The measure of system nonstationarity proposed by Macchi takes the form

$$\kappa(t) = \sqrt{\frac{\mathrm{E}[(\Delta y(t))^2]}{\sigma_v^2}} \,, \qquad (2.45)$$

where

$$\Delta y(t) = \varphi^T(t)\Delta\theta(t)$$

and

$$\Delta\theta(t) = \theta(t) - \theta(t-1)$$

is the vector of parameter increments.

Denote by

$$\widehat{y}(t|t-1;\theta(t-1)) = \varphi^T(t)\theta(t-1)$$

the one-step-ahead prediction of $y(t)$ based on a true system model frozen at instant $t-1$. Since the corresponding prediction error may be written in the form

$$y(t) - \widehat{y}(t|t-1;\theta(t-1)) = \Delta y(t) + v(t),$$

the quantity $\Delta y(t)$ can be recognized as an excess prediction error due to system time variation. Note that the measure (2.45) is scale invariant, i.e. its value does not change if both sides of the system equation are multiplied by a factor $\beta > 1$ (upscaling) or $\beta < 1$ (downscaling).

According to Macchi [117] the system can be regarded as slowly time-varying if

$$E\left[(\Delta y(t))^2\right] \ll \sigma_v^2,$$

i.e. if the relationship

$$\kappa(t) \ll 1 \qquad (2.46)$$

holds 'most of the time'. This includes the situation when system parameters are subject to occasional large changes, provided that after each jump condition (2.46) is fulfilled long enough for adaptive filter to reach its steady-state behavior.

Note that if condition (2.46) is fulfilled, the mean square value of the error forced by system nonstationarity is only a fraction of the overall mean square prediction error. On the contrary, when condition (2.46) is not met, the system undergoes persistent and significant changes over each sampling interval. Almost no adaptive system will work satisfactorily in these circumstances unless the time-varying process description is reducible to the time-invariant description.

To get some insight into the structure of (2.45) consider a nonstationary FIR system

$$\varphi(t) = [\, u(t-1), \ldots, u(t-n)\,]^T$$

with parameters varying according to the random walk model

$$\theta(t) = \theta(t-1) + w(t), \qquad \text{cov}[w(t)] = W. \qquad (2.47)$$

Denote

$$E[\varphi(t)\varphi^T(t)] = \Phi(t).$$

Then

$$E\left[(\Delta y(t))^2\right] = E\left[(\varphi^T(t)w(t))^2\right] = \text{tr}\{\Phi(t)W\},$$

leading to

$$\kappa(t) = \sqrt{\frac{\text{tr}\{\Phi(t)W\}}{\sigma_v^2}}. \qquad (2.48)$$

Note that under a stationary excitation ($\Phi(t) = \Phi_o$) the index of system nonstationarity (2.48) is constant and depends on the mean square rate of change of system parameters (W) and the signal-to-noise ratio (Φ_o/σ_v^2).

Example 2.1

For a fading telecommunication channel one can set $\Phi_o = \sigma_u^2 I_n$ (uncorrelated input symbols) and $W = \sigma_w^2 I_n$ (mutually uncorrelated channel tap changes), which results in a simple expression

$$\kappa(t) = \frac{\sigma_u \sigma_w}{\sigma_v}\sqrt{n}. \qquad (2.49)$$

The simplified expression (2.49) shows another important feature of the index (2.45), its dependence on the order of an analyzed system.

For a time-varying AR process (and generally for all systems with time-varying poles) the analysis (2.45) becomes considerably more difficult for at least two reasons.

First, it is fairly difficult to come up with 'reasonable' models of AR parameter variation, e.g. models that would guarantee stability of the analyzed system. To meet the conservative stability constraint (2.37), $\theta(t)$ should be a process with *bounded variation*. Surely neither the random walk model (2.33) nor autoregressive models such as (2.31) or (2.32) fulfill this requirement.

The second difficulty arises from the fact that for AR systems the regression vector $\varphi(t)$ and its covariance matrix depend on the parameter vector $\theta(t)$. This means that whether parameter variation can be regarded as slow or fast depends not only on the rate of parameter changes (i.e. on their relative changes) but also on which region of the parameter space is spanned. Even very small changes taking place in regions close to the stability boundary (defined for a time-invariant system) may significantly influence characteristics of the analyzed process.

The empirical approach

An interesting approach to quantification of the degree of system nonstationarity was pursued by Bohlin [29]. Assuming that parameter variation of a nonstationary autoregressive process can be approximated by the random walk model (2.47) with

$$W = \sigma_w^2 I_n,$$

Bohlin proposed the following nonstationarity index:

$$\eta(t) = \sqrt{\frac{E[y^2(t)]\sigma_w^2}{\sigma_v^2}}, \tag{2.50}$$

which resembles the quantity introduced by Macchi. Actually, observe that for a slowly varying autoregressive process

$$\varphi(t) = [y(t-1), \dots, y(t-n)]^T,$$

$$\text{tr}\{\Phi(t)\} = \sum_{i=1}^{n} E[y^2(t-i)] \cong nE[y^2(t)],$$

hence in the case considered one formally has

$$\kappa(t) \cong \sqrt{\frac{nE[y^2(t)]\sigma_w^2}{\sigma_v^2}} \cong \sqrt{n}\eta(t). \tag{2.51}$$

The main problem with this 'derivation' lies in the fact that the analysis was carried out for AR processes with parameters varying according to the random walk model, i.e. processes that are generically unstable ($E[y^2(t)] \mapsto \infty$). However, the way Bohlin handles (2.50) is somewhat different. Assuming that the model (2.47) is adequate for the analyzed data record, he finds the maximum likelihood estimate of $\eta(t)$ by estimating the root mean square value of the signal and the standard deviations of both noise components. Even though the estimation procedure is not straightforward (iterative search must be performed), the proposed approach is interesting and allows one to check the degree of process nonstationarity in an empirical way.

Note that $\eta(t)$ is defined in terms of *directly measurable* quantities, while the nonstationarity index proposed by Macchi is not. Note also that it is not crucial whether or not the system parameter changes actually obey (2.47); fitting the random walk model is regarded here as a means for evaluating the degree of system variability rather like fitting a straight line can be used to quantify the degree of collinearity of measurement points.

Applying his measure of process nonstationarity to EEG signals, Bohlin distinguished two major categories: samples that can be regarded as 'slowly changing' ($\hat{\eta} < 0.0005$) and those which can be classified as 'fast changing' ($0.0005 < \hat{\eta} < 0.005$).

2.8 Assumptions

Generally speaking, one can identify five groups of assumptions under which the properties of various parameter tracking algorithms can be analyzed.

2.8.1 Dependence among regressors

The first group consists of restrictions imposed on the asymptotic rate of dependence among regression vectors $\varphi(t)$; in statistical literature such restrictions are often called *mixing conditions*. Here are some typical assumptions falling into this category (listed in order of increasing generality):

B1.1 The process $\{\varphi(t)\}$ is a sequence of independent and identically distributed (i.i.d.) random vectors.

B1.2 The process $\{\varphi(t)\}$ is a sequence of m-dependent random vectors, i.e. such that for all t the sequences $\{\varphi(i), i \leq t\}$ and $\{\varphi(i), i \geq t + m\}$ are independent.

B1.3 The process $\{\varphi(t)\}$ is ϕ-mixing, i.e. there exists a nonincreasing function $\phi(n)$, called the mixing rate, obeying

$$\phi(n) \in [0,1], \qquad \forall n,$$

$$\phi(n) \xrightarrow[\infty]{t \to \infty} 0,$$

such that

$$|P(B|A) - P(B)| \leq \phi(n), \tag{2.52}$$

for every events $A \in \mathcal{F}_{-\infty}^t$ and $B \in \mathcal{F}_s^\infty$ where $\mathcal{F}_s^t$, $-\infty \leq s \leq t \leq \infty$ denotes the σ-algebra generated by $\{\varphi(i), s \leq i \leq t\}$ and $n = s - t$.

B1.4 The process $\{\varphi(t)\}$ is generated from a ϕ-mixing sequence $\{e(t)\}$ by a stable linear filter

$$\varphi(t) = \sum_{i=-\infty}^{\infty} A_i e(t - i) + \delta(t), \tag{2.53}$$

where $\sum_{i=-\infty}^{\infty} \|A_i\| < \infty$ and $\{\delta(t)\}$ is a bounded deterministic process.

The so-called independence assumption (B1.1) is obviously too restrictive from the practical viewpoint (although it may be fulfilled in certain adaptive array processing applications) but it played an important role in early theoretical studies and led to many insightful results which were later generalized to weaker mixing conditions.

The m-dependence assumption (B1.2) is more realistic and met, for example, by FIR systems excited by a white noise input signal. Actually, note that when $\varphi(t) = [u(t-1), \ldots, u(t-n)]^T$ and $\{u(t)\}$ is a white noise sequence the process $\{\varphi(t)\}$ is n-dependent. A white noise excitation is typical of channel equalization applications.

The ϕ-mixing concept is a pretty standard way of characterizing weakly dependent random processes. Condition (B1.3) is satisfied, for example, by sequences generated from bounded white noises via a stable linear filter as well as by certain classes of stationary Markov chains.

Finally, assumption (B1.4) admits a large class of processes other than ϕ-mixing, since the sequence $\{e(t)\}$ in (2.53) may be unbounded.

2.8.2 Dependence between system variables

Assumptions belonging to this group impose restrictions on the mutual dependence between $\varphi(t)$, $v(t)$ and $\theta(t)$. Most of the known results rest on one of the following conditions:

B2.1 The processes $\{\varphi(t)\}$, $\{v(t)\}$ and $\{\theta(t)\}$ are mutually uncorrelated.

B2.2 The processes $\{\varphi(t)\}$, $\{v(t)\}$ and $\{\theta(t)\}$ are mutually independent.

Both assumptions admit FIR systems but evidently rule out IIR systems. This is due to the dependence of the regression vector on past values of the parameter vector observed for AR(X)/ARMA(X) processes. Note that under Gaussian assumptions conditions (B2.1) and (B2.2) are equivalent.

2.8.3 Persistence of excitation

The third group of assumptions are conditions that ensure system identifiability; to guarantee that the entire parameter vector can be estimated from input/output data the sequence $\{\varphi(t)\}$ must span the entire regressor space, i.e. it should be sufficiently rich. Such conditions on *persistence of excitation* usually have the following forms:

B3.1 There exist constants $s > 0$ and $c > 0$ such that

$$\sum_{t=k+1}^{k+s} \varphi(t)\varphi^T(t) \geq cI_n, \qquad \forall k.$$

B3.2 There exist constants $\gamma > 0$, $\delta > 0$ and $x_0 > 0$ such that for all $x : x_0 > x > 0$

$$\sup_{||\beta||=1} P\left((\beta^T \varphi(t))^2 < x\right) \leq \gamma x^\delta, \qquad \forall t.$$

B3.3 There exist constants $s > 0$ and $c > 0$ such that

$$\mathrm{E}\left[\sum_{t=k+1}^{k+s} \frac{\varphi(t)\varphi^T(t)}{1 + \varphi^T(t)\varphi(t)} \bigg| \mathcal{F}^k_{-\infty}\right] \geq cI_n, \qquad \forall k.$$

B3.4 There exist constants $s > 0$ and $c > 0$ such that

$$\mathrm{E}\left[\sum_{t=k+1}^{k+s} \varphi(t)\varphi^T(t)\right] \geq cI_n, \qquad \forall k.$$

The most restrictive, deterministic persistence of excitation condition (B3.1) was the cornerstone of early stability studies of finite memory adaptive filters. Note, however, that for stochastic regressors the inequality in (B3.1) is very difficult to fulfill unless they have a persistently exciting deterministic component.

The stochastic persistence of excitation condition (B3.2) is significantly weaker than (B3.1) as it only requires that the probability of the components of $\varphi(t)$ becoming 'almost' linearly dependent should tend to zero at a sufficiently fast rate. It is known that distribution of any random vector φ can be factored (Lebesgue decomposition theorem [183]) as

$$F(\varphi) = \mu_c F_c(\varphi) + \mu_d F_d(\varphi) + \mu_s F_s(\varphi),$$

where F_c, F_d and F_s are continuous, discrete and singular (supported on hyperplanes) distributions, respectively, and μ_c, μ_d, μ_s are nonnegative constants such that

$$\mu_c + \mu_d + \mu_s = 1.$$

What (B3.2) effectively says is that the distribution of $\varphi(t)$ should be free of discrete and singular components. Additionally, it rules out 'almost discrete' and 'almost singular' components in the continuous distribution. We note that (B3.2) admits a very large class of continuous distributions, e.g. all distributions characterized by bounded probability density functions (Gaussian, uniform, etc.).

Identifiability conditions (B3.3) and (B3.4) are the least restrictive ones. Condition (B3.3) is met, for example, by stationary s-dependent sequences $\{\varphi(t)\}$ with positive definite covariance matrices and finite fourth-order moments [66]. Finally, condition (B3.4) is fulfilled by a large variety of stochastic processes, e.g. by all sequences $\{\varphi(t)\}$ characterized by positive definite covariance matrices.

2.8.4   Boundedness of system variables

Assumptions belonging to this group impose various restrictions on probability density functions of system variables:

B4.1 There exists a constant $0 < c < \infty$ such that

$$\|\varphi(t)\| < c, \qquad \forall t.$$

B4.2 For all p, $0 < p \leq P$,

$$E[||\varphi(t)||^p] < \infty, \qquad \forall t.$$

B4.3 There exist constants $\alpha > 0$ and $0 < c < \infty$ such that

$$E\left[\exp\{||e(t)||^2\}\right] < c, \qquad \forall t,$$

where $e(t)$ denotes the 'source' from which $\varphi(t)$ is formed; see (B1.4).

The boundedness assumption (B4.1) entails

$$p(\varphi(t)) = 0, \qquad \forall \varphi(t) : ||\varphi(t)|| \geq c,$$

which rules out unbounded distributions such as the Gaussian distribution. Assumption (B4.2) requires existence of all moments of $\varphi(t)$ up to the order $p = P$; in many analyses P is restricted to 4. Assumption (B4.3) imposes a restriction on the decay rate for the tails of a noise distribution (it is met, for example, by Gaussian variables).

2.8.5 Variation of system parameters

Assumptions belonging to this group specify the maximum range or the maximum speed of parameter variation:

B5.1 There exists a constant $0 < c < \infty$ such that

$$||\theta(t)|| < c, \qquad \forall t.$$

B5.2 There exists a constant $0 < c < \infty$ such that

$$E[||\theta(t)||^2] < c, \qquad \forall t.$$

B5.3 There exists a constant $0 < c < \infty$ such that

$$||\theta(t) - \theta(t-1)|| < c, \qquad \forall t.$$

B5.4 There exists a constant $0 < c < \infty$ such that

$$E[||\theta(t) - \theta(t-1)||^2] < c, \qquad \forall t.$$

Assumption (B5.1) implies boundedness of the parameter trajectory in a deterministic sense while (B5.2) implies its boundedness in the mean square sense. Since assumptions (B5.3) and (B5.4) require boundedness of parameter increments only, they relax conditions imposed by (B5.1) and (B5.2), respectively. Note that assumption (B5.2) is met if the system parameters vary according to the random walk model which yields unbounded parameter trajectories.

2.9   About computer simulations

When presenting different methods of time-varying identification one is tempted to illustrate analytical results using real data processing examples. This presents no problem in principle, since the audio signals alone provide an endless stream of nonstationary data. But all such real data experiments have an obvious drawback: since the true variation of process coefficients is not known, it is not possible to compare trajectories of parameter estimates with the true parameter trajectory.

For this reason it was decided to confine all identification results to simulated nonstationary processes. Four examples are referred to throughout the book: two second-order time-varying FIR systems governed by

$$y(t) = a_1(t)u(t-1) + a_2(t)u(t-2) + v(t) \tag{2.54}$$

and two second-order AR processes

$$y(t) = a_1(t)y(t-1) + a_2(t)y(t-2) + v(t), \tag{2.55}$$

where $u(t)$ denotes an exogenous Gaussian input signal (first-order AR process $u(t) = 0.8u(t-1) + e(t)$, $\sigma_e^2 = 1$) and $v(t)$ denotes white Gaussian noise ($\sigma_v^2 = 0.01$).

The two variants of FIR and AR processes correspond to two types of parameter changes: jump changes (Figure 2.1a) and continuous changes (Figure 2.1b). In the second case one of the process coefficients exhibits a drift-like behavior, and the second coefficient is subject to smooth deterministic (sinusoidal) changes.

For the sake of comparison identical parameter changes were simulated for FIR and AR processes. The following abbreviations will be used throughout the book

    FIR1   —   an FIR system with jump parameter changes
    FIR2   —   an FIR system with continuous parameter changes
    AR1   —   an AR system with jump parameter changes
    AR2   —   an AR system with continuous parameter changes

Typical plots of the signals generated in all four cases are shown in Figure 2.2 (for FIR systems) and in Figure 2.3 (for AR processes).

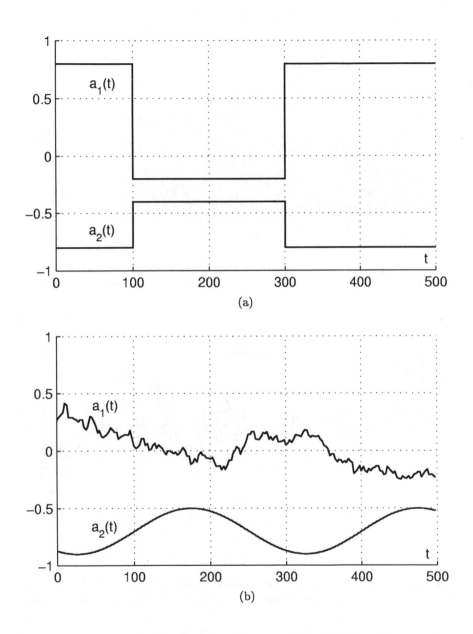

Figure 2.1 Two types of parameter changes used for simulation of nonstationary processes: jump changes (a) and continuous changes (b).

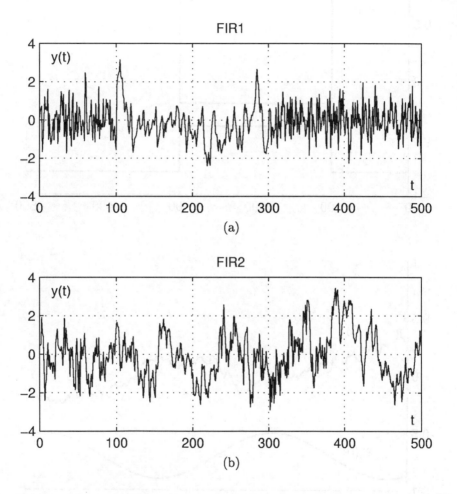

Figure 2.2 A single realization of the signal observed at the output of a nonstationary FIR system with jump parameter changes (a) and continuous parameter changes (b).

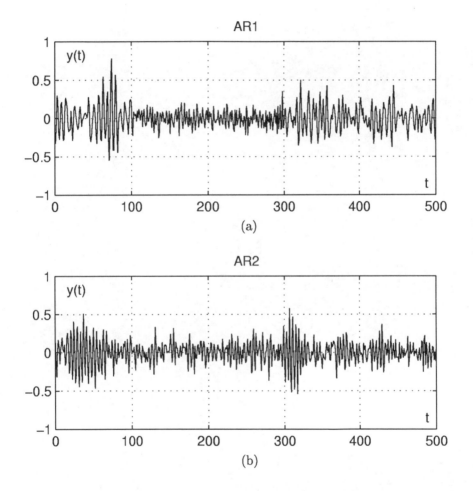

Figure 2.3 A single realization of a nonstationary AR process with jump parameter changes (a) and continuous parameter changes (b).

3

Process Segmentation

In the segmentation approach the nonstationary process is broken into a sequence of constant-length or variable-length 'frames':

$$\begin{aligned}
\Xi_i(N_i) &= \{\xi(t), t \in T_i\}, \\
T_i &= [t_{i-1} + 1, \ldots, t_i], \\
N_i &= t_i - t_{i-1},
\end{aligned}$$

each of which is analyzed separately and represented by a time-invariant model.

3.1 Nonadaptive segmentation

When the nonadaptive variant of the segmentation is used, all frames have identical lengths, i.e. the segmentation is data independent. Excellent examples of this technique are speech analysis/synthesis (in LPC systems the analysis window is typically 20–30 ms long and contains 200–300 samples taken at the telephone rate of 10 kHz) and equalization of mobile radio channels (a typical frame is 4–8 ms long and covers 100–200 complex symbols transmitted at the rate of 24 kbd).

Consider a frame of length N (for the sake of clarity the frame number will be temporarily suppressed):

$$\begin{aligned}
\Xi(N) &= \{\xi(t), t \in T\}, \\
T &= [1, \ldots, N].
\end{aligned}$$

Assuming that the process is stationary in the time interval T,

$$\theta(t) = \theta_o, \qquad t \in T,$$

the estimate of model parameters can be obtained using the method of least squares, i.e. by minimizing the sum of squared deviations of the system output $y(t)$ from its predicted output $\varphi^T(t)\theta$:

$$\widehat{\theta}(T) = \arg \min_\theta J(\theta; \Xi(0), \Xi(N)), \tag{3.1}$$

where

$$J(\theta; \Xi(0), \Xi(N)) = \sum_{t=1}^N (y(t) - \varphi^T(t)\theta)^2, \tag{3.2}$$

and

$$\Xi(0) = \{y(1), \ldots, y(1 - n_A), u(1), \ldots, y(1 - n_B)\}$$

denotes the set of initial conditions, usually derived from the preceding frame.

Setting to zero the gradient of J evaluated with respect to θ,

$$\nabla_\theta J(\theta)|_{\theta=\widehat{\theta}(T)} = 2\left(\sum_{t=1}^N \varphi(t)\varphi^T(t)\right)\widehat{\theta}(T) - 2\sum_{t=1}^N y(t)\varphi(t) = 0, \qquad (3.3)$$

and assuming that the regression matrix is positive definite and hence invertible,

$$R(T) = \sum_{t=1}^N \varphi(t)\varphi^T(t) > 0, \qquad (3.4)$$

one obtains the following explicit formula:

$$\widehat{\theta}(T) = \left(\sum_{t=1}^N \varphi(t)\varphi^T(t)\right)^{-1}\left(\sum_{t=1}^N y(t)\varphi(t)\right) = R^{-1}(T)s(T), \qquad (3.5)$$

where

$$s(T) = \sum_{t=1}^N y(t)\varphi(t). \qquad (3.6)$$

The estimate of the input noise variance $\sigma_v^2 = \text{var}[v(t)], \forall\ t \in T$ can be obtained by averaging the squared residual errors

$$\widehat{\sigma}^2(T) = \frac{1}{N}\sum_{t=1}^N (y(t) - \varphi^T(t)\widehat{\theta}(T))^2. \qquad (3.7)$$

We note that if the noise distribution is Gaussian,

$$v(t) \sim \mathcal{N}(0, \sigma_v^2), \qquad t \in T,$$

equations (3.5) and (3.7) coincide with the maximum likelihood estimators, which guarantees their efficiency provided that the analyzed process is truly stationary in the time interval T.

3.1.1   Conditions of identifiability

Note that

$$\nabla_{\theta\theta^T}^2 J(\theta)|_{\theta=\widehat{\theta}(T)} = 2R(T),$$

hence if condition (3.4) is fulfilled then the derived estimate actually minimizes the quadratic loss function (3.2) (the matrix of second-order derivatives of J is positive definite).

Since for any $x \in \mathcal{R}^n$ it holds that

$$x^T R(T)x = \sum_{t=1}^N x^T\varphi(t)\varphi^T(t)x = \sum_{t=1}^N (x^T\varphi(t))^2 \geq 0, \qquad (3.8)$$

the regression matrix must be nonnegative definite. However, it is clear from (3.8) that if there exists a nonzero vector x such that

$$x^T \varphi(t) = 0, \qquad \forall\, t \in T,$$

i.e. if the components of the regression vector $\varphi(t)$ are linearly dependent, the quadratic form (3.8) will be semidefinite and the regression matrix will be singular, making the corresponding identification problem ill-posed. In cases like this the system is referred to as *parameter nonidentifiable*. Nonidentifiability is not a property of the analyzed system but rather a consequence of defective experimental conditions under which system identification is carried out. There are two well-known sources of system nonidentifiability:

- Application of input signals that are not persistently exciting of sufficiently high order (so that the input variable $u(t-1)$ can be expressed as a linear combination of the remaining u-components of $\varphi(t)$).
- Existence of linear feedback between the input and output variables (so that the input variable $u(t-1)$ can be expressed as a linear combination of the remaining components of $\varphi(t)$).

Example 3.1

Suppose that the input signal has the form

$$u(t) = (-1)^t.$$

Observe that

$$u(t) = -u(t-1), \quad \forall t,$$

which means that if more than one u-term is present in the regression vector $\varphi(t)$ (i.e. if more than one input coefficient is estimated) the system becomes nonidentifiable. Such signals can be termed persistently exciting of order 1.

Sinusoidal signals

$$u(t) = \sin \omega t$$

are persistently exciting of order 2 since

$$u(t) = au(t-1) - u(t-2), \quad \forall t,$$

where $a = 2\cos\omega$. White noise is an example of a signal which is persistently exciting of infinite order.

Example 3.2

Suppose that a first-order plant

$$y(t) = ay(t-1) + bu(t) + v(t) \tag{3.9}$$

is identified under a linear feedback (proportional control)

$$u(t) = ky(t).$$

The regression vector associated with (3.9) has the form $\varphi(t) = [y(t), u(t)]^T$. Since

$$[y(t), u(t)] \begin{bmatrix} 1 \\ -k \end{bmatrix} = 0, \quad \forall t$$

it is clearly not possible to estimate both unknown coefficients in (3.9) unless the experimental conditions are changed.

3.1.2   Recursive least squares algorithm

The closed-form expression (3.5) requires inversion of the regression matrix $R(T)$. An alternative way of evaluating the LS estimate is via the recursive formula known as the recursive least squares (RLS) algorithm.

Derivation of the RLS algorithm

To derive an RLS algorithm we will need the following matrix inversion lemma (which can be easily proven by direct multiplication).

Lemma

Consider an $n \times n$ matrix A, an $n \times 1$ vector x and a scalar μ. Provided that all inverses below exist

$$\left(A + \mu x x^T\right)^{-1} = A^{-1} - \mu \frac{A^{-1} x x^T A^{-1}}{1 + \mu x^T A^{-1} x} . \tag{3.10}$$

∎

To start derivation of the RLS algorithm denote

$$R(t) \quad = \quad \sum_{i=1}^{t} \varphi(i) \varphi^T(i),$$

$$s(t) \quad = \quad \sum_{i=1}^{t} y(i) \varphi(i)$$

and observe that

$$\begin{aligned} R(t) &= R(t-1) + \varphi(t) \varphi^T(t), \\ s(t) &= s(t-1) + y(t) \varphi(t). \end{aligned} \tag{3.11}$$

Hence

$$\begin{aligned} \widehat{\theta}(t) &= R^{-1}(t) s(t) = R^{-1}(t) \left[R(t-1) \widehat{\theta}(t-1) + y(t) \varphi(t) \right] \\ &= R^{-1}(t) \left[R(t) \widehat{\theta}(t-1) + \varphi(t) \left(y(t) - \varphi^T(t) \widehat{\theta}(t-1) \right) \right] \tag{3.12} \\ &= \widehat{\theta}(t-1) + P(t) \varphi(t) \epsilon(t), \end{aligned}$$

where $P(t)$ denotes the inverse of the regression matrix $R(t)$,

$$P(t) = R^{-1}(t),$$

and

$$\epsilon(t) = y(t) - \varphi^T(t)\widehat{\theta}(t-1)$$

is the one-step-ahead prediction error.

Applying the matrix inversion lemma to the right-hand side of

$$P(t) = R^{-1}(t) = \left(P(t-1) + \varphi(t)\varphi^T(t)\right)^{-1},$$

one obtains $(A := P^{-1}, x := \varphi, \mu := 1)$ that

$$P(t) = P(t-1) - \frac{P(t-1)\varphi(t)\varphi^T(t)P(t-1)}{1 + \varphi^T(t)P(t-1)\varphi(t)},$$

which completes the derivation. The set of RLS recursions can be summarized as follows:

$$
\begin{aligned}
\widehat{\theta}(t) &= \widehat{\theta}(t-1) + K(t)\epsilon(t), \\
\epsilon(t) &= y(t) - \varphi^T(t)\widehat{\theta}(t-1), \\
K(t) &= \frac{P(t-1)\varphi(t)}{1 + \varphi^T(t)P(t-1)\varphi(t)}, \\
P(t) &= P(t-1) - \frac{P(t-1)\varphi(t)\varphi^T(t)P(t-1)}{1 + \varphi^T(t)P(t-1)\varphi(t)}.
\end{aligned}
\tag{3.13}
$$

Choice of initial conditions

The influence of initial conditions $\widehat{\theta}_o = \widehat{\theta}(0)$ and $P_o = P(0)$ on the RLS estimate can be easily analyzed after noting that the one-shot (nonrecursive) expression equivalent to (3.13) is

$$\widehat{\theta}(t) = [P_o^{-1} + P^{-1}(t)]^{-1}[P_o^{-1}\widehat{\theta}_o + s(t)].$$

Therefore by setting $\widehat{\theta}_o = 0$ and

$$P_o^{-1} \ll P^{-1}(T) = \sum_{t=1}^{N} \varphi(t)\varphi^T(t), \tag{3.14}$$

one can make the RLS estimate $\widehat{\theta}(N)$ practically coincide with the 'true' LS estimate (3.5) without a need to invert the regression matrix even once. The matrix P_o is frequently adopted in a diagonal form

$$P_o = \delta I_n,$$

where δ denotes a 'sufficiently large' positive constant. Note that selection of the appropriate value of δ, i.e. the value that obeys (3.14), requires some prior knowledge on the statistics of the regression vector. If the sequence $\{\varphi(t)\}$ is stationary then

$$\mathrm{E}\left[\sum_{t=1}^{N} \varphi(t)\varphi^T(t)\right] = N\Phi_o,$$

hence the recommended value of δ is

$$\delta \gg \frac{1}{N\lambda_{\min}(\Phi_o)} \,, \qquad (3.15)$$

where $\lambda_{\min}(\Phi_o)$ denotes the minimum eigenvalue of $\Phi_o = \mathrm{cov}[\varphi(t)]$. For FIR systems subject to white noise excitation, $\Phi_o = \sigma_u^2 I_n$, hence (3.15) is equivalent to

$$\delta \gg \frac{1}{N\sigma_u^2} \,.$$

Remark

Condition (3.15) guarantees that when the endpoint N of the segment is reached, the influence of P_o on $\widehat{\theta}(N)$ will be negligible. By enforcing

$$\delta \gg \frac{1}{k\lambda_{\min}(\Phi_o)} \,, \qquad n \leq k < N,$$

one can reduce the settling time of the RLS algorithm to k samples (which may be important in all applications that require online identification). The minimum settling time is equal to n, as for $t < n$ the matrix $R(t)$ is not full rank and hence it is noninvertible.

∎

When the segmentation approach is used, the proper initialization of the RLS algorithm may be difficult for the first segment only. For the second and all succeeding segments the natural choice is

$$P_i(t_{i-1}) \gg P_{i-1}(t_{i-1}), \qquad (3.16)$$

where $P_{i-1}(t_{i-1})$ is the terminal value computed for the previous data segment. To fulfill (3.16) it is sufficient to set

$$P_i(t_{i-1}) = \delta P_{i-1}(t_{i-1}),$$

where $\delta \geq 10$.

Computation of the residual sum of squares

We will show that the estimate of the input noise variance σ_v^2 is also recursively computable. Denote by $q(t)$ the residual sum of squares

$$q(t) = \sum_{s=1}^{t} \left(y(s) - \varphi^T(s)\widehat{\theta}(t)\right)^2. \qquad (3.17)$$

Hence, according to (3.7),

$$\widehat{\sigma}^2(T) = \frac{q(T)}{N} \,.$$

First of all, observe that

$$q(t) = \sum_{s=1}^{t} y^2(s) - 2\widehat{\theta}^T(t) \sum_{s=1}^{t} y(s)\varphi(s) + \widehat{\theta}^T(t) \left(\sum_{s=1}^{t} \varphi(s)\varphi^T(s) \right) \widehat{\theta}(t)$$

$$= \sum_{s=1}^{t} y^2(s) - 2\widehat{\theta}^T(t)s(t) + \widehat{\theta}^T(t)P^{-1}(t)\widehat{\theta}(t).$$

Since $\widehat{\theta}(t) = P(t)s(t)$ one gets

$$q(t) = \sum_{s=1}^{t} y^2(s) - \widehat{\theta}^T(t)P^{-1}(t)\widehat{\theta}(t).$$

Hence

$$q(t) - q(t-1) = y^2(t) + \widehat{\theta}^T(t-1)P^{-1}(t-1)\widehat{\theta}(t-1) - \widehat{\theta}^T(t)P^{-1}(t)\widehat{\theta}(t). \qquad (3.18)$$

Using

$$P^{-1}(t) = P^{-1}(t-1) + \varphi(t)\varphi^T(t),$$
$$\widehat{\theta}(t) = \widehat{\theta}(t-1) + P(t)\varphi(t)\epsilon(t),$$

one can show that

$$\widehat{\theta}^T(t)P^{-1}(t)\widehat{\theta}(t) = \widehat{\theta}^T(t-1)P^{-1}(t-1)\widehat{\theta}(t-1)$$
$$+ \widehat{\theta}^T(t-1)\varphi(t)\varphi^T(t)\widehat{\theta}(t-1)$$
$$+ \varphi^T(t)P(t)\varphi(t)\epsilon^2(t) + 2\varphi^T(t)\widehat{\theta}(t-1)\epsilon(t). \qquad (3.19)$$

Finally, after combining (3.18) with (3.19) and noting that

$$\varphi^T(t)\widehat{\theta}(t-1) = y(t) - \epsilon(t),$$

one arrives at

$$q(t) = q(t-1) + [1 - \varphi^T(t)P(t)\varphi(t)]\epsilon^2(t)$$
$$= q(t-1) + [1 - K^T(t)\varphi(t)]\epsilon^2(t)$$
$$= q(t-1) + \frac{\epsilon^2(t)}{1 + \varphi^T(t)P(t-1)\varphi(t)}, \qquad (3.20)$$

which completes our derivation.

When $t = n$ the least squares estimate $\widehat{\theta}(n)$ is the solution of the set of n linear equations

$$y(t) = \varphi^T(t)\theta, \quad t = 1, \ldots, n$$

which means that

$$y(t) - \varphi^T(t)\widehat{\theta}(n) = 0, \quad t = 1, \ldots, n$$

i.e. the corresponding residual errors are all equal to zero. This means that the recursion (3.20) should be started at instant $t = n + 1$ with the initial condition

$$q(n) = \sum_{t=1}^{n} (y(t) - \varphi^T(t)\widehat{\theta}(n))^2 = 0.$$

3.2   Adaptive segmentation

So far we have been assuming that the breakpoints which mark beginnings of consecutive analysis frames T_i are either known a priori or are chosen in a somewhat arbitrary manner. In systems exploiting adaptive process segmentation the number and localization of breakpoints depend on the analyzed data. The problem of dividing a nonstationary process into a number of stationary epochs can be formulated as follows: find a partitioning of a given set of observations $\Xi(N) = \{\xi(t), t \in T\}, T = [1, N]$ into a sequence of blocks

$$\Xi_1(N_1), \ldots, \Xi_k(N_k), \qquad N_1, \ldots, N_k > n,$$

$$\begin{aligned} \Xi_i(N_i) \cap \Xi_j(N_j) &= O, \qquad \forall\, i \neq j, \\ \cup_{i=1}^{k} \Xi_i(N_i) &= \Xi(N), \end{aligned}$$

which yields the best piecewise constant model of $\Xi(N)$ of the form

$$\left(\widehat{\theta}(t), \widehat{\sigma}^2(t)\right) = \begin{cases} \left(\widehat{\theta}(T_1), \widehat{\sigma}^2(T_1)\right) & \forall\, t \in T_1 \\ \quad\vdots & \quad\vdots \\ \left(\widehat{\theta}(T_k), \widehat{\sigma}^2(T_k)\right) & \forall\, t \in T_k \end{cases}$$

3.2.1   Segmentation based on the Akaike criterion

An interesting solution to the above problem, based on Akaike's information criterion, was proposed by Ozaki and Tong [154] and further refined by Kitagawa and Akaike [91].

Akaike's information criterion (AIC) was originally developed as a tool for model order selection [2]. The AIC statistics

$$\begin{aligned} \text{AIC} \;=\; & -2(\text{maximized log likehood of the model}) \\ & +\; 2(\text{number of parameters estimated in the model}), \qquad (3.21) \end{aligned}$$

can be interpreted as an approximately unbiased estimate of the negentropy (expected log likelihood) of the model. According to Akaike, when several competing models are considered – derived from the same data set using the method of maximum likelihood, but differing in complexity – one should select the model which minimizes the AIC index. If the compared models are 'nested' (i.e. the lower-order models are 'contained' in higer-order models) the first term on the right-hand side of (3.21) decreases with system order while the second term increases; the second term may be interpreted as a penalty for model complexity. As a result, the model selected by Akaike's criterion is always a compromise between parsimony of the adopted parameterization and the achievable degree of fit to experimental data.

Assuming the process is stationary in the time interval T and assuming the noise $\{v(t)\}$ is Gaussian,

$$v(t) \sim \mathcal{N}(0, \sigma_v^2), \qquad \forall\, t \in T,$$

then the (conditional) likelihood function of the data $\Xi(N)$ is

$$p(\Xi(N)|\Xi(0), \theta, \sigma_v^2) = (2\pi\sigma_v^2)^{-N/2} \exp\left\{-\frac{\sum_{t=1}^{N}(y(t) - \varphi^T(t)\theta)^2}{2\sigma_v^2}\right\}, \qquad (3.22)$$

and its natural logarithm, called the log likelihood function

$$L(\theta, \sigma_v^2) = \ln p(\Xi(N)|\Xi(0), \theta, \sigma_v^2),$$

attain its maximum with respect to θ and σ_v^2 for the maximum likelihood (ML) estimates $\widehat{\theta}(T)$ and $\widehat{\sigma}^2(T)$, which are identical with the LS estimates (3.5) and (3.7), respectively. Hence the maximized log likelihood of the model becomes

$$L(\widehat{\theta}(T), \widehat{\sigma}^2(T)) = -\frac{N}{2} \ln 2\pi\widehat{\sigma}^2(T) - \frac{N}{2}. \qquad (3.23)$$

When the data partitioning is imposed, the maximized log likelihood associated with the ith data segment becomes

$$L(\widehat{\theta}(T_i), \widehat{\sigma}^2(T_i)) = -\frac{N_i}{2} \ln 2\pi\widehat{\sigma}^2(T_i) - \frac{N_i}{2}, \qquad (3.24)$$

leading to the following expression for the AIC statistics:

$$\text{AIC}_i = N_i \ln 2\pi\widehat{\sigma}^2(T_i) + N_i + 2(n + 1). \qquad (3.25)$$

The AIC index for the entire data set can be evaluated by adding components obtained for each block:

$$\text{AIC} = \sum_{i=1}^{k} \text{AIC}_i = \sum_{i=1}^{k} N_i(\ln 2\pi\widehat{\sigma}^2(T_i) + 1) + 2k(n + 1). \qquad (3.26)$$

The best partitioning $\{k, t_1, \ldots, t_k\}$ of the data $\Xi(N)$ can be defined as the one that minimizes the AIC statistics (3.26). Since for every partitioning

$$\sum_{i=1}^{k} N_i = N = \text{const},$$

the comparison can be based on the modified statistics

$$\text{AIC}^\star = \sum_{i=1}^{k} N_i \ln \widehat{\sigma}^2(T_i) + 2k(n + 1). \qquad (3.27)$$

For the fixed model order n, the average value of the residual sum of squares tends to decrease with the block size, approaching zero for $N_i = n$. Therefore the value of the first, variance-dependent term on the right-hand side of (3.27) will generally decrease when $\Xi(N)$ is partitioned to a large number of short segments. On the other hand, the value of the second, debiasing term of (3.27) linearly increases with the number of data blocks, which favors parsimonious parameterizations.

Segmentation based on the aggregate/split approach

Since there are a vast number of different segmentation variants associated with various selections of the number and localization of breakpoints, the problem of adaptive segmentation is computationally prohibitive in its general formulation. To reduce the number of degrees of freedom consider the following procedure based on aggregation of fixed-length segments $(t_i - t_{i-1} = N, \forall i)$. Each time a new potential breakpoint is reached, two hypotheses are considered:

H_0 (homogeneity) The new data segment is statistically consistent with the preceding $k, k \geq 1$ segments, i.e. the new observations should be used to update and improve the existing mathematical model of the process.

H_1 (inhomogeneity) The new data segment is statistically inconsistent with the preceding segment(s), i.e. the new observations should be used to build a new mathematical model of the process.

To verify the homogeneity hypothesis, identification of two competitive models is carried out in each time interval T_i: an *aggregated* model $\hat{\theta}(T_i, \ldots, T_{i-k})$ combining both the 'old' and the 'new' data into one parameter estimate and a *split* model $\hat{\theta}(T_i)$ based on the 'new' data only (Figure 3.1). Denote by $\hat{\sigma}^2(T_i, \ldots, T_{i-k})$ and $\hat{\sigma}^2(T_i)$

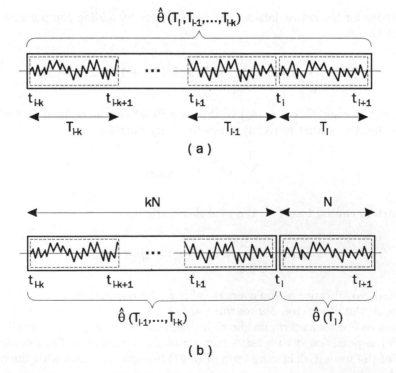

Figure 3.1   The aggregated model (a) versus the split model (b).

the residual noise variance estimates obtained for the aggregated and split models, respectively. When the end of the ith frame is reached, the AIC statistic is evaluated for both models:

$$
\begin{aligned}
\text{AIC}_0^\star &= (k+1)N \ln \widehat{\sigma}^2(T_i, \dots, T_{i-k}) + 2(n+1), \\
\text{AIC}_1^\star &= kN \ln \widehat{\sigma}^2(T_{i-1}, \dots, T_{i-k}) + N \ln \widehat{\sigma}^2(T_i) + 4(n+1),
\end{aligned}
\qquad (3.28)
$$

and the following segmentation rule is applied:

$$\text{if } \text{AIC}_0^\star < \text{AIC}_1^\star:$$

$$\widehat{\theta}(t) = \widehat{\theta}(T_i, \dots, T_{i-k}), \quad \forall t \in T_i \cup \dots \cup T_{i-k},$$

$$\text{if } \text{AIC}_0^\star \geq \text{AIC}_1^\star:$$

$$
\widehat{\theta}(t) = \begin{cases} \widehat{\theta}(T_{i-1}, \dots, T_{i-k}) & \forall t \in T_{i-1} \cup \dots \cup T_{i-k} \\ \widehat{\theta}(T_i) & \forall t \in T_i \end{cases}
$$

or equivalently

$$
\text{AIC}_1^\star - \text{AIC}_0^\star = \begin{matrix} H_0 \\ \gtrless \\ H_1 \end{matrix} \ 0.
\qquad (3.29)
$$

Segmentation based on the global/local approach

The main drawback of the preceding scheme is that the positions of the breakpoints are fixed a priori. A more flexible solution can be obtained by considering a growing-memory *global* model, combining all samples starting from the preceding breakpoint, and a fixed-memory *local* model combining a prescribed number of past samples only. Suppose that the local analysis interval covers $N > n$ samples and that N is the minimum allowable segment length (Figure 3.2). Furthermore, denote by $\widehat{\theta}(t, N)$ the LS estimate based on the N most recent process observations $\xi(t), \dots, \xi(t - N + 1)$, and let $\widehat{\sigma}^2(t, N)$ be the corresponding estimate of the noise variance.

Then for every t, starting from $t_{i-1} + 2N$, the following two hypotheses are considered:

H_0 (homogeneity) The last N process observations are statistically consistent with the preceding $N_i = t - t_{i-1} - N$ observations, i.e. a single global model $\widehat{\theta}(t, N_i + N)$ can be used to represent the data from the analysis interval $[t_{i-1} + 1, t]$.

H_1 (inhomogeneity) The last N process observations are statistically inconsistent with the preceding $N_i = t - t_{i-1} - N$ observations, i.e. the analysis interval should be divided into two subintervals: subinterval $[t_{i-1} + 1, t_i]$, $t_i = t - N$, represented by the model $\widehat{\theta}(t - N, N_i)$, and subinterval $[t_i + 1, t]$, represented by the model $\widehat{\theta}(t, N)$.

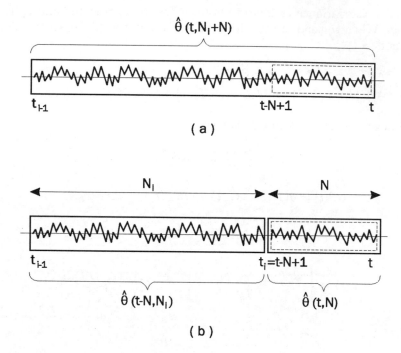

Figure 3.2   The global model (a) versus the local model (b).

The decision rule of form (3.29) can be based on comparing of the corresponding AIC statistic in a similar way as before:

$$\begin{aligned}
\text{AIC}_0^\star &= (N + N_i) \ln \widehat{\sigma}^2(t, N_i + N) + 2(n + 1), \\
\text{AIC}_1^\star &= N_i \ln \widehat{\sigma}^2(t - N, N_i) + N \ln \widehat{\sigma}^2(t, N) + 4(n + 1)
\end{aligned} \qquad (3.30)$$

Then the new rule reads as follows:

$$\text{if } \text{AIC}_0^\star < \text{AIC}_1^\star:$$

$$\widehat{\theta}(s) = \widehat{\theta}(t, N_i + N), \quad \forall s \in [t_{i-1} + 1, t],$$

$$\text{if } \text{AIC}_0^\star \geq \text{AIC}_1^\star:$$

$$\widehat{\theta}(s) = \begin{cases} \widehat{\theta}(t - N, N_i) & \forall s \in [t_{i-1} + 1, t - N] \\ \widehat{\theta}(t, N) & \forall s \in [t - N + 1, t] \end{cases}$$

Each time a new breakpoint $t_i = t - N$ is introduced ($\text{AIC}_0^\star \geq \text{AIC}_1^\star$) a new global model is initialized, identical with the local model, and the search for the next potential breakpoint is continued.

Computational aspects

The parameter estimates for the aggregated/split models can be updated recursively using the RLS algorithm (3.13), provided that the following initial conditions are imposed:

$$\widehat{\theta}_a(t_{i-1}) = \widehat{\theta}(t_{i-1})$$
$$P_a(t_{i-1}) = P(t_{i-1})$$

(which insures continuation of the estimation process) and

$$\widehat{\theta}_s(t_{i-1}) = 0$$
$$P_s(t_{i-1}) = 10P(t_{i-1})$$

(which effectively resets the estimation algorithm at instant t_i), where $\widehat{\theta}(t_{i-1})$, equal to $\widehat{\theta}(T_{i-1}, \ldots, T_{i-k})$ or $\widehat{\theta}(T_{i-1})$, is the estimate that won the competition at the previous stage and $P(t_{i-1})$ denotes the corresponding covariance matrix updated by the RLS filter.

The estimates of the residual noise variances, needed to evaluate the AIC statistics, are also recursively computable. Denote by $q(T_i, \ldots, T_{i-k})$ and $q(T_i)$ the residual sums of squares corresponding to the aggregated and split models, respectively. Both quantities can be updated using formula (3.20) provided the initial conditions are set to

$$q_a(t_{i-1}) = q(t_{i-1})$$

and

$$q_s(t_{i-1} + n) = 0,$$

where $q(t_{i-1})$ denotes the residual sum of squares obtained for the winning model, equal to $q(T_{i-1}, \ldots, T_{i-k})$ or $q(T_{i-1})$. The actual variance estimates can be obtained from

$$\widehat{\sigma}^2(T_i, \ldots, T_{i-k}) = \frac{q(T_i, \ldots, T_{i-k})}{(k+1)N},$$
$$\widehat{\sigma}^2(T_i) = \frac{q(T_i)}{N}.$$

The recursive algorithm for computation of parameters of the local model (in the global/local approach),

$$\widehat{\theta}(t, N) = \left(\sum_{i=t-N+1}^{t} \varphi(i)\varphi^T(i) \right)^{-1} \left(\sum_{i=t-N+1}^{t} \varphi(i)y(i) \right) = R^{-1}(t, N)s(t, N)$$

is a two-step algorithm. In the first step the analysis interval is extended by one sample to include a new observation $\xi(t)$. Since

$$R(t, N+1) = R(t-1, N) + \varphi(t)\varphi^T(t),$$
$$s(t, N+1) = s(t-1, N) + \varphi(t)y(t)$$

this step can be carried out by using the standard RLS algorithm:

$$\hat{\theta}(t, N+1) = \hat{\theta}(t-1, N) + K(t, N+1)\epsilon(t),$$

$$\epsilon(t) = y(t) - \varphi^T(t)\hat{\theta}(t-1, N),$$

$$K(t, N+1) = \frac{P(t-1, N)\varphi(t)}{1 + \varphi^T(t)P(t-1, N)\varphi(t)},$$

$$P(t, N+1) = P(t-1, N) - \frac{P(t-1, N)\varphi(t)\varphi^T(t)P(t-1, N)}{1 + \varphi^T(t)P(t-1, N)\varphi(t)}. \qquad (3.31)$$

In the second step the analysis window is contracted by one sample to remove dependence of the parameter estimates on the observation $\xi(t - N)$ collected $N + 1$ instants away. Note that

$$R(t, N) = R(t, N+1) - \varphi(t-N)\varphi^T(t-N),$$

$$s(t, N) = s(t, N+1) - \varphi(t-N)y(t-N).$$

Applying the matrix inversion lemma to

$$P(t, N) = R^{-1}(t, N) = \left(P^{-1}(t, N+1) - \varphi(t-N)\varphi^T(t-N)\right)^{-1},$$

one obtains $(A := P^{-1}, x := \varphi, \mu := -1)$ that

$$P(t, N) = P(t, N+1) + \frac{P(t, N+1)\varphi(t-N)\varphi^T(t-N)P(t, N+1)}{1 - \varphi^T(t-N)P(t, N+1)\varphi(t-N)},$$

leading to the following recursions:

$$\hat{\theta}(t, N) = \hat{\theta}(t, N+1) - K(t, N)\eta(t),$$

$$K(t, N) = \frac{P(t, N+1)\varphi(t-N)}{1 - \varphi^T(t-N)P(t, N+1)\varphi(t-N)},$$

$$\eta(t) = y(t-N) - \varphi^T(t-N)\hat{\theta}(t, N+1),$$

$$P(t, N) = P(t, N+1) + \frac{P(t, N+1)\varphi(t-N)\varphi^T(t-N)P(t, N+1)}{1 - \varphi^T(t-N)P(t, N+1)\varphi(t-N)}. \quad (3.32)$$

Note that, when combined together, equations (3.31) and (3.32) constitute a recursive algorithm for computation of the local estimate $\hat{\theta}(t, N)$.

The recursive algorithm for computation of $\hat{\sigma}^2(t, N)$ can be derived analogously. Denote

$$q(t, N) = \sum_{i=t-N+1}^{t} \left(y(i) - \varphi^T(i)\hat{\theta}(t, N)\right)^2$$

and observe that

$$\hat{\sigma}^2(t, N) = \frac{q(t, N)}{N}.$$

The first part of the algorithm, based on window expansion, is identical with the one derived in Section 3.1.2:

$$q(t, N+1) = q(t-1, N) + \left(1 - K^T(t, N+1)\varphi(t)\right)\epsilon^2(t)$$

$$= q(t-1, N) + \frac{\epsilon^2(t)}{1 + \varphi^T(t)P(t-1, N)\varphi(t)}, \qquad (3.33)$$

while the second part, based on window contraction, can be shown to obey

$$
\begin{aligned}
q(t, N) &= q(t, N+1) - \left(1 + K^T(t, N)\varphi(t - N)\right) \eta^2(t) \\
&= q(t, N+1) - \frac{\eta^2(t)}{1 - \varphi^T(t - N)P(t, N+1)\varphi(t - N)} .
\end{aligned}
\tag{3.34}
$$

Remark

Both local estimation algorithms derived above are reminiscent of the following two-step recursive algorithm:

$$
\begin{aligned}
p(t, N+1) &= p(t-1, N) + x(t), \\
p(t, N) &= p(t, N+1) - x(t - N).
\end{aligned}
\tag{3.35}
$$

for evaluation of the local sum of N signal samples

$$
p(t, N) = \sum_{i=0}^{N-1} x(t - i).
\tag{3.36}
$$

It is worth noticing that when computations are carried out in finite precision arithmetics, the quantity $p(t, N)$ computed recursively via (3.35) will tend to diverge from the analogous quantity evaluated by means of direct summation (3.36). Divergence is a direct consequence of *rounding* performed by digital computers. Due to finite wordlength effects, operations such as addition or multiplication are irreversible in a sense that in general

$$
[[x + y]_R - y]_R \neq x
$$

and

$$
[[x \cdot y]_R : y]_R \neq x,
$$

where $[\cdot]_R$ denotes the rounding operation.

To keep divergence under control, one should reinforce the recursive algorithm (3.35) from time to time by direct computation of (3.36). Quite obviously the same remark concerns both local estimation algorithms (3.31), (3.32) and (3.33), (3.34).

3.2.2 *Segmentation based on the generalized likelihood ratio test*

The generalized likelihood ratio (GLR) test proposed by Willsky and Jones [194], [195] is a tool for detecting additive changes in linear dynamic systems described by state space equations. The method can be used to choose between two hypotheses:

H_0 (homogeneity) All data observed in the analysis interval T are statistically consistent i.e. they can be represented by a single mathematical model.

H_1 (inhomogeneity) Due to data inconsistency the analysis interval T, of length N, should be divided into two subintervals T_1 (of length N_1) and T_2 (of length N_2) represented by two different mathematical models.

The decision is based on comparing the maximized likelihoods of the corresponding models:

$$\eta(\Xi(N)) = 2\ln \frac{\max_{\theta(T),\sigma^2(T)} p(\Xi(N)|\Xi(0),H_0)}{\max_{\theta(T_1),\sigma^2(T_1),\theta(T_2),\sigma^2(T_2)} p(\Xi(N)|\Xi(0),H_1)}$$

$$= 2 \max_{\theta(T),\sigma^2(T)} \ln p(\Xi(N)|\Xi(0),H_0)$$

$$- 2 \max_{\theta(T_1),\sigma^2(T_1),\theta(T_2),\sigma^2(T_2)} \ln p(\Xi(N)|\Xi(0),H_1) \underset{H_1}{\overset{H_0}{\gtrless}} \eta_0, \qquad (3.37)$$

where η_0 is a threshold value chosen so as to provide a reasonable trade-off between two kinds of detection errors: false alarms (detection of nonexistent system changes) and missed alarms (failure to detect true system changes).

Under the null hypothesis and Gaussian assumptions about $\{v(t)\}$, the maximized conditional log likelihood function of the entire data set is

$$\max_{\theta(T),\sigma^2(T)} \ln p(\Xi(N)|\Xi(0),H_0) = \ln p(\Xi(N)|\Xi(0),\widehat{\theta}(T),\widehat{\sigma}^2(T))$$

$$= -\frac{N}{2}\ln 2\pi\widehat{\sigma}^2(T) - \frac{N}{2}, \qquad (3.38)$$

where $\widehat{\theta}(T), \widehat{\sigma}^2(T)$ are the (conditional) ML estimates of θ and σ_v^2, identical with (3.5) and (3.7), respectively.

Under the alternative hypothesis,

$$\max_{\theta(T_1),\sigma^2(T_1),\theta(T_2),\sigma^2(T_2)} \ln p(\Xi(N)|\Xi(0),H_1)$$

$$= \ln p(\Xi(N)|\Xi(0),\widehat{\theta}(T_1),\widehat{\sigma}^2(T_1),\widehat{\theta}(T_2),\widehat{\sigma}^2(T_2))$$

$$= -\frac{N_1}{2}\ln 2\pi\widehat{\sigma}^2(T_1) - \frac{N_2}{2}\ln 2\pi\widehat{\sigma}^2(T_2) - \frac{N}{2}, \qquad (3.39)$$

where $\widehat{\theta}(T_1), \widehat{\sigma}^2(T_1)$ and $\widehat{\theta}(T_2), \widehat{\sigma}^2(T_2)$ are the corresponding local ML estimates.

Replacing the full likelihood functions in (3.37) with conditional likelihoods and taking advantage of (3.38) and (3.39), one arrives at the following decision rule for segmentation of an ARX process:

$$N_1\ln\widehat{\sigma}^2(T_1) + N_2\ln\widehat{\sigma}^2(T_2) - N\ln\widehat{\sigma}^2(T) \underset{H_1}{\overset{H_0}{\gtrless}} \eta_0. \qquad (3.40)$$

Since the classical GLR test was designed to choose between two hypotheses only, it cannot be used to resolve the general unconstrained segmentation problem described in Section 3.2. It can be used, however, for sequential segmentation replacing the AIC-based decision rules in the aggregate/split and global/local segmentation procedures.

Remark

Note that when only two partitioning variants are considered at a time – a global model versus two local models – the rule based on the Akaike criterion is identical with (3.40) provided that the decision threshold η_0 is set to $\eta_0 = 2(n + 1)$.

3.3 Extension to ARMAX processes

3.3.1 Iterative estimation algorithms

Since the noise components of the regression vector associated with the ARMAX model,

$$\varphi(t) = [y(t-1), \ldots, y(t-n_A), u(t-1), \ldots, u(t-n_B), v(t-1), \ldots, v(t-n_C)]^T,$$

are not directly observable, estimation of the corresponding model coefficients is less straightforward than in the ARX case.

Consider a finite-length data segment

$$\Xi(N) = \{\xi(t), t \in T\},$$
$$T = [1, \ldots, N].$$

Assuming that the process is stationary in the time interval T, the prediction error measure of fit, analogous to (3.2), corresponding to a hypothetical parameter vector $\theta \in D_I$ where

$$D_I = \{\theta : \text{all zeros of } C(q^{-1}, \theta) = 1 + \sum_{i=1}^{n_C} c_i q^{-i}$$

lie inside the unit circle in the complex plane$\}$

can be defined in the following *implicit* form:

$$J(\theta; \Xi(0), \Xi(N)) = \sum_{t=1}^{N} v^2(t, \theta), \qquad (3.41)$$

where

$$v(t, \theta) = y(t) - \varphi^T(t, \theta)\theta, \qquad (3.42)$$

and $\varphi(t, \theta)$ is the so-called pseudolinear regression vector

$$\varphi(t, \theta) = [y(t-1), \ldots, y(t-n_A), u(t-1), \ldots, u(t-n_B), v(t-1, \theta), \ldots, v(t-n_C, \theta)]^T.$$

Note that, unlike the ARX case, the components of $\varphi(t, \theta)$ have to be computed recursively. The invertibility condition $\theta \in D_I$ must be fulfilled in order to guarantee stability of the inverse filtering scheme (3.42). For notational convenience let

$$J_N(\theta) = \frac{1}{N} J(\theta; \Xi(0), \Xi(N)).$$

The minimum prediction error (PE) estimate of the unknown vector of system parameters,

$$\hat{\theta}(T) = \arg\min_{\theta \in D_I} J(\theta; \Xi(0), \Xi(N)) = \arg\min_{\theta \in D_I} J_N(\theta), \qquad (3.43)$$

can be searched iteratively using the following Newton–Raphson scheme:

$$\widehat{\theta}^{(i)}(T) = \widehat{\theta}^{(i-1)}(T) - \gamma \left[\nabla^2_{\theta\theta^T} J_N(\widehat{\theta}^{(i-1)}(T)) \right] \left[\nabla_\theta J_N(\widehat{\theta}^{(i-1)}(T)) \right], \tag{3.44}$$

where $\widehat{\theta}^{(i)}(T)$ denotes the ith iteration point in the search and γ is the positive gain coefficient, usually set in the range $[0.5, 1]$. To complete each step of the iterative search procedure, one should evaluate the gradient

$$\nabla_\theta J_N(\theta) = \frac{1}{N} \sum_{t=1}^{N} v(t,\theta) \nabla_\theta v(t,\theta) \tag{3.45}$$

and the Hessian

$$\nabla^2_{\theta\theta^T} J_N(\theta) = \frac{1}{N} \sum_{t=1}^{N} \nabla_\theta v(t,\theta) \nabla_\theta^T v(t,\theta) + \frac{1}{N} \sum_{t=1}^{N} v(t,\theta) \nabla^2_{\theta\theta^T} v(t,\theta). \tag{3.46}$$

Both quantities can be computed using the following recursions for $\nabla_\theta v(t,\theta)$ and $\nabla^2_{\theta\theta^T} v(t,\theta)$:

$$\nabla_\theta v(t,\theta) = -\sum_{i=1}^{n_C} c_i \nabla_\theta v(t-i,\theta) - \varphi(t,\theta), \tag{3.47}$$

$$\nabla^2_{\theta\theta^T} v(t,\theta) = -\sum_{i=1}^{n_C} c_i \nabla^2_{\theta\theta^T} v(t-i,\theta) - G(t,\theta) - G^T(t,\theta), \tag{3.48}$$

$$G(t,\theta) = [0,\ldots,0, \nabla_\theta v(t-1,\theta),\ldots,\nabla_\theta v(t-n_C,\theta)],$$

which can easily be obtained by differentiating both sides of (3.42). Note that computation of (3.45) and (3.46) requires iteration of (3.47) and (3.48) from $t=1$ to $t=N$ for each value of i in (3.44).

Example 3.3

Consider the first-order ARMAX plant

$$y(t) = ay(t-1) + bu(t-1) + v(t) + cv(t-1) = \varphi^T(t)\theta + v(t).$$

In the case considered $\theta = [a,b,c]^T$ and

$$v(t,\theta) = y(t) - ay(t-1) - bu(t-1) - cv(t-1,\theta). \tag{3.49}$$

Differentiating both sides of (3.49) with respect to a, b and c, one obtains

$$\frac{\partial v(t,\theta)}{\partial a} = -c\frac{\partial v(t-1,\theta)}{\partial a} - y(t-1),$$
$$\frac{\partial v(t,\theta)}{\partial b} = -c\frac{\partial v(t-1,\theta)}{\partial b} - u(t-1),$$
$$\frac{\partial v(t,\theta)}{\partial c} = -c\frac{\partial v(t-1,\theta)}{\partial c} - v(t-1,\theta). \tag{3.50}$$

Note that (3.50) can be written in the more compact form

$$\nabla_\theta v(t, \theta) = -c \, \nabla_\theta v(t-1, \theta) - \varphi(t-1, \theta), \qquad (3.51)$$

where

$$\varphi(t, \theta) = [\, y(t-1), u(t-1), v(t-1, \theta) \,]^T.$$

A similar evaluation gives a matrix of second-order derivatives of $v(t, \theta)$ (i.e. the first-order derivatives of (3.51)), resulting in

$$\nabla^2_{\theta\theta^T} v(t, \theta) = -c \, \nabla^2_{\theta\theta^T} v(t-1, \theta) - G(t, \theta) - G^T(t, \theta), \qquad (3.52)$$

where

$$G(t, \theta) = \begin{bmatrix} 0 & 0 & \frac{\partial v(t-1,\theta)}{\partial a} \\ 0 & 0 & \frac{\partial v(t-1,\theta)}{\partial b} \\ 0 & 0 & \frac{\partial v(t-1,\theta)}{\partial c} \end{bmatrix}.$$

∎

Since the quantities $\nabla_\theta v(t, \theta)$ and $\nabla^2_{\theta\theta^T} v(t, \theta)$ evaluated according to (3.47) and (3.48) will diverge if $\theta \notin D_{\mathrm{I}}$, at the end of each cycle of the iterative search procedure the invertibility of the obtained approximation must be checked by examining the stability of the polynomial

$$C(q^{-1}, \widehat{\theta}^{(i)}(T)) = 1 + \sum_{j=1}^{n_C} \widehat{c}_j^{(i)} q^{-j}.$$

If the invertibility condition is not fulfilled, the estimate obtained from (3.44) must be projected back into the invertibility subspace D_{I}, e.g. by ignoring the corresponding c-updates – the portion of $\widehat{\theta}^{(i)}(T)$ which corresponds to the moving average part of the model is retained from $\widehat{\theta}^{(i-1)}(T)$. Some more advanced invertibility enforcement schemes can also be applied [113], [61].

The iteration is stopped if the absolute values of the differences between the estimates obtained in successive iterations are less than a small prescribed value and/or if the norm of the gradient $\nabla_\theta J_N(\theta)$ is less than a small prespecified value.

Remark 1

As with all gradient-based procedures, the iterative algorithm (3.44) is sensitive to the choice of the starting value for parameter estimates $\widehat{\theta}^{(0)}(T)$; convergence is guaranteed only when the starting values are not too far from the true values (for small values of N the error surface can be shown to have multiple local minima). The least squares estimates are often recommended as a good starting point for an iterative search based on (3.44).

Remark 2

The computational complexity of the iterative algorithm (3.44) can be significantly reduced by neglecting the second term on the right-hand side of (3.46). The resulting Gauss–Newton scheme, which allows one to avoid computation of the second-order derivatives of $v(t, \theta)$, has the same asymptotic convergence rate as the original Newton–Raphson algorithm [179], [87].

3.3.2 *Recursive estimation algorithms*

When it comes to identification of time-varying systems, the practical usefulness of the iterative algorithms, such as (3.44), is limited to nonadaptive segmentation. To obtain a computationally efficient solution of the adaptive segmentation problem, one needs a recursive estimation algorithm.

When the vector of parameter estimates is close to its true value one can set

$$v(t, \widehat{\theta}(t-1)) \cong \epsilon(t),$$

where

$$\epsilon(t) = y(t) - \widetilde{\varphi}^T(t)\widehat{\theta}(t-1)$$

denotes the one-step-ahead prediction error and

$$\widetilde{\varphi}(t) = [\, y(t-1), \ldots, y(t-n_A), u(t-1), \ldots, u(t-n_B), \epsilon(t-1), \ldots, \epsilon(t-n_C)\,]^T$$

is an approximation of the pseudolinear regression vector $\varphi(t, \theta)$.

Based on similar arguments one can set

$$\nabla_\theta v(t, \widehat{\theta}(t-1)) \cong \psi(t),$$

where

$$\psi(t) = [\, y^F(t-1), \ldots, y^F(t-n_A), u^F(t-1), \ldots, u^F(t-n_B), \epsilon^F(t-1), \ldots, \epsilon^F(t-n_C)\,]^T,$$

and $y^F(t), u^F(t), \epsilon^F(t)$ denote the filtered components of the regression vector $\widetilde{\varphi}(t)$:

$$y^F(t) = -\sum_{i=1}^{n_C} \widehat{c}_i(t-1)y^F(t-i) - y(t),$$

$$u^F(t) = -\sum_{i=1}^{n_C} \widehat{c}_i(t-1)u^F(t-i) - u(t),$$

$$\epsilon^F(t) = -\sum_{i=1}^{n_C} \widehat{c}_i(t-1)\epsilon^F(t-i) - \epsilon(t). \tag{3.53}$$

These relationships can be regarded as an 'instantaneous' version of (3.47).

The two approximations described above are the cornerstone of the following recursive prediction error (RPE) algorithm [179]:

$$\widehat{\theta}(t) = \widehat{\theta}(t-1) + K(t)\epsilon(t),$$

$$\epsilon(t) = y(t) - \widetilde{\varphi}^T(t)\widehat{\theta}(t-1),$$

$$K(t) = \frac{P(t-1)\psi(t)}{1 + \psi^T(t)P(t-1)\psi(t)},$$

$$P(t) = P(t-1) - \frac{P(t-1)\psi(t)\psi^T(t)P(t-1)}{1 + \psi^T(t)P(t-1)\psi(t)}, \tag{3.54}$$

which can be considered as a recursively computable variant of the Gauss–Newton procedure mentioned in Remark 2 of Section 3.3.1. Quite obviously, for the RPE algorithm to converge, the invertibility condition has to be checked at the end of each computational cycle; if $\widehat{\theta}(t) \notin D_I$ the parameter estimates should be modified so as to meet the identifiability constraint.

The residual sum of squares $q(t)$ can be evaluated (approximately) using a recursive algorithm (3.20). Since the starting value $q(n)$ is in general nonzero (it would have been zero had $\widehat{\theta}(n)$ been the 'exact' PE estimate) it can be evaluated by direct summation of the corresponding squared prediction errors.

Another recursive scheme for identification of ARMAX systems is known as the extended least squares (ELS) algorithm. The ELS scheme is basically identical with the RLS scheme except that the regression vector $\varphi(t)$ is replaced with $\widetilde{\varphi}(t)$. The resulting algorithm

$$
\begin{aligned}
\widehat{\theta}(t) &= \widehat{\theta}(t-1) + K(t)\epsilon(t), \\
\epsilon(t) &= y(t) - \widetilde{\varphi}^T(t)\widehat{\theta}(t-1), \\
K(t) &= \frac{P(t-1)\widetilde{\varphi}(t)}{1 + \widetilde{\varphi}^T(t)P(t-1)\widetilde{\varphi}(t)}, \\
P(t) &= P(t-1) - \frac{P(t-1)\widetilde{\varphi}(t)\widetilde{\varphi}^T(t)P(t-1)}{1 + \widetilde{\varphi}^T(t)P(t-1)\widetilde{\varphi}(t)}
\end{aligned} \tag{3.55}
$$

is simpler than the RPE algorithm (3.54) and, since inverse filtering of regression variables is not performed, the invertibility checks and interventions are not required. On the negative side, the convergence analysis reveals that the ELS estimates may not converge to the true parameter values if the identified system does not obey the following strict positive realness (SPR) condition [178]

$$
\theta \in D_{\text{SPR}}, \quad D_{\text{SPR}} = \left\{ \theta : \text{Re}\left\{ \frac{1}{C(e^{j\omega}, \theta)} - \frac{1}{2} \right\} > 0, \quad \forall \omega \right\}. \tag{3.56}
$$

When $n_C > 1$, i.e. the number of estimated moving average coefficients exceeds one, D_{SPR} is a subset of D_I, which means that the ELS algorithm may fail to converge in certain areas of the admissible parameter space.

Remark 1

It should be stressed that, despite notational similarities, the quantities $P(t)$ and $K(t)$ updated in the RPE and ELS algorithms are *not* identical.

Remark 2

To improve the rate of convergence of the RPE and ELS schemes, the regression vector $\widetilde{\varphi}(t)$ can be replaced with

$$
\widetilde{\varphi}(t) = [y(t-1), \ldots, y(t-n_A), u(t-1), \ldots, u(t-n_B), \epsilon_+(t-1), \ldots, \epsilon_+(t-n_C)]^T,
$$

where

$$
\epsilon_+(t) = y(t) - \widetilde{\varphi}^T(t)\widehat{\theta}(t)
$$

are the so-called a posteriori errors (used instead of the prediction errors $\epsilon(t)$).

3.3.3   Conditions of identifiability

In the ARMA(X) case, linear feedback or insufficiently exciting input signals are not the only potential sources of system nonidentifiability. Linear dependence among regression variables may also occur if the polynomial degrees of the system are overestimated, i.e. the model is *overparameterized*.

Example 3.4

Suppose that the second-order ARMA model

$$y(t) = a_1 y(t-1) + a_2 y(t-2) + v(t) + c_1 v(t-1) + c_2 v(t-2) \qquad (3.57)$$

is fitted to the first-order process

$$y(t) = a y(t-1) + v(t) + c v(t-1). \qquad (3.58)$$

Note that the regression vector associated with (3.57),

$$\varphi(t) = [\, y(t-1), y(t-2), v(t-1), v(t-2) \,]^T,$$

has linearly dependent components since, according to (3.58)

$$[\, 1, -a, -1, -c \,]\, \varphi(t) = 0, \quad \forall t.$$

Quite obviously, the rank deficiency of $\varphi(t)$ is inherited, at least asymptotically, by the pseudolinear regression vector $\varphi(t, \theta)$ used in the identification procedures described in Sections 3.3.1 and 3.3.2.

∎

In order to guarantee parameter identifiability of a general ARMAX system, one has to assume that the polynomials $A(q^{-1})$, $B(q^{-1})$ and $C(q^{-1})$ have no common factors. This coprimeness requirement is obviously not fulfilled in the example above. Note that the process (3.58) has an infinite number of representations of the form (3.57) with

$$a_1 = a + \beta, \quad a_2 = -a\beta, \quad c_1 = c - \beta, \quad c_2 = -c\beta,$$

and arbitrary β, all of which can be obtained by multiplying both sides of (3.58) with $1 - \beta q^{-1}$.

3.3.4   Adaptive segmentation

Under Gaussian assumptions the prediction error estimators are identical with the maximum likelihood estimators. Hence all results based on the AIC statistics and the GLR test (both derived for ML estimators) remain valid for ARMAX systems. Realization of the aggregate/split scheme is straightforward. Realization of the global/local scheme is technically difficult since, due to the approximate nature of the RPE approach, there are no natural extensions of the two-step algorithms (3.31), (3.32) and (3.33), (3.34) to the ARMAX case. This means it is not possible to estimate the parameters of the local model in a computationally efficient way.

Comments and extensions

Section 3.2

- When process characteristics change abruptly due to the failure of some system components, switching of the operating mode, external disturbances, etc., segmentation can be achieved by means of change detection. A survey of different techniques for change detection can be found in Basseville [15] and Basseville and Nikiforov [16].

4

Weighted Least Squares

4.1   Estimation principles

Denote by $\{w(i)\}$ the weighting sequence, a sequence of nonnegative coefficients obeying

$$w(0) = 1 \geq w(1) \geq \ldots \geq 0, \tag{4.1}$$

$$\sum_{i=0}^{\infty} w(i) < \infty, \tag{4.2}$$

which will be used to 'localize' estimation, namely to make its results insensitive (or at least less sensitive) to observations collected in the remote past. To achieve the effect of forgetting or discounting 'old' data, the sum of squares minimized in the method of least squares is replaced with the weighted sum of squares, resulting in the following weighted least squares (WLS) estimator:

$$\widehat{\theta}(t) = \arg\min_{\theta} J_w(\theta, \Xi(t), \Xi(0)), \tag{4.3}$$

where

$$J_w(\theta, \Xi(t), \Xi(0)) = \sum_{i=0}^{t-1} w(i)[y(t-i) - \varphi^T(t-i)\theta]^2. \tag{4.4}$$

Minimization of (4.4) can be carried out analogously as minimization of an ordinary sum of squares (3.2), leading to the following explicit expression for the WLS estimator:

$$\begin{aligned}
\widehat{\theta}(t) &= \left(\sum_{i=0}^{t-1} w(i)\varphi(t-i)\varphi^T(t-i) \right)^{-1} \left(\sum_{i=0}^{t-1} w(i)y(t-i)\varphi(t-i) \right) \\
&= R^{-1}(t)s(t),
\end{aligned} \tag{4.5}$$

subject to the identifiability condition

$$R(t) = \sum_{i=0}^{t-1} w(i)\varphi(t-i)\varphi^T(t-i) > 0.$$

4.2   Estimation windows

The two windows most frequently used in practice are a rectangular window

$$w(i) = \begin{cases} 1 & 0 \le i < N \\ 0 & \text{elsewhere} \end{cases} \tag{4.6}$$

and an exponential window

$$w(i) = \lambda^i, \qquad 0 < \lambda < 1. \tag{4.7}$$

In the first case only the N most recent observations (equally weighted) are used for parameter estimation. This approach is often called a sliding window approach as it corresponds to moving a constant-length analysis window along the data.

In the second case the weights decay at an exponential rate, gradually decreasing the influence of old data on the current parameter estimates. This approach is frequently termed exponential forgetting. The constant λ in (4.7) is called a forgetting constant.

The recursive algorithm for computation of the sliding window least squares (SWLS) estimate was derived in Section 3.2.1. Note that under (4.6) the WLS estimator (4.5) is identical with the local LS estimator $\hat{\theta}(t, N)$ introduced for adaptive process segmentation. Hence computation of the sliding window estimates can be carried out by the two-step recursive algorithm (3.31) and (3.32).

Derivation of the recursive version of the exponentially weighted least squares (EWLS) estimator is a simple modification of the derivation for an RLS algorithm. Observe that for (4.7)

$$\begin{aligned} R(t) &= \lambda R(t-1) + \varphi(t)\varphi^T(t), \\ s(t) &= \lambda s(t) + y(t)\varphi(t), \end{aligned}$$

leading to

$$\begin{aligned} \hat{\theta}(t) &= R^{-1}(t)s(t) = R^{-1}(t)\left[\lambda R(t-1)\hat{\theta}(t-1) + y(t)\varphi(t)\right] \\ &= R^{-1}(t)\left[R(t)\hat{\theta}(t-1) + \varphi(t)(y(t) - \varphi^T(t)\hat{\theta}(t-1))\right] \\ &= \hat{\theta}(t-1) + P(t)\varphi(t)\epsilon(t), \end{aligned}$$

where

$$P(t) = R^{-1}(t) = \left(\lambda P^{-1}(t-1) + \varphi(t)\varphi^T(t)\right)^{-1}.$$

After applying the matrix inversion lemma ($A := \lambda P^{-1}$, $x := \varphi$, $\mu := 1$) one arrives at the following algorithm:

$$\begin{aligned} \hat{\theta}(t) &= \hat{\theta}(t-1) + K(t)\epsilon(t), \\ \epsilon(t) &= y(t) - \varphi^T(t)\hat{\theta}(t-1), \\ K(t) &= \frac{P(t-1)\varphi(t)}{\lambda + \varphi^T(t)P(t-1)\varphi(t)}, \\ P(t) &= \frac{1}{\lambda}\left[P(t-1) - \frac{P(t-1)\varphi(t)\varphi^T(t)P(t-1)}{\lambda + \varphi^T(t)P(t-1)\varphi(t)}\right], \end{aligned} \tag{4.8}$$

which closely resembles (3.13); it is identical with (3.13) when $\lambda = 1$, i.e. there is no forgetting.

Another, frequently encountered form of the recursive EWLS algorithm can be obtained by replacing $P(t)$ in (4.8) with $\tilde{P}(t) = P(t)/\lambda$:

$$
\begin{aligned}
\widehat{\theta}(t) &= \widehat{\theta}(t-1) + K(t)\epsilon(t), \\
\epsilon(t) &= y(t) - \varphi^T(t)\widehat{\theta}(t-1), \\
K(t) &= \frac{\tilde{P}(t-1)\varphi(t)}{1 + \varphi^T(t)\tilde{P}(t-1)\varphi(t)}, \\
\tilde{P}(t) &= \frac{1}{\lambda}\left[\tilde{P}(t-1) - \frac{\tilde{P}(t-1)\varphi(t)\varphi^T(t)\tilde{P}(t-1)}{1 + \varphi^T(t)\tilde{P}(t-1)\varphi(t)}\right]
\end{aligned}
\tag{4.9}
$$

Remark

When the conventional EWLS algorithm is used, the rate of forgetting is the same in all directions of the parameter space, i.e. it is implicitly assumed that all process coefficients vary with the same speed. Replacing the single forgetting factor λ in the covariance update equation of (4.9) with the set of forgetting factors $\lambda_1, \ldots, \lambda_n$, assigned individually for each of n process coefficients, Saelid and Foss [166] arrived at the following EWLS algorithm with vector forgetting :

$$
\begin{aligned}
\widehat{\theta}(t) &= \widehat{\theta}(t-1) + K(t)\epsilon(t), \\
\epsilon(t) &= y(t) - \varphi^T(t)\widehat{\theta}(t-1), \\
K(t) &= \frac{\tilde{P}(t-1)\varphi(t)}{1 + \varphi^T(t)\tilde{P}(t-1)\varphi(t)}, \\
\tilde{P}(t) &= L(t)\left[\tilde{P}(t-1) - \frac{\tilde{P}(t-1)\varphi(t)\varphi^T(t)\tilde{P}(t-1)}{1 + \varphi^T(t)\tilde{P}(t-1)\varphi(t)}\right]L(t),
\end{aligned}
\tag{4.10}
$$

where

$$
L(t) = \text{diag}\{\lambda_1^{-1/2}, \ldots, \lambda_n^{-1/2}\},
$$
$$
0 < \lambda_i \leq 1, \quad i = 1, \ldots, n.
$$

Vector forgetting allows one to better incorporate the prior knowledge about the speed of variation of different components of θ. Generally, the faster a given process parameter varies with time, the smaller the forgetting factor to guarantee a good tracking performance.

4.3   Static characteristics of WLS estimators

We will start by examining the behavior of the WLS estimators under time-invariant conditions, i.e. in the case where $\theta(t) = \theta$, $\forall t$.

4.3.1 Effective window width

Suppose that the process made up of regression vectors is stationary:

C1 $\{\varphi(t)\}$ is a zero-mean, stationary process with positive definite covariance matrix $E[\varphi(t)\varphi^T(t)] = \Phi_o$.

Then the expected value of the weighted regression matrix obeys

$$E[R(t)] = E\left[\sum_{i=0}^{t-1} w(i)\varphi(t-i)\varphi^T(t-i)\right] = k_t\Phi_o, \qquad (4.11)$$

where the quantity

$$k_t = \sum_{i=0}^{t-1} w(i) \qquad (4.12)$$

can be interpreted as the 'window area'. According to (4.11) the coefficient $k_t \leq t$ characterizes the amount of energy extracted from the input/output data as a result of applying the weighting sequence $\{w(i)\}$. Since, with weighting, the average energy content of the set of t observations is the same as the energy content of the set of k_t observations without weighting, the quantity k_t is usually called the effective number of observations or the effective window width. Note that

$$k_t^{\mathrm{SWLS}} = \min(t, N), \quad k_t^{\mathrm{EWLS}} = \frac{1 - \lambda^t}{1 - \lambda}$$

for the rectangular and exponential windows, respectively.

4.3.2 Equivalent window width

In order to quantify the accuracy of WLS estimators under stationary conditions, consider the following *simplified* WLS estimator:

$$\widehat{\widehat{\theta}}(t) = \frac{1}{k_t}\Phi_o^{-1}\sum_{i=0}^{t-1} w(i)y(t-i)\varphi(t-i), \qquad (4.13)$$

which can be obtained from (4.5) after applying the following approximation:

$$R^{-1}(t) \cong (E[R(t)])^{-1} = [k_t\Phi_o]^{-1}, \qquad (4.14)$$

valid for sufficiently large values of k_t. The mathematical foundations of this approximation will be discussed in Section 4.8.

Substituting $y(t-i) = \varphi^T(t-i)\theta + v(t-i)$ into (4.13) and using orthogonality of $\varphi(t-i)$ and $v(t-i)$, one obtains

$$E[\widehat{\theta}(t)] \cong E[\widehat{\widehat{\theta}}(t)]$$

$$= \frac{1}{k_t}\Phi_o^{-1}E\left\{\sum_{i=0}^{t-1} w(i)\varphi(t-i)\varphi^T(t-i)\theta + \sum_{i=0}^{t-1} w(i)\varphi(t-i)v(t-i)\right\} = \theta. \qquad (4.15)$$

Similarly,

$$\text{cov}[\widehat{\theta}(t)] \cong \text{cov}[\widehat{\widehat{\theta}}(t)] = \frac{1}{k_t^2}\Phi_o^{-1}Q(t)\Phi_o^{-1}, \tag{4.16}$$

where

$$Q(t) = \{\sum_{\substack{i=0 \\ i<j}}^{t-1} + \sum_{\substack{i=0 \\ i=j}}^{t-1} + \sum_{\substack{i=0 \\ i>j}}^{t-1}\} \; \mathbf{E}\left[w(i)w(j)\varphi(t-i)\varphi^T(t-j)v(t-i)v(t-j)\right]$$

$$= Q_1(t) + Q_2(t) + Q_3(t). \tag{4.17}$$

Since, for $i < j$, the random variable $v(t-i)$ is independent of $v(t-j)$, $\varphi(t-i)$ and $\varphi(t-j)$, and since $\mathbf{E}[v(t-i)] = 0$, the term $Q_1(t)$ in (4.17) is zero. Similarly, the term $Q_2(t)$ is zero. Finally, due to independence of $v(t-i)$ and $\varphi(t-i)$,

$$Q_3(t) = \sum_{i=0}^{t-1} \mathbf{E}[w^2(i)\varphi(t-i)\varphi^T(t-i)]\mathbf{E}[v^2(t-i)] = \Phi_o\sigma_v^2\left(\sum_{i=0}^{t-1}w^2(i)\right),$$

which after substitution into (4.17) yields

$$\text{cov}[\widehat{\theta}_{\text{WLS}}(t)] \cong \frac{\sigma_v^2\Phi_o^{-1}}{l_t^{\text{WLS}}}, \tag{4.18}$$

where

$$l_t^{\text{WLS}} = \frac{\left(\sum_{i=0}^{t-1}w(i)\right)^2}{\sum_{i=0}^{t-1}w^2(i)}. \tag{4.19}$$

According to (4.18) in the stationary case all WLS estimators characterized by the same value of l_t are equivalent from the viewpoint of estimation accuracy, irrespective of the shape of the corresponding windows. Consequently, l_t characterizes the amount of information about θ which is extracted from the input/output data as a result of applying the method of weighted least squares. Note that the average information content of the set of t data samples in the case of weighting is the same as the information content of l_t samples in the case of no weighting (the covariance matrices of the corresponding WLS and LS estimators are identical). Following (4.18) we shall call l_t the equivalent number of observations or the equivalent window width. Note that for the rectangular window

$$l_t^{\text{SWLS}} = \min(t, N),$$

and for the exponential window

$$l_t^{\text{EWLS}} = \frac{(1-\lambda^t)}{(1+\lambda^t)}\frac{(1+\lambda)}{(1-\lambda)}.$$

Due to (4.1)

$$l_t \geq k_t$$

(rectangular windows are the only ones for which equality holds), which demonstrates some peculiarity of weighted estimation: the amount of information extracted from the data by applying a nonrectangular window is always greater than the corresponding amount of energy.

The discrepancy between the effective and equivalent number of observations can be substantial, e.g. for an exponential window

$$\frac{l_t^{\text{EWLS}}}{k_t^{\text{EWLS}}} = \frac{1+\lambda}{1+\lambda^t} \xrightarrow[t\to\infty]{} 1+\lambda \cong 2,$$

for typical (close to 1) values of λ. This discrepancy may be a source of much confusion when comparing WLS estimators characterized by different windows.

4.3.3   Degree of window concentration

The last group of static characteristics, which tend to explain some properties of WLS estimators are related to the degree of concentration of the window $\{w(i)\}$. The median-based concentration measure

$$c_t = \frac{m_t}{l_t} \,, \tag{4.20}$$

where

$$m_t \quad : \quad \sum_{i=0}^{m_t} w(i) \cong 0.5 \sum_{i=0}^{t} w(i),$$

is one possibility. Another useful measure can be based on comparing k_t and l_t:

$$e_t = \frac{k_t}{l_t} \,. \tag{4.21}$$

Note that *small* values of the coefficients c_t and e_t correspond to a high degree of concentration of $\{w(i)\}$ around $i = 0$.

4.4   Dynamic time-domain characteristics of WLS estimators

Now we will try to address a more difficult problem: how to assess the parameter tracking capabilities of WLS estimators.

Consider any parameter trajectory

$$\Theta(t) = \{\theta(1), \dots, \theta(t)\}$$

characterizing the time variation of an analyzed process and suppose the mean square parameter tracking error

$$\mathcal{T}(t) = \mathcal{T}[\widehat{\theta}(t)] = \mathrm{E}\left[\|\widehat{\theta}(t) - \theta(t)\|^2\right] \tag{4.22}$$

is adopted as a measure of tracking quality. Introducing the notion of an averaged parameter trajectory

$$\bar{\theta}(t) = \mathrm{E}[\widehat{\theta}(t)],$$

one can write $\mathcal{T}(t)$ in the form

$$\mathcal{T}(t) = \mathcal{T}_b(t) + \mathcal{T}_v(t), \tag{4.23}$$

where

$$\mathcal{T}_b(t) = \| \bar{\theta}(t) - \theta(t) \|^2 \tag{4.24}$$

is a bias component of the mean square tracking error and

$$\mathcal{T}_v(t) = \mathrm{E}\left[\| \widehat{\theta}(t) - \bar{\theta}(t) \|^2 \right] \tag{4.25}$$

denotes its variance component (fluctuations of parameter estimates around their mean values).

Evaluation of both components of the mean square parameter tracking error is a difficult task. The source of the main technical difficulty lies in the fact that in the general case characteristics of the regression vector are parameter dependent and hence also time dependent. An interesting special case in which the problem remains analytical is identification of a time-varying FIR system

$$y(t) = \varphi^T(t)\theta(t) + v(t), \qquad \varphi(t) = [\, u(t-1), \ldots, u(t-n_B)\,]^T,$$

arising in adaptive equalization of communication channels, for example. Note that for an FIR system the characteristics of $\varphi(t)$ are independent of parameter variations. This allows one to derive pretty insightful results on parameter tracking.

4.4.1 Impulse response associated with WLS estimators

Consider a time-varying FIR system obeying the stationary condition (C1) (met when the input signal is stationary and persistently exciting of sufficiently high order). Since the approximation (4.14) remains valid in this case (under some mild additional constraints discussed in Section 4.8), one gets

$$
\begin{aligned}
\bar{\theta}(t) \;=\; & \mathrm{E}[\widehat{\theta}(t)] \cong \frac{1}{k_t}\Phi_o^{-1}\mathrm{E}\left[\sum_{i=0}^{t-1} w(i)y(t-i)\varphi(t-i)\right] \\
\;=\; & \frac{1}{k_t}\Phi_o^{-1}\mathrm{E}\left[\sum_{i=0}^{t-1} w(i)\varphi(t-i)\varphi^T(t-i)\theta(t-i)\right] \\
& + \frac{1}{k_t}\Phi_o^{-1}\mathrm{E}\left[\sum_{i=0}^{t-1} w(i)v(t-i)\varphi(t-i)\right] \\
\;=\; & \frac{1}{k_t}\sum_{i=0}^{t-1} w(i)\theta(t-i).
\end{aligned}
\tag{4.26}
$$

Most of our further developments will be based on this strikingly simple result. According to (4.26) the process $\{\bar{\theta}(t)\}$ can be approximately viewed as a result of passing the process $\{\theta(t)\}$ through a linear filter (filter *associated* with the WLS estimator) of window-dependent, i.e. generally also time-dependent, coefficients.

The result obtained has interesting implications. First, observe that (4.26) can be rewritten as a set of independent relationships for different coefficients comprising the parameter vector $\theta(t)$. This means that the bias errors are (approximately) uncoupled. Second, note that, in the mean, the evolution of parameter estimates depends on the applied window but *does not* depend on the statistics of the input signal. In particular, according to (4.26) the tracking speed of WLS estimators does not depend on the eigenvalue structure of the regression matrix Φ_o, i.e. it is exactly the same in all directions of the parameter space.

When examining the tracking capabilities of the WLS algorithm, one is interested primarily in its long-term or steady-state behavior, i.e. the behavior observed when effective and equivalent numbers of observations approach their limiting (steady-state) values:

$$k_\infty = \lim_{t \to \infty} k_t, \qquad l_\infty = \lim_{t \to \infty} l_t.$$

The long-term counterpart of (4.26) can be obtained by setting $t \mapsto \infty$ or by assuming that the infinite observation history is available at instant t. In the latter case one obtains

$$\bar{\theta}_{\mathrm{WLS}}(t) \cong \sum_{i=0}^{\infty} h_{\mathrm{WLS}}(i)\theta(t-i), \tag{4.27}$$

where

$$h_{\mathrm{WLS}}(i) = \frac{w(i)}{k_\infty}. \tag{4.28}$$

According to (4.27), after a sufficiently long time, $\{\bar{\theta}(t)\}$ can be approximately viewed as an output of a linear causal time-invariant filter with impulse response $\{h_{\mathrm{WLS}}(i)\}$ excited by the process $\{\theta(t)\}$. The sequence $\{h_{\mathrm{WLS}}(i), i = 0, 1, \ldots\}$ will be further called the impulse response associated with the WLS estimator. Note that

$$\sum_{i=0}^{\infty} h_{\mathrm{WLS}}(i) = 1 \tag{4.29}$$

and

$$\sum_{i=0}^{\infty} h_{\mathrm{WLS}}^2(i) = \frac{1}{l_\infty^{\mathrm{WLS}}}. \tag{4.30}$$

Note also that for EWLS and SWLS approaches, the corresponding associated impulse responses are

$$h_{\mathrm{SWLS}}(i) = \begin{cases} 1/N & 0 \leq i \leq N \\ 0 & \text{elsewhere} \end{cases}$$

$$h_{\mathrm{EWLS}}(i) = \begin{cases} (1-\lambda)\lambda^i & i \geq 0 \\ 0 & i < 0 \end{cases}$$

4.4.2 Variability of WLS estimators

It is straightforward to check that

$$\widehat{\theta}(t) - \overline{\theta}(t)$$

$$= R^{-1}(t) \left[\sum_{i=0}^{t-1} w(i)\varphi(t-i)\varphi^T(t-i)\gamma(i,t) + \sum_{i=0}^{t-1} w(i)\varphi(t-i)v(t-i) \right], \qquad (4.31)$$

where

$$\gamma(i,t) = \theta(t-i) - \overline{\theta}(t).$$

Applying approximation (4.14) and exploiting the mutual independence of $\{\varphi(t)\}$ and $\{v(t)\}$, one arrives at the following expression characterizing fluctuations of the estimate $\widehat{\theta}(t)$ around its mean trajectory $\overline{\theta}(t)$:

$$V(t) = \mathrm{E}\left[(\widehat{\theta}(t) - \overline{\theta}(t))(\widehat{\theta}(t) - \overline{\theta}(t))^T\right] = V_1(t) + V_2(t) \qquad (4.32)$$

where

$$V_1(t) = \frac{1}{k_t^2}\Phi_o^{-1}\mathrm{E}\left[\sum_{i=0}^{t-1}\sum_{j=0}^{t-1} w(i)w(j)\varphi(t-i)\varphi^T(t-j)v(t-i)v(t-j)\right]\Phi_o^{-1} \qquad (4.33)$$

and

$$V_2(t) =$$

$$\frac{1}{k_t^2}\Phi_o^{-1}\mathrm{E}\left[\sum_{i=0}^{t-1}\sum_{j=0}^{t-1} w(i)w(j)\varphi(t-i)\varphi^T(t-i)\gamma(i,t)\gamma^T(j,t)\varphi(t-j)\varphi^T(t-j)\right]\Phi_o^{-1}.$$

$$(4.34)$$

The matrix $V_1(t)$, identical with (4.17), can be recognized as the estimation error covariance matrix of the time-invariant system. Hence the *variability* matrix $V(t)$ can be split in two components. The first one (cf. (4.18)) is

$$V_1(t) \cong \frac{\sigma_v^2\Phi_o^{-1}}{l_t}$$

and describes the estimation accuracy under stationary conditions. The second one, $V_2(t)$, reflects the additional degradation of the estimation accuracy caused by parameter variation. By inspection the matrix $V_2(t)$ is positive semidefinite, so

$$V(t) \geq V_1(t),$$

i.e. $V_1(t)$ can be interpreted as a lower bound on the variability matrix under nonstationary conditions.

As far as the matrix $V_2(t)$ is concerned, its value depends on the actual parameter trajectory and on the rate of decorrelation of the input signal. The above statement can be easily verified in the case where both the input signal $u(t)$ and the noise $v(t)$ are zero-mean and Gaussian. It is known that expectation of a product of four zero-mean Gaussian variables $x_1, \ldots, x_4$ can be written in the following form [179]:

$$\mathrm{E}[x_1 x_2 x_3 x_4] = \mathrm{E}[x_1 x_2]\mathrm{E}[x_3 x_4] + \mathrm{E}[x_1 x_3]\mathrm{E}[x_2 x_4] + \mathrm{E}[x_1 x_4]\mathrm{E}[x_2 x_3].$$

Applying this rule to (4.34) one arrives at

$$V_2(t) = V_{21}(t) + V_{22}(t) + V_{23}(t),$$

where

$$V_{21}(t) = \frac{1}{k_t^2} \sum_{i=0}^{t-1} \sum_{j=0}^{t-1} w(i)w(j)\Phi_o^{-1}\Phi_{i-j}\gamma(i,t)\gamma^T(j,t)\Phi_{i-j}\Phi_o^{-1}, \qquad (4.35)$$

$$V_{22}(t) = \frac{1}{k_t^2} \sum_{i=0}^{t-1} \sum_{j=0}^{t-1} w(i)w(j)\Phi_o^{-1}\Phi_{i-j}\Phi_o^{-1}\gamma^T(i,t)\Phi_{i-j}\gamma(j,t), \qquad (4.36)$$

$$V_{23}(t) = \frac{1}{k_t^2} \sum_{i=0}^{t-1} \sum_{j=0}^{t-1} w(i)w(j)\gamma(i,t)\gamma^T(j,t)$$

$$= \left(\frac{1}{k_t} \sum_{i=0}^{t-1} w(i)\theta(t-i) - \bar{\theta}(t) \right) \left(\frac{1}{k_t} \sum_{i=0}^{t-1} w(i)\theta(t-i) - \bar{\theta}(t) \right)^T = 0, \qquad (4.37)$$

and

$$\Phi_{i-j} = \mathrm{E}[\varphi(i)\varphi^T(j)].$$

Generally speaking, $V_1(t)$ is a dominant term in (4.32) for 'slow' parameter variations (for a time-invariant system $V_2(t) = O$) while $V_2(t)$ plays the crucial role for 'fast' variations. Note that the variance component of the mean square parameter tracking error $\mathcal{T}(t)$ can easily be obtained from the variability matrix $V(t)$:

$$\mathcal{T}_v(t) = \mathrm{tr}\{V(t)\}.$$

4.5   Dynamic frequency-domain characteristics of WLS estimators

The analysis of distortion effects in parameter tracking can be carried out using the concept of frequency characteristics associated with the WLS estimator

4.5.1   Frequency characteristics associated with WLS estimators

Recall from (4.27) that

$$\bar{\theta}(t) \cong \sum_{i=0}^{\infty} h_{\mathrm{WLS}}(i)\theta(t-i).$$

Make the following assumption:

C2 $\{\theta(t)\}$ is a zero-mean wide-sense stationary process with a cross-spectral density matrix $S_\theta(\omega)$.

Then the bias component of the mean square parameter tracking error $\mathcal{T}_b(t)$, averaged over different realizations of the parameter trajectory $\Theta(t) = \{\theta(1), \ldots, \theta(t)\}$, can be expressed in the following form:

$$\mathcal{T}_b = \underset{\Theta(t)}{\mathrm{E}} \left[\| \bar{\theta}(t) - \theta(t) \|^2 \right] \cong \frac{1}{\pi} \int_0^\pi E_{\mathrm{WLS}}(\omega) \, \mathrm{tr}\{S_\theta(\omega)\} \, d\omega, \tag{4.38}$$

where

$$E_{\mathrm{WLS}}(\omega) = |1 - H_{\mathrm{WLS}}(\omega)|^2 \tag{4.39}$$

and

$$H_{\mathrm{WLS}}(\omega) = \sum_{i=0}^\infty h_{\mathrm{WLS}}(i) e^{-j\omega i} = A_{\mathrm{WLS}}(\omega) e^{j\phi_{\mathrm{WLS}}(\omega)}. \tag{4.40}$$

We will call $H_{\mathrm{WLS}}(\omega)$, the Fourier transform of $h_{\mathrm{WLS}}(t)$, the *frequency response* associated with the WLS estimator, and we will call $E_{\mathrm{WLS}}(\omega)$ its *parameter tracking characteristic*. The quantities $A_{\mathrm{WLS}}(\omega)$ and $\phi_{\mathrm{WLS}}(\omega)$ are the associated amplitude and phase characteristics, respectively.

Note that (4.38) specifies the frequency distribution of the bias error in a particularly useful form as it clearly separates the true parameter variation, reflected by the spectral density function $S_\theta(\omega)$, from the tracking capability of the algorithm, reflected by the *parameter tracking characteristic* $E_{\mathrm{WLS}}(\omega)$. It is obvious that good tracking performance can be achieved only in the case where the spectral density function $S_\theta(\omega)$ matches the passband region of $E_{\mathrm{WLS}}(\omega)$. Hence the shape of the tracking characteristic should be a good guideline to evaluate and compare the tracking capabilities of different estimators.

4.5.2 Properties of associated frequency characteristics

First of all, observe that

$$A_{\mathrm{WLS}}(\omega) \leq \sum_{i=0}^\infty |h_{\mathrm{WLS}}(i) e^{-j\omega i}| \leq \sum_{i=0}^\infty h_{\mathrm{WLS}}(i) = A_{\mathrm{WLS}}(0) = 1,$$

which means the associated filter must be a lowpass filter, i.e. the corresponding WLS algorithms are capable of tracking the slow parameter variations only.

Using the Parseval's theorem one obtains

$$\frac{1}{\pi} \int_0^\pi |H_{\mathrm{WLS}}(\omega)|^2 d\omega = \frac{1}{\pi} \int_0^\pi A_{\mathrm{WLS}}(\omega)^2 d\omega = \sum_{i=0}^\infty h_{\mathrm{WLS}}^2(i) = \frac{1}{l_\infty},$$

showing that the equivalent estimation bandwidth of WLS estimators is inversely proportional to the equivalent width of the applied window.

To get more insights into the tracking characteristics of WLS estimators, we will use the technique of integral approximations. Denote by $\tilde{w}(\cdot) \in L_1[0, \infty]$, $\tilde{w}(0) = 1$ a 'prototype' analog window (any positive, nonincreasing and integrable function defined on the interval $[0, \infty]$). Setting

$$w(i) = \tilde{w}(t)|_{t=i}, \qquad i = 0, 1, \ldots,$$

one obtains the following integral approximations, valid for 'sufficiently wide' windows:

$$k_\infty = \sum_{i=0}^{\infty} w(i) \cong \int_0^\infty \tilde{w}(s)\,ds = \tilde{k}_\infty$$

and

$$l_\infty = \frac{1}{\sum_{i=0}^{\infty} h_{\mathrm{WLS}}^2(i)} \cong \frac{1}{\int_0^\infty \tilde{h}_{\mathrm{WLS}}^2(s)\,ds} = \tilde{l}_\infty,$$

where

$$\tilde{h}_{\mathrm{WLS}}(s) = \frac{\tilde{w}(s)}{\tilde{k}_\infty}$$

is a continuous-time analog of the associated impulse response $h_{\mathrm{WLS}}(i)$.

The same technique also leads to another useful approximation:

$$H_{\mathrm{WLS}}(\omega) = \sum_{i=0}^{\infty} h_{\mathrm{WLS}}(i)e^{-j\omega i} \cong \int_0^\infty \tilde{h}_{\mathrm{WLS}}(s)e^{-j\omega s}\,ds = \tilde{H}_{\mathrm{WLS}}(\omega),$$

valid in the range of small values of ω.

Consider now an adjustable-size window

$$w^\eta(i) = \tilde{w}(\eta i), \qquad i = 0, 1, \ldots.$$

Note that by changing the coefficient $\eta > 0$ of $w^\eta(i)$, one can increase ($\eta < 1$) or decrease ($\eta > 1$) the effective width of the window without changing its shape, namely

$$k_\infty^\eta = \sum_{i=0}^{\infty} w^\eta(i) \cong \int_0^\infty \tilde{w}(\eta s)\,ds = \frac{\tilde{k}_\infty}{\eta} \cong \frac{k_\infty}{\eta}. \tag{4.41}$$

In an analogous way one can show that

$$l_\infty^\eta \cong \frac{l_\infty}{\eta} \tag{4.42}$$

and

$$H_{\mathrm{WLS}}^\eta(\omega) \cong H_{\mathrm{WLS}}(\omega/\eta). \tag{4.43}$$

As a direct consequence of (4.41) to (4.43) one arrives at the following conclusion: for small ω the frequency characteristics $H_{\mathrm{WLS}}(\omega)$ and $E_{\mathrm{WLS}}(\omega)$ can be approximately expressed as functions of a *normalized frequency*

$$\Omega = l_\infty \omega.$$

The corresponding quantities $H_{\mathrm{WLS}}(\Omega)$ and $E_{\mathrm{WLS}}(\Omega)$ will be called the *normalized frequency characteristics*. Comparing the plots of the normalized tracking characteristics amounts to comparing tracking properties of estimators characterized by the same l_∞, i.e. estimators yielding the same estimation accuracy under time-invariant conditions.

Example 4.1

Consider the exponential window. By direct calculation

$$k_\infty = \frac{1}{1-\lambda}, \qquad l_\infty = \frac{1+\lambda}{1-\lambda},$$

$$h_{\mathrm{EWLS}}(i) = (1-\lambda)\lambda^i, \qquad H_{\mathrm{EWLS}}(\omega) = \frac{1-\lambda}{1-\lambda e^{-j\omega}}.$$

The 'prototype' analog counterpart of an exponential window is

$$\tilde{w}(s) = e^{-\gamma s}, \qquad \gamma = -\ln \lambda > 0,$$

leading to the following integral approximations:

$$\tilde{k}_\infty = \frac{1}{\gamma}, \qquad \tilde{l}_\infty = \frac{2}{\gamma},$$

$$\tilde{h}_{\mathrm{EWLS}}(i) = \gamma e^{-\gamma s}, \qquad \tilde{H}_{\mathrm{EWLS}}(\omega) = \frac{\gamma}{\gamma + j\omega} = \frac{1}{1 + j\frac{\tilde\Omega}{2}},$$

where $\tilde\Omega = \tilde{l}_\infty \omega$. For forgetting constants close to one, $1 + \lambda \cong 2$ and $\ln \lambda \cong \lambda - 1$, leading to (with an error smaller than 5% for $0.9 < \lambda < 1$)

$$l_\infty \cong \frac{2}{1-\lambda} \cong \tilde{l}_\infty, \qquad \Omega \cong \tilde\Omega.$$

Similarly, using the approximations $\sin\omega \cong \omega$, $\cos\omega \cong 1$ (with an error smaller than 1% for $|\omega| \le 0.14$) one obtains

$$H_{\mathrm{EWLS}}(\omega) \cong \frac{1-\lambda}{(1-\lambda) + j\lambda\omega} \cong \frac{1}{1 + j\frac{\Omega}{2}} = \tilde{H}_{\mathrm{EWLS}}(\omega).$$

If the same analysis is carried out for a rectangular window, one arrives at

$$H_{\mathrm{SWLS}}(\omega) = \frac{\sin\frac{N\omega}{2}}{N\sin\frac{\omega}{2}} e^{-j\omega\frac{N-1}{2}} \cong \frac{\sin\frac{\Omega}{2}}{\frac{\Omega}{2}} e^{-j\frac{\Omega}{2}} = \tilde{H}_{\mathrm{SWLS}}(\omega).$$

It is easy to check that in both cases the approximations obtained are very accurate over the entire passband of the associated filters, for all reasonable equivalent window widths ($l_\infty \ge 20$).

4.5.3 Estimation delay of WLS estimators

Bias errors observed for WLS estimators are caused by two kinds of distortions: amplitude distortions, governed by the amplitude characteristic $A_{\mathrm{WLS}}(\omega)$ of the associated filter, and phase distortions, governed by the corresponding phase characteristic $\phi_{\mathrm{WLS}}(\omega)$. Shifting the phase of a sinusoidal signal is equivalent to delaying it in time. Hence the expected trajectory of WLS filters can be regarded, to some extent, as a delayed version of the true trajectory. The two popular measures of delay can be used to characterize the time-shifting properties of the associated filter. The phase delay

$$\tau_p(\omega) = -\frac{\phi_{\text{WLS}}(\omega)}{\omega} \tag{4.44}$$

is useful as a measure of delay at specific frequencies, and the group delay

$$\tau_g(\omega) = -\frac{d\phi_{\text{WLS}}(\omega)}{d\omega} \tag{4.45}$$

is a measure of delay over a band (group) of frequencies. Although the time-shifting properties of $H_{\text{WLS}}(\omega)$ are completely characterized by its phase and group delay functions, in many applications it may be convenient to have a single quantity reflecting the mean or dominant delay introduced by the associated filter, called the *estimation delay* τ_e of the WLS algorithm. Since the averaged trajectory of parameter estimates $\{\bar{\theta}(t)\}$ is a corrupted version of the true parameter trajectory $\{\theta(t)\}$, the definition of the estimation delay is not obvious. We will discuss this problem from two different standpoints.

The time-domain approach

Estimation delay τ_e can be defined as an average delay of $\bar{\theta}(t)$ with respect to $\theta(t)$:

$$\tau_{\text{ave}} = \arg\min_{\tau} \mathop{\mathrm{E}}_{\Theta(t)} \left[\| \bar{\theta}(t) - \theta(t - \tau) \|^2 \right], \tag{4.46}$$

assuming that the corresponding expectation exists.

The expressions for τ_{ave} derived in this way obviously depend on the assumed model of parameter changes. For sinusoidal changes

$$\theta(t) = \theta_o \sin(\omega t + \phi)$$

one obtains $\tau_{\text{ave}} = \tau_p(\omega)$. Suppose in turn that evolution of system parameters can be described by the random walk model

$$\theta(t - i) = \sum_{j=0}^{i} w(t), \qquad \text{cov}\{w(t)\} = W,$$

where $\{w(t)\}$ is a white noise process independent of $\{v(t)\}$; to enable the steady-state analysis a *reverse-time* model is adopted. Then

$$\mathop{\mathrm{E}}_{\Theta(t)} \left[\| \bar{\theta}(t) - \theta(t - \tau) \|^2 \right] = \mathop{\mathrm{E}}_{\Theta(t)} \left[\| \sum_{i=0}^{\infty} h_{\text{WLS}}(i) \left(\theta(t - i) - \theta(t - \tau) \right) \|^2 \right]$$

$$= \text{tr}\{W\} \sum_{i=0}^{\infty} \sum_{j=0}^{\infty} h_{\text{WLS}}(i) h_{\text{WLS}}(j) m(\tau, i, j), \tag{4.47}$$

where

$$m(\tau, i, j) = \tau + \min(i, j) - \min(\tau, i) - \min(\tau, j).$$

The problem of minimizing the double sum in (4.47) is equivalent to minimizing the following expression:

$$\tau - 2 \sum_{i=0}^{\infty} h_{\text{WLS}}(i) \min(\tau, i) \cong \tau - 2 \int_{0}^{\infty} \tilde{h}_{\text{WLS}}(s) \min(\tau, s) \, ds,$$

leading to the condition

$$\int_{0}^{\tau} \tilde{h}_{\text{WLS}}(s) \, ds = 0.5.$$

Hence, for sufficiently wide windows, the average delay can be computed as a median of the associated impulse response

$$\tau_{\text{med}} : \sum_{i=0}^{\tau_{\text{med}}} h_{\text{WLS}}(i) \cong 0.5.$$

The frequency-domain approach

Since the associated filter $h_{\text{WLS}}(i)$ is lowpass, its time-shifting properties can be roughly characterized by the nominal (low-frequency) delay time

$$\tau_{\text{nom}} = \lim_{\omega \mapsto 0} \tau_p(\omega) = \lim_{\omega \mapsto 0} \tau_g(\omega) = \sum_{i=0}^{\infty} iw(i).$$

For a typical window the nominal delay is approximately half its equivalent width,

$$\tau_{\text{nom}} \cong \frac{l_{\infty}}{2},$$

e.g. for the rectangular window $\tau_{\text{nom}} = (N-1)/2$ and for the exponential window $\tau_{\text{nom}} = \lambda/(1-\lambda)$.

4.5.4 Matching characteristics of WLS estimators

After adopting one of the definitions for the estimation delay τ_e, the mean square parameter matching error

$$\mathcal{M}(t) = \text{E}[\|\widehat{\theta}(t) - \theta(t - \tau_e)\|^2] \tag{4.48}$$

can be written in a form analogous to (4.23):

$$\mathcal{M}(t) = \mathcal{M}_b(t) + \mathcal{M}_v(t), \tag{4.49}$$

where

$$\mathcal{M}_b(t) = \|\bar{\theta}(t) - \theta(t - \tau_e)\|^2 \tag{4.50}$$

is a bias component of $\mathcal{M}(t)$ and

$$\mathcal{M}_v(t) = \text{E}[\|\widehat{\theta}(t) - \bar{\theta}(t)\|^2] = \mathcal{T}_v(t) \tag{4.51}$$

denotes its variance component, identical with the corresponding component of $\mathcal{T}(t)$. Under (C2) the frequency decomposition of the average bias error becomes

$$\overline{\mathcal{M}_b} = \mathop{\mathrm{E}}_{\Theta(t)} [\|\bar{\theta}(t) - \theta(t - \tau_e)\|^2] \cong \frac{1}{\pi} \int_0^\pi \tilde{E}_{\mathrm{WLS}}(\omega) \, \mathrm{tr}\{S_\theta(\omega)\} \, d\omega, \qquad (4.52)$$

where the quantity

$$\tilde{E}_{\mathrm{WLS}}(\omega) = |1 - e^{-j\omega\tau_e} H_{\mathrm{WLS}}(\omega)|^2, \qquad (4.53)$$

will be called the *parameter matching characteristic* of the WLS estimator.

4.6   The principle of uncertainty

As we already know, WLS estimators are capable of tracking slow parameter variations only. To make this observation more precise, we will introduce a notion of an *estimation bandwidth* $[0, \omega_\beta]$, i.e. the frequency range in which parameter changes can be tracked 'successfully'. The limiting frequency ω_β can be defined in several ways based on the corresponding parameter tracking or matching characteristics of the WLS estimator. For tracking, one can set

$$\omega_\beta : \begin{cases} \displaystyle\sup_{\omega \in [0, \omega_\beta]} E(\omega) \ \leq \beta \\ \text{or} \\ \int_0^{\omega_\beta} E(\omega) d\omega \ \leq \beta \end{cases} \qquad (4.54)$$

where $\beta > 0$ is a constant which governes the desired tracking accuracy. Using the technique of integral approximations one can show that for an adjustable-size window $w^\eta(i)$ (Section 4.5.2)

$$\omega_\beta^\eta \cong \eta\omega_\beta,$$

which means that the estimation bandwidth is inversely proportional to the equivalent window width:

$$\omega_\beta \propto \frac{1}{l_\infty}. \qquad (4.55)$$

Recall that the limiting steady-state estimation accuracy of WLS estimators is also inversely proportional to l_∞:

$$\inf_{\Theta(t)} \mathcal{T}_v(\infty) = \mathrm{tr}\{V_1(\infty)\} = \sigma_v^2 \frac{\mathrm{tr}\{\Phi_o^{-1}\}}{l_\infty}. \qquad (4.56)$$

Combining (4.55) and (4.56) one arrives at the following 'principle of uncertainty':

$$\frac{\mathcal{T}_v(l_\infty)}{\omega_\beta(l_\infty)} \geq c, \qquad (4.57)$$

where c is a constant depending on the process and noise covariances and on the window shape. The inequality derived above is a precise formulation of one of the basic principles of nonstationary system identification, saying that the choice of the window width is always a compromise between the acceptable estimation accuracy and acceptable tracking ability of the corresponding estimation algorithm. Actually, it is evident from (4.57) that the two requirements are contradictory to a certain extent: the greater the equivalent window width, the greater the attainable estimation accuracy, reflected by the decrease in $T_v(l_\infty)$, but the smaller the estimation bandwidth $\omega_\beta(l_\infty)$, i.e. the frequency range in which the time-varying parameters can be tracked 'successfully'.

Figure 4.1 provides good insight into the core of the memory optimization problem. The plots show evolution of an autoregressive coefficient of a first-order AR process

$$y(t) = ay(t-1) + v(t),$$

subject to a jump change at instant $t = 200$, and its exponentially weighted least squares estimates obtained for two values of the forgetting constant, $\lambda = 0.99$ and $\lambda = 0.9$. Note that for $\lambda = 0.99$ (long memory) the estimation algorithm provides high estimation accuracy in periods where parameters do not change or change slowly, but is pretty insensitive to parameter changes. On the other hand, $\lambda = 0.9$ (short memory) guarantees fast parameter tracking at the expense of low estimation accuracy attainable under time-invariant conditions. The forgetting constant should be chosen so as to provide a reasonable balance between the bias component (tracking) and the variance component (accuracy) of the mean square parameter tracking error. We will discuss this problem in some detail in Section 5.5.2. Even though derived for WLS estimators, the uncertainty principle seems to be a fundamental limitation valid for all finite-memory identification algorithms and all kinds of time-varying systems, not necessarily FIR systems.

4.7 Comparison of the EWLS and SWLS approaches

We have shown that WLS estimators characterized by the same equivalent window width behave identically under time-invariant conditions, namely, the corresponding error covariance matrices are approximately the same. This is a very good basis for comparing parameter tracking or matching abilities of estimators characterized by different window *shapes*. Comparing estimators characterized by different values of l_∞ barely makes any sense and it resembles comparing runners that specialize in different distances.

Figure 4.2 shows plots of normalized parameter tracking and matching characteristics for EWLS and SWLS estimators; note that comparing the plots of normalized characteristics, i.e. plots expressed in terms of a scaled frequency $\Omega = \omega l_\infty$, amounts to comparing parameter tracking or matching properties for estimators having the same equivalent memory span. Their interpretation is pretty straightforward. Since the plot of the normalized tracking characteristic of the EWLS estimator lies below that corresponding to the SWLS estimator, it is obvious that the parameter tracking properties of the EWLS approach are better. This effect can be attributed to a higher degree of concentration of the exponential window around $i = 0$, which results in a smaller estimation delay hence smaller bias errors.

On the other hand, for parameter matching applications, rectangular windows should be preferred to exponential windows. The better parameter matching properties of the SWLS approach can be explained by linearity of the corresponding associated phase characteristic. Note that the parameter matching bandwidth is significantly larger than the parameter tracking bandwidth, irrespective of how the estimation

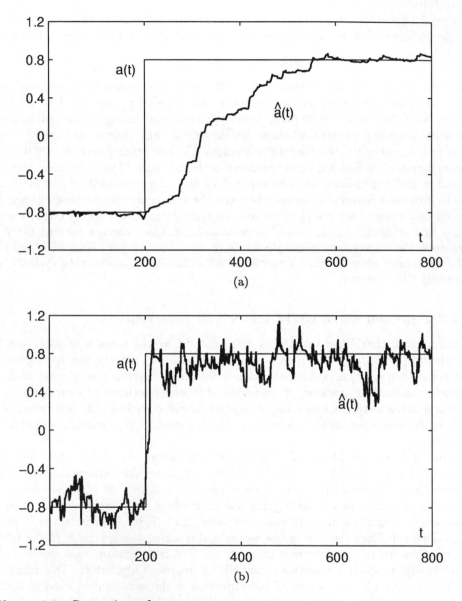

Figure 4.1 Comparison of a true parameter trajectory and its exponentially weighted least squares estimates obtained for $\lambda = 0.99$ (a) and $\lambda = 0.9$ (b).

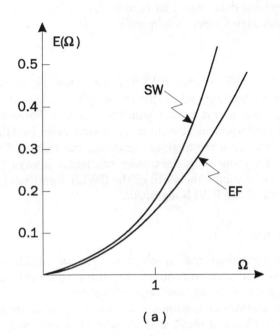

(a)

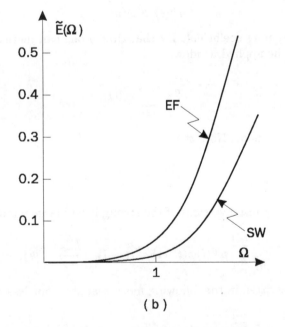

(b)

Figure 4.2 Comparison of normalized parameter tracking and matching characteristics for WLS estimators based on the exponential forgetting (EF) and sliding window (SW) approach.

bandwidth is actually defined. This means that compensation of the estimation delay by introducing a decision delay equal to τ_e may dramatically improve the estimation results, provided that such a delay is admissible.

Remark

When comparing the EWLS and SWLS approaches some authors considered estimation algorithms characterized by the same *effective* window width, arriving at the false conclusion that a rectangular window provides faster tracking. Since the equivalent width of the exponential window is almost twice its effective width (both quantities are identical for a rectangular window), the result of this comparison is not surprising: the longer-memory parameter tracker is always slower. In the case considered here the estimation bandwidth of the SWLS algorithm is roughly twice the estimation bandwidth of the EWLS algorithm.

4.8   Technical issues

Since most of the analytical results obtained for the WLS estimator were de facto derived for its simplified version (4.13), some comments are needed on the mathematical legitimacy of the applied approximation.

We will start from asymptotic arguments. Consider a prototype analog window $\widetilde{w}(\cdot)$ introduced in Section 4.5.2 and recall that an adjustable-size discrete window can be defined in the form

$$w^\eta(i) = \widetilde{w}(\eta i),$$

where $\eta > 0$ is a scaling coefficient. By changing η one can increase or decrease the effective width of the applied window,

$$k_t^\eta = \sum_{i=0}^{t-1} w^\eta(i),$$

without changing its shape. Note that

$$\lim_{\substack{t \to \infty \\ \eta \to 0}} k_t^\eta = \infty.$$

Using one of the generalized versions of the strong law of large numbers, one can show that

$$\frac{1}{k_t^\eta} \sum_{i=0}^{t-1} w^\eta(i) \varphi(t-i) \varphi^T(t-i) \xrightarrow[\substack{\eta \mapsto \infty \\ t \mapsto \infty \\ a.s.}]{} \Phi_o, \qquad (4.58)$$

which can also be written in the following more readable, but less rigorous form:

$$\frac{1}{k_t} \sum_{i=0}^{t-1} w(i) \varphi(t-i) \varphi^T(t-i) \xrightarrow[\substack{k_t \mapsto \infty \\ a.s.}]{} \Phi_o. \qquad (4.59)$$

The derivation of (4.58) given in [137] was based on the m-dependence assumption (B1.2) and the moment condition (B4.2) (with $P = 4$). Extension of (4.58) to weaker mixing conditions imposed on $\{\varphi(t)\}$ is also possible. According to (4.59), when the effective window width k_t tends to infinity, the inverse of the normalized weighted regression matrix

$$\Phi(t) = \frac{1}{k_t} \sum_{i=0}^{t-1} w(i)\varphi(t-i)\varphi^T(t-i)$$

can be approximately replaced with Φ_o^{-1}, which results in (4.14). The problem with the justification given above is that it is based on asymptotic arguments ($k_t \mapsto \infty$). Since in practice the window size is finite and fixed, a more careful analysis is needed to substantiate (4.14).

First, we will show that under assumption (B2.2) on independence of system variables, assumption (B5.2) on mean square parameter boundedness, and assumption (C1) on stationarity of regressors, the discrepancy between $\widehat{\theta}(t)$ and $\widehat{\widehat{\theta}}(t)$ can be quantified in terms of the difference

$$\Delta(t) = \Phi^{-1}(t) - \Phi_o^{-1},$$

namely that

$$E[\|\widehat{\theta}(t) - \widehat{\widehat{\theta}}(t)\|^2] = O\left(\mathrm{tr}\left\{E[\Delta(t)]\right\}\right). \tag{4.60}$$

First of all, observe that

$$\delta(t) = \widehat{\theta}(t) - \widehat{\widehat{\theta}}(t) = \sum_{i=0}^{t-1} x_1(i)x_2(i), \tag{4.61}$$

where

$$x_1(i) = \sqrt{\frac{w(i)}{k_t}}\Delta(t)\varphi(t-i),$$

$$x_2(i) = \sqrt{\frac{w(i)}{k_t}}y(t-i).$$

Using the Schwartz inequality one gets

$$E[\|\delta(t)\|^2] \leq E\left[\sum_{i=0}^{t-1}\|x_1(i)\|^2\right]E\left[\sum_{i=0}^{t-1}\|x_2(i)\|^2\right]. \tag{4.62}$$

Observe that

$$E\left[\sum_{i=0}^{t-1}\|x_1(i)\|^2\right] = E\left[\mathrm{tr}\left\{\sum_{i=0}^{t-1}x_1(i)x_1^T(i)\right\}\right]$$

$$= \mathrm{tr}\left\{E\left[\Delta(t)\Phi(t)\Delta^T(t)\right]\right\}. \tag{4.63}$$

Furthermore, since

$$y(t-i) = \varphi^T(t-i)\theta(t-i) + v(t-i),$$

it holds that

$$\mathrm{E}\left[\sum_{i=0}^{t-1} \|x_2(i)\|^2\right] \leq \frac{1}{k_t} \sum_{i=0}^{t-1} \mathrm{E}[v^2(t-i)]$$

$$+ \frac{1}{k_t} \sum_{i=0}^{t-1} \mathrm{E}[\|\theta(t-i)\|^2]\mathrm{E}[\mathrm{tr}\{w(i)\varphi(t-i)\varphi^T(t-i)\}]$$

$$\leq \sigma_v^2 + C\,\mathrm{tr}\Phi_o = O(1). \qquad (4.64)$$

Combining (4.62) with (4.63) and (4.64) one arrives at (4.60). For a rectangular window of length N and i.i.d. Gaussian regressors (assumption B1.1), $\Phi(t)$ is Wishart distributed with N degrees of freedom, leading [144] to

$$\mathrm{E}\left[\Phi^{-1}(t)\right] = \frac{\Phi_o^{-1}}{N-n-1}$$

and hence

$$\mathrm{E}\left[\Delta(t)\right] = \frac{(n+1)\Phi_o^{-1}}{N-n-1}\ .$$

Therefore

$$\mathrm{E}[\|\widehat{\theta}(t) - \widehat{\widehat{\theta}}(t)\|^2] = O(1/N). \qquad (4.65)$$

A generalization of (4.65) to m-dependent Gaussian regressors can be found in [144]. Given that the regression matrix is invertible in the mean sense, i.e.

$$\mathrm{E}\left[\Phi^{-1}(t)\right] < \infty, \qquad \forall t, \qquad (4.66)$$

a more general (but also more conservative) bound can be derived for non-Gaussian regressors and arbitrary weighting sequences

$$\mathrm{E}[\|\widehat{\theta}(t) - \widehat{\widehat{\theta}}(t)\|^2] < \frac{c}{\sqrt{l_t}}\ , \qquad (4.67)$$

where c is a deterministic constant.

Conditions under which the uniform identifiability assumption (4.66) is fulfilled for WLS estimators with finite-duration windows $w(i) = 0, \forall i \geq i_0$ were examined in [144]. For independent regressors (assumption B1.1) with finite moments of sufficiently high order (assumption B4.2) the necessary and sufficient invertibility condition takes the following form (repeated here for convenience):

B3.2 There exist constants $\gamma > 0$, $\delta > 0$ and $x_0 > 0$ such that for all $x : x_0 > x > 0$,

$$\sup_{\|\beta\|=1} P\left((\beta^T\varphi(t))^2 < x\right) \leq \gamma x^\delta, \qquad \forall t.$$

See Section 2.8.3 for the interpretation of this condition.

Condition (B3.2) is *not necessary* for infinite-duration windows, such as the exponential window. The invertibility results summarized above can easily be extended to weaker mixing conditions, such as m-dependent regressors (B1.2), ϕ-mixing regressors (B1.3) as well as to sequences formed from white noise by linear filters [144].

Remark 1

To prove invertibility of $\Phi(t)$ in the mean sense it suffices to find conditions under which the determinant of $\Phi(t)$ is invertible in the kth moment

$$E\left[\left(\det \Phi(t)\right)^{-k}\right] < \infty \tag{4.68}$$

for sufficiently large k. Quite obviously, in order to satisfy (4.68) one needs

$$P\left(\det \Phi(t) = 0\right) = 0, \tag{4.69}$$

which cannot be guaranteed by imposing on $\varphi(t)$ conditions such as

$$E[\varphi(t)\varphi^T(t)] > 0, \tag{4.70}$$

or its extensions, such as (B4.3). Consider, for example, identification of the FIR system in the case where $\{u(t)\}$ is an i.i.d. sequence taking only the two values $+1$ and -1, with probabilities p $(0 < p < 1)$ and $1 - p$, respectively. The distribution of $\varphi(t)$ is *discrete*, so it is straightforward to show that, for any finite-duration window, condition (4.68) is not fulfilled even though condition (4.70) is fulfilled. ∎

Remark 2

Boundedness of $\Delta(t)$ is sufficient to prove boundedness of the mean square parameter tracking error for WLS estimators. Actually, note that

$$E[\|\widehat{\theta}(t) - \theta(t)\|^2] \leq 2E[\|\widehat{\theta}(t) - \overline{\widehat{\theta}}(t)\|^2] + 2E[\|\overline{\widehat{\theta}}(t) - \theta(t)\|^2]. \tag{4.71}$$

Since boundedness of the second term on the right-hand side of (4.71) can be easily shown for mean square bounded parameter trajectories (B5.2) and regressors with bounded fourth-order moments (B4.2), the inequality (4.71) implies

$$E[\|\widehat{\theta}(t) - \theta(t)\|^2] < \infty.$$

4.9   Computer simulations

Figures 4.3 to 4.12 show identification results obtained using two WLS algorithms (EWLS and SWLS) for the time-varying finite impulse response systems (FIR1 and FIR2) and autoregressive processes (AR1 and AR2) described in Section 2.9.

Figures 4.3 and 4.8 show the output of linear filters associated with the EWLS and SWLS estimators. Note that for both FIR systems these 'theoretical' plots stay in very good agreement with the results of averaging (over 100 simulation runs) the estimated parameter trajectories (Figures 4.5 and 4.10). Estimation results obtained for a single realization of the identified process are shown in Figures 4.4 and 4.9.

Figures 4.6, 4.7 and 4.11, 4.12 show identification results obtained for the time-varying autoregressive processes AR1 and AR2. Even though all theoretical results were derived for FIR systems and cannot be extended to autoregressive processes

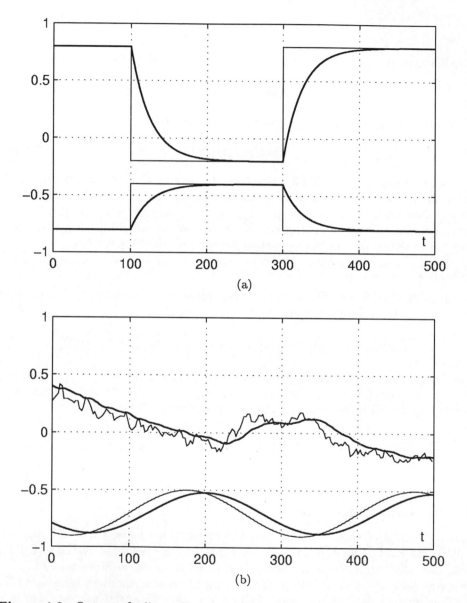

Figure 4.3 Output of a linear filter associated with the exponentially weighted least squares (EWLS) estimator for a system with jump parameter changes (a) and continuous parameter changes (b); the equivalent memory of the estimation algorithm was set to $l_{\infty} = 50$ ($\lambda = 0.96$).

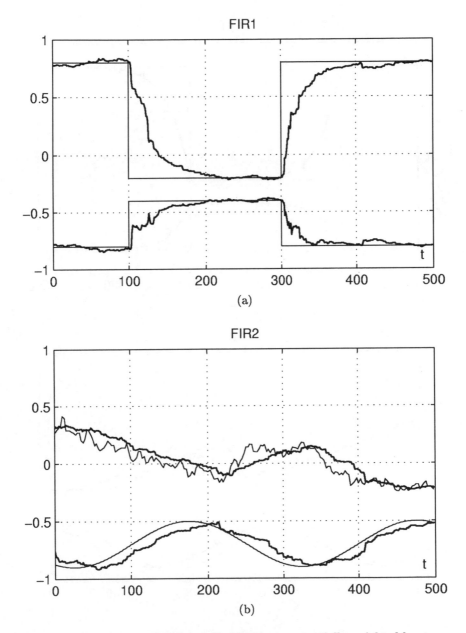

Figure 4.4 Parameter estimates yielded by the exponentially weighted least squares (EWLS) algorithm for a single realization of an FIR system with jump parameter changes (a) and continuous parameter changes (b); the equivalent memory of the estimation algorithm was set to $l_\infty = 50$ ($\lambda = 0.96$).

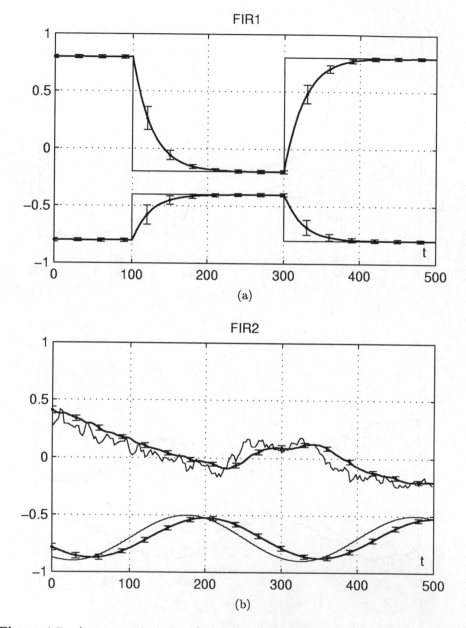

Figure 4.5 Average trajectories of parameter estimates yielded by the exponentially weighted least squares (EWLS) algorithm for an FIR system with jump parameter changes (a) and continuous parameter changes (b); vertical bars show standard deviation of the estimates.

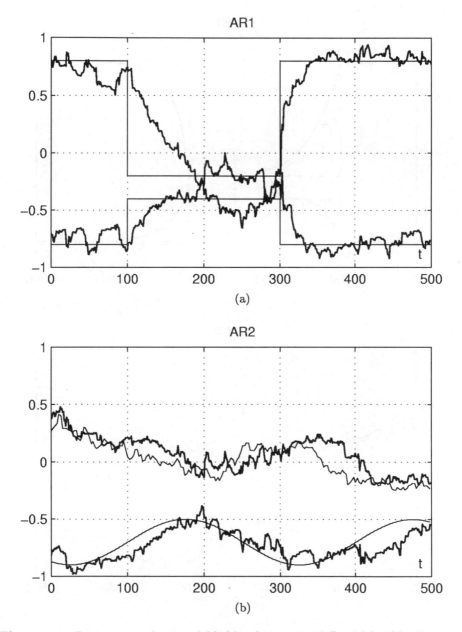

Figure 4.6 Parameter estimates yielded by the exponentially weighted least squares (EWLS) algorithm for a single realization of an AR system with jump parameter changes (a) and continuous parameter changes (b); the equivalent memory of the estimation algorithm was set to $l_\infty = 50$ ($\lambda = 0.96$).

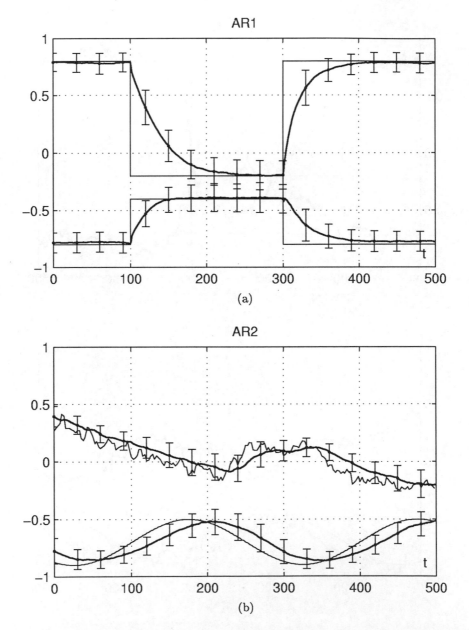

Figure 4.7 Average trajectories of parameter estimates yielded by the exponentially weighted least squares (EWLS) algorithm for an AR system with jump parameter changes (a) and continuous parameter changes (b); vertical bars show standard deviation of the estimates.

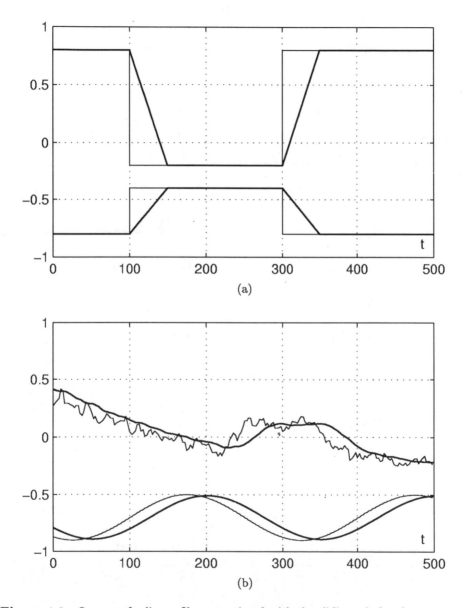

Figure 4.8 Output of a linear filter associated with the sliding window least squares (SWLS) estimator for a system with jump parameter changes (a) and continuous parameter changes (b); the equivalent memory of the estimation algorithm was set to $l_\infty = 50$ ($N = 50$).

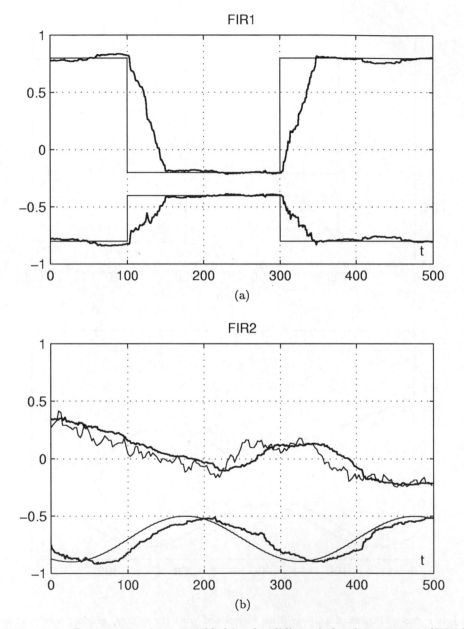

Figure 4.9   Parameter estimates yielded by the sliding window least squares (SWLS) algorithm for a single realization of an FIR system with jump parameter changes (a) and continuous parameter changes (b); the equivalent memory of the estimation algorithm was set to $l_\infty = 50$ ($N = 50$).

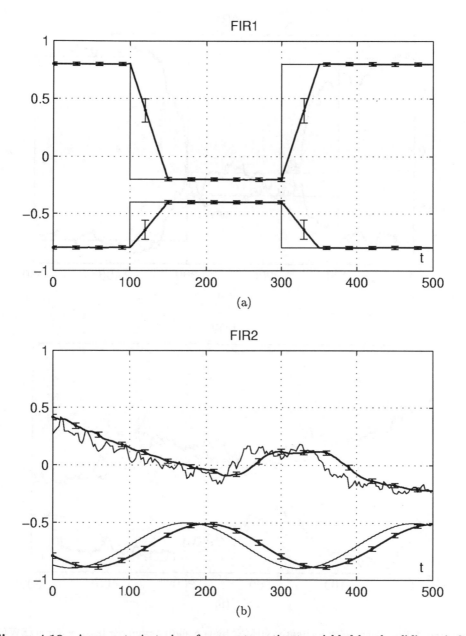

Figure 4.10 Average trajectories of parameter estimates yielded by the sliding window least squares (SWLS) algorithm for an FIR system with jump parameter changes (a) and continuous parameter changes (b); vertical bars show standard deviation of the estimates.

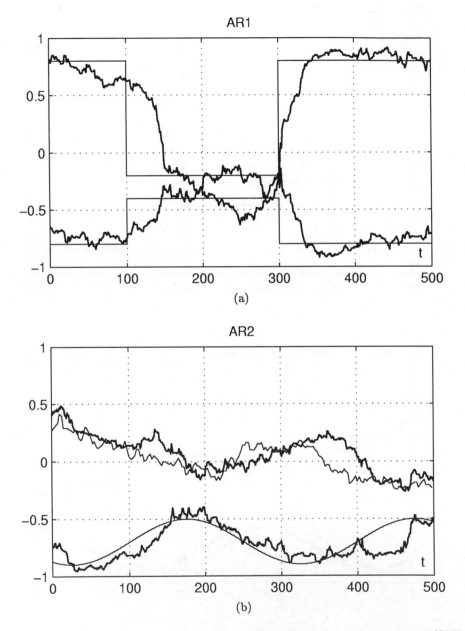

Figure 4.11 Parameter estimates yielded by the sliding window least squares (SWLS) algorithm for a single realization of an AR system with jump parameter changes (a) and continuous parameter changes (b); the equivalent memory of the estimation algorithm was set to $l_\infty = 50$ ($N = 50$).

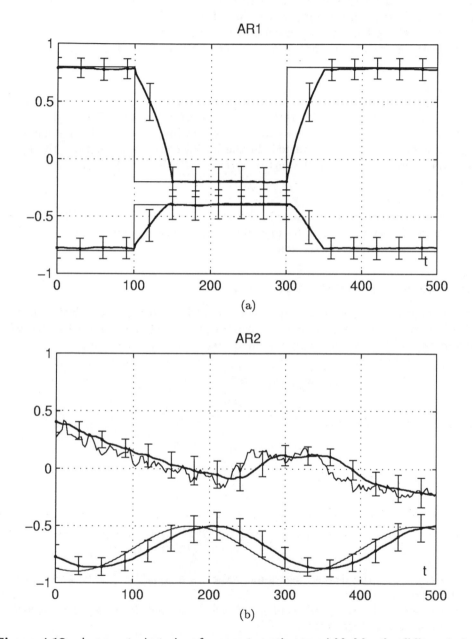

Figure 4.12 Average trajectories of parameter estimates yielded by the sliding window least squares (SWLS) algorithm for an AR system with jump parameter changes (a) and continuous parameter changes (b); vertical bars show standard deviation of the estimates.

(for an autoregressive process the regression matrix $\Phi_o(t) = \text{cov}[\varphi(t)]$ is parameter-dependent and hence also time-dependent), the plots shown in Figures 4.6, 4.11 (single realizations) and Figures 4.7, 4.12 (ensemble averages) very much resemble those obtained for the finite impulse systems FIR1 and FIR2. This shows that at the *qualitative* level the concept of associated time and frequency characteristics can be used to describe tracking behavior of WLS filters for a wider class of processes than FIR.

Remark 1

All plots show the steady-state tracking behavior of WLS algorithms. To reach the steady state, parameter estimation was initialized at instant $t = -500$, i.e. before the analysis interval $T = [0, 500]$ started. In the initial convergence period, the system parameters were constant $(a_1(t) = a_1(0), a_2(t) = a_2(0)$ for $t < 0)$.

Remark 2

The greater variance of the estimates in the AR case, compared to the FIR case, is the consequence of a smaller signal-to-noise ratio. The tracking capabilities of the WLS algorithm are approximately the same in both cases.

4.10   Extension to ARMAX processes

The exponential forgetting technique can be readily extended to identification of ARMAX systems. The exponentially weighted prediction error (EWPE) algorithm, which minimizes the exponentially weighted prediction error criterion

$$J_\lambda(\theta, \Xi(t), \Xi(0)) = \sum_{i=0}^{t-1} \lambda^i v^2(t - i, \theta),$$

can be summarized as follows:

$$
\begin{aligned}
\widehat{\theta}(t) &= \widehat{\theta}(t-1) + K(t)\epsilon(t), \\
\epsilon(t) &= y(t) - \widetilde{\varphi}^T(t)\widehat{\theta}(t-1), \\
K(t) &= \frac{P(t-1)\psi(t)}{\lambda + \psi^T(t)P(t-1)\psi(t)}, \\
P(t) &= \frac{1}{\lambda}\left[P(t-1) - \frac{P(t-1)\psi(t)\psi^T(t)P(t-1)}{\lambda + \psi^T(t)P(t-1)\psi(t)}\right],
\end{aligned}
\tag{4.72}
$$

where $\psi(t)$ denotes the filtered regression vector (see equation (3.53)) and $\widehat{\theta}(t)$ must pass the invertibility test.

The exponentially weighted version of the extended least squares (EWELS) algorithm has an analogous form:

$$
\begin{aligned}
\widehat{\theta}(t) &= \widehat{\theta}(t-1) + K(t)\epsilon(t), \\
\epsilon(t) &= y(t) - \widetilde{\varphi}^T(t)\widehat{\theta}(t-1), \\
K(t) &= \frac{P(t-1)\widetilde{\varphi}(t)}{\lambda + \widetilde{\varphi}^T(t)P(t-1)\widetilde{\varphi}(t)},
\end{aligned}
$$

$$P(t) = \frac{1}{\lambda}\left[P(t-1) - \frac{P(t-1)\tilde{\varphi}(t)\tilde{\varphi}^T(t)P(t-1)}{1+\tilde{\varphi}^T(t)P(t-1)\tilde{\varphi}(t)}\right]. \qquad (4.73)$$

Due to analytical difficulties the properties of both algorithms listed above were examined for time-invariant systems only. It can be shown that the true parameter vector is a stationary point of the EWELS algorithm provided that the identified system obeys the strict positive realness condition (3.56). The EWPE algorithm is free from this limitation.

Comments and extensions

Section 4.2

- The initial convergence of the EWLS algorithm was studied in some detail by Moustakides [127]. As shown there, the settling time of the EWLS filter depends not only on the initialization matrix but also on the level of the measurement noise. If the signal-to-noise ratio (for an FIR system) is 'large' or 'medium', the norm of the initialization matrix P_o should be 'large' to guarantee quick initial convergence, exactly as for the RLS algorithm (Section 3.1.2). On the other hand, in the low-SNR environment it is preferable to initialize the algorithm using a matrix with a 'small' norm.

Section 4.3

- The equivalent width of an exponential window was established independently by many authors; see Niedźwiecki [131], Medaugh and Griffiths [124], Ling and Proakis [109] and Porat [158]. An expression describing the equivalent width of an arbitrarily shaped window was derived by Niedźwiecki [132]. Note that an identical measure of window width (given without any theoretical justification) was proposed by Makhoul and Cosell in an interesting study into how window shape affects the quality of speech in LPC systems [120].

- When self-tuning minimum variance regulators are designed, one of the system coefficients, preferably the high-frequency gain b_1 in (1.67) or (1.68), is usually fixed at a preselected value β and not estimated. Such gain freezing improves numerical conditioning of the identification algorithm and prevents parameter estimates from drifting (see the remarks in Section 8.1). Even though in general the assumed gain β differs from b_1, for a wide range of ratios $\beta/b_1 > 0$ a self-tuning regulator converges to the optimal regulator (i.e. a true minimum variance regulator). Since identification is carried out in a closed loop, the estimation bias is automatically compensated by feedback [9].
  A strange effect was pointed out by Niedźwiecki [140]. According to [140], when a finite-memory identification is combined with gain freezing, the equivalent 'closed-loop' memory of the estimator depends on the ratio β/b_1, namely $l_\infty^c = (\beta/b_1)l_\infty$, where l_∞ denotes the 'open-loop memory' of an adaptive filter. The practical consequences of this are quite sound. Since the true gain coefficient b_1 is usually not known and β can take a wide range of values without affecting the stability of the closed-loop system (e.g. between 1 and 100 in the ore crushing experiment described by Borisson and Syding [30]), one has to be cautious in choosing the

memory settings, such as the forgetting constant λ in the EWLS algorithm. Certain values of λ are often regarded as 'reasonable': $\lambda = 0.99$ or $\lambda = 0.98$ are the typical values recommended for practical applications. The results mentioned above indicate that when the gain freezing technique is used, such advice may be misleading unless the true value of b_1 (or the range of values) is known a priori. If $b_1 = 1$, $\beta = 0.1$ and $\lambda = 0.99$ the 'closed-loop' equivalent memory of the EWLS filter is only 20 instead of 200; conversely, if $b_1 = 1$, $\beta = 10$ and $\lambda = 0.90$ the equivalent memory span is 200! This also means that if the true b_1 varies with time, the algorithm's memory is time-varying.

Section 4.4

- Analysis of the tracking capabilities of WLS estimators reveals intriguing dualities between time-varying identification and signal processing. This duality, pointed out by Willsky [196] for time-invariant systems, is further explored in Niedźwiecki [142]. The two-stage estimation scheme proposed in [142] combines the standard WLS identification with lowpass filtering of parameter estimates. It is shown that *explicit* filtering of parameter estimates is, to some extent, equivalent to *implicit* filtering imposed by the WLS approach; namely, the parameter tracking properties of such filtered weighted least squares (FWLS) estimators are approximately the same as the tracking capabilities of the WLS estimator characterized by the appropriately defined weighting sequence. The main advantage of this approach is that it allows for efficient implementation of banks of adaptive filters characterized by different memory lengths but without compromising the good tracking capabilities of WLS estimators. It also provides the designer with greater flexibility in shaping the window.

Section 4.5

- The concept of associated frequency characteristics of WLS estimators is due to Niedźwiecki [133], [137], [138]. For EWLS estimators some elements of this analysis can be found in the paper of Eleftheriou and Falconer [44]. Frequency-domain analysis of the bias errors was presented by Niedźwiecki [143] (some of the results presented in [143] were later 'rediscovered' in [107]).

Section 4.6

- The 'uncertainty principle' formulated in Section 4.6 resembles in many ways the analogous principle established for nonparametric spectrum estimators; there is an excellent discussion by Priestley [159]. Note, however, that the problems solved in both cases are entirely *different* in nature. For this reason, one should be cautious not to generalize the known results on design of spectral windows, used to estimate the spectral density function of a *stationary* signal, to the estimation of time-varying coefficients of a *nonstationary* system.

5

Least Mean Squares

5.1   Estimation principles

The least mean squares (LMS) approach to system identification or tracking is based on a simple gradient-based error minimization strategy. Denote by $J(\theta)$ the performance measure parameterized in terms of θ. Assuming that $J(\theta)$ is differentiable with respect to θ, a simple iterative algorithm searching for its minimum can be constructed, using the method of steepest descent:

$$\widehat{\theta}^{(i)} = \widehat{\theta}^{(i-1)} - \frac{1}{2}\mu\nabla_{\theta}J(\widehat{\theta}^{(i-1)}),\qquad(5.1)$$

where μ denotes a small positive stepsize parameter. The minus sign in (5.1) assures that we move *down* the local gradient, i.e. towards the minimum of $J(\cdot)$ and the scaling factor $1/2$ is introduced to simplify further results.

Let

$$\theta^{\star} = \arg\min_{\theta} J(\theta).$$

If $J(\theta)$ is quadratic around $\theta^{\star}$ and μ is sufficiently small, one can easily show that

$$\widehat{\theta}^{(i)} \underset{i\mapsto\infty}{\longmapsto} \theta^{\star},$$

provided that the initial estimate $\widehat{\theta}^{(0)}$ is close enough to $\theta^{\star}$ (Figure 5.1).

If the analyzed system is time invariant and the joint second-order statistic of $(y(t), \varphi(t))$ is known a priori one can set

$$J(\theta) = \mathrm{E}\left[v^2(t,\theta)\right]\qquad(5.2)$$

where

$$v(t,\theta) = y(t) - \varphi^T(t)\theta.$$

If no prior knowledge is available and/or the system is time-varying, one can use the instantaneous quality measure

$$J_t(\theta) = v^2(t,\theta),\qquad(5.3)$$

obtained by neglecting the expectation sign in (5.2). By analogy with (5.1), the recursive algorithm capable of following $\theta(t)$ can be written in the form

$$\widehat{\theta}(t) = \widehat{\theta}(t-1) - \frac{1}{2}\mu\nabla_{\theta}J_t(\widehat{\theta}(t-1)).\qquad(5.4)$$

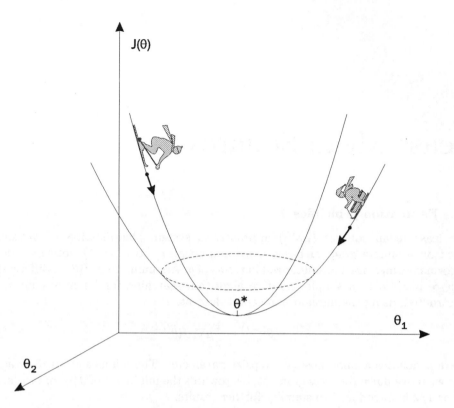

Figure 5.1   The steepest descent minimization strategy: always move down the hill.

Observe that

$$\nabla_\theta J_t(\theta) = -2\varphi(t)v(t,\theta),$$

allowing one to rewrite (5.4) in the form known as a standard LMS algorithm:

$$\widehat{\theta}(t) = \widehat{\theta}(t-1) + \mu\varphi(t)\epsilon(t), \tag{5.5}$$

where

$$\epsilon(t) = v(t,\widehat{\theta}(t-1)) = y(t) - \varphi^T(t)\widehat{\theta}(t-1)$$

denotes the one-step-ahead prediction error.

There are some important differences between (5.1) and (5.5). The first algorithm is iterative and deterministic. Note that if the cost function $J(\cdot)$ is convex and differentiable around $\theta^\star$ then

$$\lim_{\theta \mapsto \theta^\star} \nabla_\theta J(\theta) = 0,$$

which guarantees that the correction term in (5.1) will eventually die out to zero as the iterative search continues.

The second algorithm is recursive and stochastic. Observe that even if the true parameter vector is time invariant, i.e.

$$\theta(t) = \theta^\star, \qquad \forall t,$$

the gradient of $J_t(\cdot)$ will not vanish for $t \mapsto \infty$ unless the input noise $\{v(t)\}$ is zero. Actually, observe that in the case considered

$$v(t, \theta^\star) = v(t),$$

which means that the correction term in (5.4),

$$\nabla_\theta J_t(\widehat{\theta}(t-1)) = -2\varphi(t)\left[\varphi^T(t)(\theta^\star - \widehat{\theta}(t-1)) + v(t)\right],$$

is nonzero even if $\widehat{\theta}(t)$ approaches $\theta^\star$. This means in turn that, for a constant step size μ, the LMS recursion will never converge to a constant parameter vector. Even in a stationary case the parameter estimates will fluctuate around the true parameter values (note that $\mathrm{E}[\nabla_\theta J_t(\theta^\star)] = 0$), which is an obvious sign of the finite-memory property of the algorithm.

5.2   Convergence and stability of LMS algorithms

We will start with the following example.

Example 5.1

Consider a scalar time-invariant system governed by

$$y(t) = \varphi(t)\theta + v(t), \tag{5.6}$$

where θ denotes an unknown (but constant) coefficient and $\varphi(t)$ is a measurable regression variable. Assume that $\{\varphi(t)\}$ and $\{v(t)\}$ are independent sequences of independent random variables with zero mean and variances σ_φ^2 and σ_v^2, respectively. Finally, assume that

$$\mathrm{E}[\varphi^4(t)] = c_\varphi^4 < \infty. \tag{5.7}$$

Denote by

$$\widetilde{\theta}(t) = \theta - \widehat{\theta}(t)$$

the parameter estimation error yielded by the LMS algorithm

$$\widehat{\theta}(t) = \widehat{\theta}(t-1) + \mu\varphi(t)(y(t) - \varphi(t)\widehat{\theta}(t-1)). \tag{5.8}$$

Combining (5.6), (5.7) and (5.8) one arrives at the following error equation:

$$\widetilde{\theta}(t) = (1 - \mu\varphi^2(t))\widetilde{\theta}(t-1) - \mu\varphi(t)v(t). \tag{5.9}$$

To study convergence in the mean of the obtained recursion, one can evaluate expectations of both sides of (5.9):

$$\mathrm{E}[\widetilde{\theta}(t)] = (1 - \mu\sigma_\varphi^2)\mathrm{E}[\widetilde{\theta}(t-1)]. \tag{5.10}$$

According to (5.10), if

$$|1 - \mu\sigma_\varphi^2| < 1,$$

which is equivalent to

$$0 < \mu < \frac{2}{\sigma_\varphi^2},$$ (5.11)

the mean value of the estimation error will asymptotically decay to zero irrespective of the imposed initial conditions, i.e.

$$E[\widehat{\theta}(t)] \xrightarrow[t \to \infty]{} \theta.$$

To study the mean square convergence of (5.8) we will square and take expectations of both sides of (5.9):

$$\sigma_\theta^2(t) = f(\mu)\sigma_\theta^2(t-1) + \mu^2 \sigma_\varphi^2 \sigma_v^2,$$ (5.12)

where

$$f(\mu) = 1 - 2\mu\sigma_\varphi^2 + \mu^2 c_\varphi^4.$$

Since (5.11) is a linear time-invariant difference equation for $\sigma_\theta^2(t)$, it can be solved explicitly. To guarantee stability of (5.12) in the mean square sense, one should choose the stepsize μ so as to fulfill

$$|f(\mu)| < 1,$$

which is equivalent to

$$f(\mu) < 1$$

since $f(\mu) = E[(1 - \mu\varphi(t))^2] \geq 0$.

Solution of (5.12) is straightforward and leads to the following constraint on μ:

$$0 < \mu < \frac{2\sigma_\varphi^2}{c_\varphi^4}.$$ (5.13)

Note that the mean square stability condition (5.13) is more stringent than the mean stability condition (5.11) since, according to the Schwartz inequality,

$$c_\varphi^4 = E[\varphi^4(t)] \geq (E[\varphi^2(t)])^2 = \sigma_\varphi^4.$$

If the distribution of $\varphi(t)$ is Gaussian then

$$c_\varphi^4 = 3\sigma_\varphi^4,$$

leading to

$$0 < \mu < \frac{2}{3\sigma_\varphi^2}.$$

Here the mean square stability bound is three times smaller than the mean stability bound (5.11).

If condition (5.13) is met, the asymptotic solution of (5.12) can be obtained in the form

$$\sigma_\theta^2(\infty) = \lim_{t \to \infty} \sigma_\theta^2(t) = \frac{1}{1 - \mu\frac{c_\varphi^4}{2\sigma_\varphi^2}} \frac{\mu}{2}\sigma_v^2.$$ (5.14)

■

Example 5.1, despite its simplicity, clearly illustrates some basic properties of the LMS algorithm:

1. The LMS filter is stable provided that the stepsize parameter μ is sufficiently small.
2. Under time-invariant conditions the mean square parameter estimation error approaches a constant value proportional to μ.

5.2.1 Analysis for independent regressors

Convergence in the mean

The error equation associated with the general LMS algorithm (5.5),

$$\tilde{\theta}(t) = \left(I_n - \mu\varphi(t)\varphi^T(t)\right)\tilde{\theta}(t-1) - \mu\varphi(t)v(t), \tag{5.15}$$

can be analyzed along the same lines as the scalar equation (5.9). Note that the expectation of the second term on the right-hand side of (5.15) is zero. Hence, given that $\{\varphi(t)\}$ is a stationary i.i.d. sequence,

$$\mathrm{E}[\tilde{\theta}(t)] = \mathrm{E}[\left(I_n - \mu\varphi(t)\varphi^T(t)\right)\tilde{\theta}(t-1)]$$

$$= \mathrm{E}[(I_n - \mu\varphi(t)\varphi^T(t))]\mathrm{E}[\tilde{\theta}(t-1)] = (I_n - \mu\Phi_o)\mathrm{E}[\tilde{\theta}(t-1)] \tag{5.16}$$

To guarantee convergence in the mean of the LMS algorithm, that is

$$\mathrm{E}[\tilde{\theta}(t)] \xrightarrow[t \to \infty]{} 0,$$

or equivalently

$$\mathrm{E}[\hat{\theta}(t)] \xrightarrow[t \to \infty]{} \theta,$$

the absolute values of all eigenvalues of the matrix $(I_n - \mu\Phi_o)$ should be smaller than 1. To get more insights into the conditions and geometry of convergence, we shall analyze (5.15) in a different system of coordinates. Denote by Q a unitary matrix, made up of the eigenvectors of Φ_o, converting Φ_o into a diagonal form:

$$Q^T Q = QQ^T = I_n, \qquad Q^T \Phi_o Q = \Lambda_o \tag{5.17}$$

where

$$\Lambda_o = \mathrm{diag}\{\lambda_1(\Phi_o), \dots, \lambda_n(\Phi_o)\}$$

is a diagonal matrix made up of the eigenvalues of Φ_o. Multiplying both sides of (5.15) with Q^T and denoting

$$\gamma(t) = Q^T \tilde{\theta}(t)$$

one arrives at

$$\mathrm{E}[\gamma(t)] = (I_n - \mu\Lambda_o)\mathrm{E}[\gamma(t-1)]. \tag{5.18}$$

Denoting by $\gamma_i(\tau)$ the ith component of $\gamma(t)$, one can rewrite (5.18) in the following explicit, decoupled form:

$$\mathrm{E}[\gamma_i(t)] = (1 - \mu\lambda_i(\Phi_o))^t \mathrm{E}[\gamma_i(0)], \qquad i = 1, \dots, n. \tag{5.19}$$

Since both the stepsize μ and the eigenvalues of the covariance matrix Φ_o must be positive, the stability/convergence condition of (5.18), i.e. the stability/convergence condition of (5.16), can be put in the form

$$|1 - \mu\lambda_{\max}(\Phi_o)| < 1, \tag{5.20}$$

where $\lambda_{\max}(\Phi_o)$ denotes the maximum eigenvalue of Φ_o:

$$\lambda_{\max}(\Phi_o) = \max\{\lambda_1(\Phi_o), \ldots, \lambda_n(\Phi_o)\}.$$

Solving (5.20) one obtains the following constraint:

$$0 < \mu < \frac{2}{\lambda_{\max}(\Phi_o)}, \tag{5.21}$$

which is an extension of (5.11).

Remark

The mean convergence condition stated above may be difficult to verify in practice. Note that

$$\lambda_{\max}(\Phi_o) \le \sum_{i=1}^{n} \lambda_i(\Phi_o) = \mathrm{tr}\{\Phi_o\}.$$

Hence (5.21) can be replaced with the more conservative constraint

$$0 < \mu < \frac{2}{\mathrm{tr}\{\Phi_o\}}, \tag{5.22}$$

which is much easier to check since

$$\mathrm{tr}\{\Phi_o\} = \mathrm{tr}\left\{\mathrm{E}[\varphi(t)\varphi^T(t)]\right\} = \mathrm{E}[\|\varphi(t)\|^2]. \tag{5.23}$$

Note that, for FIR systems, equation (5.23) can be interpreted as the total average input signal power (tap input power), a quantity which can be easily estimated from the received signal.

∎

Analysis of the diagonalized error equation (5.18) reveals another important feature of the LMS algorithm – its sensitivity to the covariance structure of the regression vector $\varphi(t)$. Even though under (5.20) all components of $\gamma(t)$ converge exponentially to zero, the actual speed of convergence depends on the eigenvalue spread of the covariance matrix Φ_o; convergence takes place faster in those directions in the parameter space which are associated with large-eigenvalue eigenvectors of Φ_o and slower in directions determined by small-eigenvalue eigenvectors. If the *eigenvalue disparity* index of Φ_o is large,

$$\delta(\Phi_o) = \frac{\lambda_{\max}(\Phi_o)}{\lambda_{\min}(\Phi_o)} \gg 1,$$

the variation of the convergence speed over different components of γ_i may be significant.

Mean square convergence

Convergence of $\widehat{\theta}(t)$ to θ 'in the mean' ensures that under stationary conditions the LMS parameter estimates will oscillate around the true parameter values, i.e. there will be no estimation bias. Convergence in the mean, however, is not sufficient to ensure the mean square stability of the LMS algorithm. To prove stochastic stability of (5.15) one should show that variations of $\hat{\theta}(t)$ around θ are bounded in some probabilistic sense. To derive conditions of the mean square boundedness of the estimation error $\tilde{\theta}(t)$, we will square both sides of (5.15) and take the expectation:

$$E[\|\tilde{\theta}(t)\|^2] = E[\|\left(I_n - \mu\varphi(t)\varphi^T(t)\right)\tilde{\theta}(t-1)\|^2]$$

$$- 2\mu E\left[\tilde{\theta}^T(t-1)\left(I_n - \mu\varphi(t)\varphi^T(t)\right)\varphi(t)v(t)\right] + \mu^2 E[\|\varphi(t)v(t)\|^2]$$

$$= I_1 + I_2 + I_3. \tag{5.24}$$

Assuming that $\varphi(t)$ and $v(t)$ have finite fourth-order moments and that regression vectors are independent, one obtains

$$I_1 = E\left[\tilde{\theta}^T(t-1)\left(I_n - \mu\varphi(t)\varphi^T(t)\right)^2\tilde{\theta}(t-1)\right]$$

$$= E\left[\tilde{\theta}^T(t-1)\Psi_o\tilde{\theta}(t-1)\right], \tag{5.25}$$

where

$$\Psi_o = E[\left(I_n - \mu\varphi(t)\varphi^T(t)\right)^2] = I_n - 2\mu\Phi_o + \mu^2 C_o$$

and

$$C_o = E[\left(\varphi(t)\varphi^T(t)\right)^2].$$

Observe that

$$I_1 \leq \lambda_{\max}(\Psi_o)E[\|\tilde{\theta}(t-1)\|^2]. \tag{5.26}$$

Furthermore, since μ is positive,

$$\lambda_{\max}(\Psi_o) \leq 1 - 2\mu\lambda_{\min}(\Phi_o) + \mu^2\lambda_{\max}(C_o) = f_{\max}(\mu). \tag{5.27}$$

Since $v(t)$ has zero mean and is independent of $\varphi(t)$ and $\tilde{\theta}(t-1)$,

$$I_2 = 0. \tag{5.28}$$

Finally,

$$I_3 = \mu^2 E[\|\varphi(t)\|^2]E[v^2(t)] = \mu^2 c, \tag{5.29}$$

where

$$c = \sigma_v^2 \text{tr}\{\Phi_o\}.$$

Combining (5.24) with (5.26), (5.28) and (5.29) one arrives at

$$E[\|\tilde{\theta}(t)\|^2] \leq f_{\max}(\mu)E[\|\tilde{\theta}(t-1)\|^2] + \mu^2 c, \tag{5.30}$$

leading to the following *sufficient* condition for the mean square stability of (5.15):

$$f_{\max}(\mu) < 1,$$

resulting in (cf. (5.27))

$$0 < \mu < 2\frac{\lambda_{\min}(\Phi_o)}{\lambda_{\max}(C_o)}. \tag{5.31}$$

One can show that

$$\lambda_{\max}(C_o) \geq \lambda_{\max}^2(\Phi_o),$$

which is a straightforward consequence of the fact that $C_o \geq \Phi_o^2$. Hence

$$2\frac{\lambda_{\min}(\Phi_o)}{\lambda_{\max}(C_o)} \leq \frac{1}{\lambda_{\max}(\Phi_o)} .$$

The mean square stability constraint (5.31) is therefore more stringent than the mean convergence constraint (5.21).

5.2.2   Analysis for dependent regressors

Conditions of stability

Note that the error equation (5.15) can be rewritten in the form

$$\tilde{\theta}(t) = F(t,0)\tilde{\theta}(0) + \mu\sum_{k=1}^{t} F(t,k)\varphi(t)v(t), \tag{5.32}$$

where

$$F(t,k) = \prod_{i=k+1}^{t}\left(I_n - \mu\varphi(i)\varphi^T(i)\right), \qquad F(t,t) = I_n, \tag{5.33}$$

is the fundamental matrix of the homogeneous error equation

$$\tilde{\theta}(t) = \left(I_n - \mu\varphi(i)\varphi^T(i)\right)\tilde{\theta}(t-1). \tag{5.34}$$

For stationary independent regressors the analysis carried out in the previous subsection was pretty straightforward since

$$\mathrm{E}\left[F(t,k)\right] = \prod_{i=k}^{t}\mathrm{E}\left[\left(I_n - \mu\varphi(i)\varphi^T(i)\right)\right] = \left(I_n - \mu\Phi_o\right)^{t-k}.$$

Note that if the convergence condition (5.21) is fulfilled, the expectation of $F(t,k)$ tends to zero exponentially fast (the dominant time constant is associated with the largest eigenvalue of Φ_o), ensuring exponential stability of (5.34) and hence also of (5.32). Unfortunately, as already remarked, the independence assumption is almost never fulfilled in practice. The analysis of (5.32) for correlated regressors is considerably more difficult, but the approach is basically the same: the key to the analysis of (5.32) is proving exponential stability of (5.34), i.e. establishing conditions under which for any $p \geq 1$ there exist positive constants c_p, φ_p and μ_p such that

$$\left(\mathrm{E}\left[\| F(t,k) \|^p\right]\right)^{1/p} \leq c_p(1 - \mu\varphi_p)^{t-k}, \qquad \forall t \geq k, \forall \mu \in (0, \mu_p); \tag{5.35}$$

see Guo and Ljung [69] and Macchi [117].

Investigation into products of random matrices and the corresponding stochastic linear difference equations has a long history; see Guo and Ljung [69] and the references therein. Assumptions allowing one to prove exponential stability of (5.34) can be grouped into four categories:

1. Persistence of excitation: the sequence $\{\varphi(t)\}$ must span the entire regression space.
2. Slow adaptation: the stepsize μ has to be sufficiently small.
3. Boundedness: the sequence $\{\varphi(t)\}$ must be bounded in some deterministic or stochastic sense.
4. Mixing: the sequence $\{\varphi(t)\}$ must obey some sort of asymptotic independence condition.

The persistence of the excitation condition is an essential part of all stability proofs. It ensures that the matrix

$$\mu\varphi(i)\varphi^T(i),$$

subtracted from the identity matrix in (5.33), is 'on the average' full rank, otherwise at least one eigenvalue of $F(t, k)$ could remain close to 1, contradicting (5.35). The deterministic (hard) identifiability condition (B3.1), see e.g. Bitmead and Anderson [25] and [26], dominated the early stability studies but was recently replaced with a stochastic (soft) identifiability condition (B3.4); see Guo and Ljung [69]. Moreover, as shown in [69], under some mild boundedness and mixing conditions imposed on $\{\varphi(t)\}$, (B3.4) is a sufficient and *necessary* condition for (5.35) to hold.

The slow adaptation condition is usually expressed in the following forms [117]

$$\mu \ll \frac{2}{\lambda_{\max}(\Phi_o)} \tag{5.36}$$

or

$$\mu \ll \frac{2}{\operatorname{tr}\{\Phi_o\}}. \tag{5.37}$$

It is closely related to the principle of parsimony discussed in Chapter 1; see the next subsection for more details. On a more technical level, it allows one to use the averaging technique when analyzing $F(t, k)$.

The degree of restrictiveness of boundedness conditions imposed on $\{\varphi(t)\}$ is usually adversely affected by the restrictiveness of mixing conditions and vice versa. Strong boundedness conditions, such as the deterministic boundedness assumption (B4.1), allow one to considerably relax mixing conditions [180], [67]; strong mixing conditions, such as the m-dependence assumption (B1.2), can be used to relax boundedness conditions [117]. Guo and Ljung [69] have formulated the set of pretty weak boundedness and mixing conditions, assumptions (B4.3) and (B1.4), which allow one to treat the case of unbounded (e.g. Gaussian) and correlated random regressors.

Analysis based on averaging

Let $\Phi(t) = \mathrm{E}[\varphi(t)\varphi^T(t)]$. If the LMS algorithm is exponentially stable in the sense described above it is possible to show, using the averaging technique, that the covariance matrix of the estimation error

$$\operatorname{cov}[\widehat{\theta}(t)] = \mathrm{E}[\widetilde{\theta}(t)\widetilde{\theta}^T(t)]$$

is approximately equal to $\Pi(t)$, a solution of the following linear deterministic difference equation:

$$\Pi(t+1) = (I_n - \mu\Phi(t))\,\Pi(t)\,(I_n - \mu\Phi(t)) + \mu^2\Phi(t)\sigma_v^2 \tag{5.38}$$

associated with (5.32); see Guo and Ljung [67]. The approximation holds in the sense that

$$\| \mathrm{E}[\tilde{\theta}(t)\tilde{\theta}^T(t)] - \Pi(t) \| \le \sigma(\mu) \, \| \Pi(t) \|, \tag{5.39}$$

where $\sigma(\mu) \mapsto 0$ as $\mu \mapsto 0$, i.e. the approximation becomes tight for small stepsizes (the slow adaptation case).

5.3   Static characteristics of LMS estimators

Suppose that the LMS algorithm is exponentially stable in the sense described in the previous section. Under stationary conditions $(\Phi(t) = \Phi_o)$ the matrix $\Pi(t)$ tends to its steady-state value Π_∞ obtained by solving

$$\Pi_\infty = (I_n - \mu\Phi_o)\Pi_\infty(I_n - \mu\Phi_o) + \mu^2\Phi_o\sigma_v^2, \tag{5.40}$$

or equivalently

$$\Phi_o\Pi_\infty + \Pi_\infty\Phi_o - \mu\Phi_o\Pi_\infty\Phi_o = \mu\Phi_o\sigma_v^2. \tag{5.41}$$

It turns out that equation (5.41) can be easily solved in a different set of coordinates. After postmultiplying equation (5.41) with a unitary matrix Q, defined in (5.17), and premultiplying it with Q^T one obtains

$$\Lambda_o\tilde{\Pi}_\infty + \tilde{\Pi}_\infty\Lambda_o - \mu\Lambda_o\tilde{\Pi}_\infty\Lambda_o = \mu\Lambda_o\sigma_v^2, \tag{5.42}$$

where

$$\tilde{\Pi}_\infty = Q^T\Pi_\infty Q.$$

A careful analysis of (5.42) leads to the conclusion that $\tilde{\Pi}_\infty$ must be a diagonal matrix. Actually, denote by $[A]_{ij}$ the element of the matrix A with coordinates i and j and observe that

$$[\Lambda_o\tilde{\Pi}_\infty]_{ij} = \lambda_i(\Phi_o)\tilde{\pi}_{ij},$$

$$[\tilde{\Pi}_\infty\Lambda_o]_{ij} = \lambda_j(\Phi_o)\tilde{\pi}_{ji},$$

$$[\Lambda_o\tilde{\Pi}_\infty\Lambda_o]_{ij} = \mu\lambda_i(\Phi_o)\lambda_j(\Phi_o)\tilde{\pi}_{ij},$$

where, owing to the symmetry of $\tilde{\Pi}_\infty$,

$$\tilde{\pi}_{ij} = [\tilde{\Pi}_\infty]_{ij} = \tilde{\pi}_{ji}.$$

Since

$$[\mu\Lambda_o\sigma_v^2]_{ij} = 0, \qquad \forall i \ne j,$$

the off-diagonal elements of $\tilde{\Pi}_\infty$ must, according to (5.42), obey the following set of equations:

$$c_{ij}\tilde{\pi}_{ij} = 0, \qquad i,j = 1,\ldots,n, \quad i \ne j,$$

where

$$c_{ij} = \lambda_i(\Phi_o) + \lambda_j(\Phi_o) - \mu\lambda_i(\Phi_o)\lambda_j(\Phi_o).$$

Under the mean stability constraint (5.21) $c_{ij} > 0$ for all $i \ne j$, which entails

$$\tilde{\pi}_{ij} = 0, \qquad \forall i \ne j,$$

i.e. the matrix $\widetilde{\Pi}_\infty$ must be diagonal.

Exploiting the diagonality of $\widetilde{\Pi}_\infty$ one can rewrite (5.42) in the form

$$(2\Lambda_o - \mu\Lambda_o^2)\widetilde{\Pi}_\infty = \mu\Lambda_o\sigma_v^2, \tag{5.43}$$

leading to

$$\widetilde{\Pi}_\infty = \mu(2I_n - \mu\Lambda_o)^{-1} = \text{diag}\left\{\frac{\mu}{2 - \mu\lambda_1(\Phi_o)}, \ldots, \frac{\mu}{2 - \mu\lambda_n(\Phi_o)}\right\}\sigma_v^2.$$

Finally,

$$\Pi_\infty = Q\widetilde{\Pi}_\infty Q^T.$$

Note that in the so-called slow adaptation case (5.36) the following approximation holds:

$$\frac{\mu}{2 - \mu\lambda_i(\Phi_o)} \cong \frac{\mu}{2},$$

resulting in

$$\lim_{t\to\infty} \text{cov}[\widehat{\theta}(t)] \cong \frac{\mu}{2}\sigma_v^2 I_n. \tag{5.44}$$

Remark

Note that (5.44) is an exact solution of a simplified difference equation which can be obtained by neglecting the term proportional to $\mu^2\Pi_\infty$ in (5.40).

5.3.1   Equivalent memory of LMS estimators

Recall that an equivalent memory of a WLS estimator was quantified by comparing the corresponding estimation error covariance matrix with the matrix characterizing the LS estimator; a meaningful comparison was facilitated by the fact that both matrices are identical up to the memory-dependent scaling factor. Unfortunately, extension of this procedure to LMS estimators is not possible, since the matrix Π_∞ has an entirely different structure than its LS/WLS counterparts. Comparison has to be based on a different principle.

Prediction-oriented measure

As already argued in Section 2.4, due to the asymptotic independence of the vector of parameter estimates and the regression vector, holding under the slow adaptation condition, the asymptotic value of the mean square one-step-ahead prediction error for an FIR system subject to a stationary excitation can be expressed in the form

$$\text{E}[\epsilon_{ex}^2(t)] \cong \mathcal{P}[\widehat{\theta}(t)] = \text{E}[\|\,\theta(t) - \widehat{\theta}(t)\,\|_{\Phi_o}^2]$$

$$= \text{tr}\left\{\text{cov}\left[\widehat{\theta}(t)\right]\Phi_o\right\}. \tag{5.45}$$

Since

$$\lim_{t\to\infty} \text{cov}[\widehat{\theta}_{\text{LMS}}(t)] \cong \frac{\mu}{2}\sigma_v^2 I_n,$$

one arrives at

$$P_\infty^{LMS} = \lim_{t \to \infty} P[\widehat{\theta}_{LMS}(t)] \cong \frac{\mu}{2} \text{tr}\{\Phi_o\}\sigma_v^2. \tag{5.46}$$

The analogous quantities for WLS estimators (4.18) are

$$\lim_{t \to \infty} \text{cov}[\widehat{\theta}_{WLS}(t)] \cong \frac{\Phi_o^{-1}}{l_\infty}\sigma_v^2,$$

and

$$P_\infty^{WLS} = \lim_{t \to \infty} P[\widehat{\theta}_{WLS}(t)] \cong \frac{n}{l_\infty}\sigma_v^2. \tag{5.47}$$

If the stepsize μ is chosen so as to satisfy the condition

$$P_\infty^{LMS} = P_\infty^{WLS}, \tag{5.48}$$

both estimation algorithms are characterized by the same mean square prediction errors, i.e. they have identical prediction capabilities under stationary conditions. Based on the observations made above, one can define the asymptotic memory span of the LMS algorithm by solving (5.48) with respect to l_∞:

$$l_\infty^{LMS} = \frac{2n}{\mu \, \text{tr}\{\Phi_o\}} . \tag{5.49}$$

According to (5.49), the equivalent memory of the LMS estimator is inversely proportional to the stepsize μ, an intuitively obvious property as small adaptation gain decreases the potential influence of a single measurement on parameter estimates and vice versa.

As it stems from (5.49), the memory of LMS is also inversely proportional to the magnitude of the regression vector

$$\text{tr}\{\Phi_o\} = \text{E}\left[\|\varphi(t)\|^2\right],$$

which may be pretty troublesome in practice, since it means that the user-dependent quantities (μ) are mixed up in (5.49) with the signal-dependent quantities (Φ_o). In particular, observe that for a stationary FIR system,

$$\text{tr}\{\Phi_o\} = n\sigma_u^2,$$

allowing one to rewrite (5.49) in the form

$$l_\infty = \frac{2}{\mu\sigma_u^2} , \tag{5.50}$$

which means that the memory of the LMS algorithm depends on the stepsize and on the power of the input signal.

Finally, note that (5.49) can be expressed in terms of the 'average' eigenvalue of the matrix Φ_o:

$$l_\infty = \frac{2}{\mu\lambda_{ave}(\Phi_o)} , \tag{5.51}$$

where

$$\lambda_{\text{ave}}(\Phi_o) = \frac{1}{n} \sum_{i=1}^{n} \lambda_i(\Phi_o).$$

The notion of an equivalent memory for the LMS estimator allows one to make an interesting interpretation of the slow adaptation condition (5.37). Using (5.49) one can rewrite (5.37) in the form

$$l_\infty \gg n, \tag{5.52}$$

which is nothing but a precise formulation of the celebrated principle of parsimony: the memory span of the estimation algorithm should be much larger than the number of estimated coefficients.

Remark

The fractional amount by which the steady-state mean square prediction error exceeds its minimum attainable value σ_v^2,

$$\Delta_\infty = \lim_{t \to \infty} \frac{\sigma_\epsilon^2(t) - \sigma_v^2}{\sigma_v^2},$$

is usually called the *final misadjustment*. The corresponding formulas for the WLS and LMS estimators are

$$\Delta_\infty^{\text{WLS}} = \frac{n}{l_\infty}, \qquad \Delta_\infty^{\text{LMS}} = \frac{\mu \operatorname{tr}\{\Phi_o\}}{2}.$$

Tracking-oriented measure

If the primary objective of the estimation algorithm is to minimize the mean square parameter tracking error

$$\mathcal{T}[\widehat{\theta}(t)] = \mathrm{E}[\|\theta(t) - \widehat{\theta}(t)\|^2] = \operatorname{tr}\{\operatorname{cov}[\widehat{\theta}(t)]\},$$

evaluation of the 'equivalent' memory of the LMS filter should be based on a different principle, namely on comparing the parameter tracking abilities of the corresponding algorithms. After solving

$$\mathcal{T}_\infty^{\text{LMS}} = \mathcal{T}_\infty^{\text{WLS}}, \tag{5.53}$$

where

$$\mathcal{T}_\infty^{\text{LMS}} = \lim_{t \to \infty} \mathcal{T}[\widehat{\theta}_{\text{LMS}}(t)] \cong \frac{n\mu}{2} \sigma_v^2,$$

$$\mathcal{T}_\infty^{\text{WLS}} = \lim_{t \to \infty} \mathcal{T}[\widehat{\theta}_{\text{WLS}}(t)] \cong \frac{\operatorname{tr}\{\Phi_o^{-1}\}}{l_\infty} \sigma_v^2,$$

one obtains

$$l_\infty^{\star\text{LMS}} = \frac{2 \operatorname{tr}\{\Phi_o^{-1}\}}{n\mu}, \tag{5.54}$$

where the star was introduced to distinguish between the prediction-oriented and tracking-oriented results (for WLS estimators such distinction is not necessary, i.e. $l_\infty = l_\infty^\star$).

Note that (5.54) can be rewritten as

$$l_\infty^\star = \frac{2\lambda_{\text{ave}}(\Phi_o^{-1})}{\mu}, \qquad (5.55)$$

where $\lambda_{\text{ave}}(\Phi_o^{-1})$ denotes the 'average' eigenvalue of Φ_o^{-1}:

$$\lambda_{\text{ave}}(\Phi_o^{-1}) = \frac{1}{n}\sum_{i=1}^{n}\lambda_i(\Phi_o^{-1}).$$

To compare the two memory measures, observe that

$$\text{tr}\{\Phi_o\} = \sum_{i=1}^{n}\lambda_i(\Phi_o)$$

and

$$\text{tr}\{\Phi_o^{-1}\} = \sum_{i=1}^{n}\lambda_i^{-1}(\Phi_o).$$

Hence, using the inequality

$$\left(\sum_{i=1}^{n}x_i\right)\left(\sum_{i=1}^{n}\frac{1}{x_i}\right) \geq n^2,$$

valid for any sequence of positive numbers $\{x_i\}$ and following in a straightforward way from the Cauchy–Schwartz inequality, one obtains

$$\frac{l_\infty^\star}{l_\infty} = \frac{\text{tr}\{\Phi_o\}\text{tr}\{\Phi_o^{-1}\}}{n^2} \geq 1, \qquad (5.56)$$

where equality holds if and only if $\lambda_1(\Phi_o) = \cdots = \lambda_n(\Phi_o)$, i.e. if Φ_o is similar to an identity matrix. Since

$$\begin{aligned}
\text{tr}\{\Phi_o\} &\leq n\lambda_{\max}(\Phi_o), \\
\text{tr}\{\Phi_o^{-1}\} &\leq \frac{n}{\lambda_{\min}(\Phi_o)},
\end{aligned}$$

the upper bound on the discrepancy between $l_\infty^\star$ and l_∞ can be expressed in terms of the eigenvalue disparity index $\delta(\Phi_o)$:

$$\frac{l_\infty^\star}{l_\infty} \leq \delta(\Phi_o).$$

Since almost all applications of adaptive filters fall into the prediction-oriented category, the predictive measure of the equivalent memory span (5.49) has greater practical significance than the tracking measure (5.54).

5.3.2 Normalized LMS estimators

The memory of an LMS algorithm depends on the magnitude of the regression vector, i.e. it is not scale invariant. Actually, note that the LMS algorithm with stepsize μ, running on a set of measurements scaled by a factor η,

$$y'(t) = \eta y(t), \qquad \varphi'(t) = \eta \varphi(t),$$

will yield identical results as an LMS algorithm with a stepsize μ/η^2 and fed with the original measurements. Quite obviously, this is an undesirable property. Since scaling does not affect the information content of the data, one has the right to expect identification results to be identical for the same stepsize μ.

The scale-invariant version of the LMS algorithm can be obtained by replacing the stepsize μ in (5.5) with a normalized stepsize

$$\frac{\bar{\mu}}{\|\varphi(t)\|^2}. \tag{5.57}$$

The resulting algorithm,

$$\widehat{\theta}(t) = \widehat{\theta}(t-1) + \frac{\bar{\mu}}{\|\varphi(t)\|^2}\varphi(t)\epsilon(t), \tag{5.58}$$

is usually called the normalized LMS (NLMS).

To avoid numerical problems, arising when the squared norm $\|\varphi(t)\|^2$ becomes very small or zero, the denominator in (5.57) is usually replaced with

$$\varepsilon + \|\varphi(t)\|^2,$$

where ε is a small positive constant – a 'safety valve' preventing the normalized stepsize from getting too large.

The mean convergence condition [191], [77] for NLMS is

$$0 < \bar{\mu} < 2, \tag{5.59}$$

and its equivalent memory span can be obtained from

$$l_\infty^{\text{NLMS}} = \frac{2n}{\bar{\mu}}. \tag{5.60}$$

Another useful variant of a scale-invariant LMS algorithm, employing the concept of exponential forgetting, is

$$\widehat{\theta}(t) = \widehat{\theta}(t-1) + \frac{\tilde{\mu}}{r(t)}\varphi(t)\epsilon(t), \tag{5.61}$$

$$r(t) = \lambda r(t-1) + \|\varphi(t)\|^2, \quad 0 < \lambda < 1,$$

and will be referred to as the trace LMS algorithm (TLMS). The name stems from the fact that $r(t)$, the normalizing factor in (5.61), is identical with the trace of the exponentially weighted regression matrix $R(t)$ in the EWLS approach:

$$r(t) = \sum_{i=0}^{t-1} \lambda^i \|\varphi(t-i)\|^2 = \text{tr}\left\{\sum_{i=0}^{t-1} \lambda^i \varphi(t-i)\varphi^T(t-i)\right\} = \text{tr}\left\{R(t)\right\}.$$

Note that in the scalar case ($n = 1$) the TLMS algorithm coincides with the EWLS algorithm provided that $\tilde{\mu}$ is set to 1. The mean stability condition for TLMS reads

$$0 < \tilde{\mu} < \frac{2}{1 - \lambda}, \tag{5.62}$$

and its equivalent estimation memory can be obtained from

$$l_{\infty}^{\text{TLMS}} = \frac{2n}{\tilde{\mu}(1 - \lambda)}. \tag{5.63}$$

Unfortunately, only the prediction-oriented equivalent memory l_{∞} is 'stabilized' if the normalized versions of the LMS algorithm are used. The corresponding tracking-oriented memory spans

$$l_{\infty}^{\star \text{NLMS}} = \frac{2 \operatorname{tr}\{\Phi_o^{-1}\} \operatorname{tr}\{\Phi_o\}}{n\bar{\mu}}, \quad l_{\infty}^{\star \text{TLMS}} = \frac{2 \operatorname{tr}\{\Phi_o^{-1}\} \operatorname{tr}\{\Phi_o\}}{n\tilde{\mu}(1 - \lambda)}, \tag{5.64}$$

still depend on the covariance structure of regressors.

5.4 Dynamic characteristics of LMS estimators

5.4.1 *Impulse response associated with LMS estimators*

In order to gain insights into the parameter tracking capabilities of the LMS algorithm, consider a time-varying system obeying the conditions on stationarity and independence of regressors (B1.1) and on mutual independence of $\{\varphi(t)\}$, $\{v(t)\}$ and $\{\theta(t)\}$ (B2.2).

Combining the system equation with (5.8) one arrives at

$$\widehat{\theta}(t) = \left(I_n - \mu\varphi(t)\varphi^T(t)\right)\widehat{\theta}(t - 1) + \mu\varphi(t)\varphi^T(t)\theta(t) + \mu\varphi(t)v(t), \tag{5.65}$$

leading to

$$\bar{\theta}(t) = \mathrm{E}[\widehat{\theta}(t)] = (I_n - \mu\Phi_o)\bar{\theta}(t - 1) + \mu\Phi_o\theta(t). \tag{5.66}$$

Note that in the steady state the recursive relationship between $\bar{\theta}(t)$ and $\theta(t)$ can be written in the explicit form

$$\bar{\theta}_{\text{LMS}}(t) = \sum_{i=0}^{\infty} H_{\text{LMS}}(i)\theta(t - i), \tag{5.67}$$

where the quantity

$$H_{\text{LMS}}(i) = \mu \left(I_n - \mu\Phi_o\right)^i \Phi_o \tag{5.68}$$

can be interpreted as an impulse response associated with the LMS estimator. Unlike the WLS case, where a sequence of scalar coefficients $\{h_{\text{WLS}}(i)\}$ was sufficient to characterize evolution 'in the mean' of all components of $\widehat{\theta}(t)$, the impulse response $\{H_{\text{LMS}}(i)\}$ is made up of matrices depending both on the stepsize μ (which is a user-dependent quantity) and on Φ_o (which depends on the input data).
Using the orthogonal transformation Q (cf. (5.17)) one gets

$$H_{\text{LMS}}(i) = \mu Q \left(I_n - \mu\Lambda_o\right)^i \Lambda_o Q^T \tag{5.69}$$

$$= \mu Q \, \operatorname{diag}\left\{\lambda_1(\Phi_o)\left(1 - \mu\lambda_1(\Phi_o)\right)^i, \dots, \lambda_n(\Phi_o)\left(1 - \mu\lambda_n(\Phi_o)\right)^i\right\} Q^T.$$

Observe that the relationship

$$\sum_{i=0}^{\infty} H_{\text{LMS}}(i) = \mu Q \left[\sum_{i=0}^{\infty} (I_n - \mu \Lambda_o)^i \right] \Lambda_o Q^T = \mu Q \, (\mu \Lambda_o)^{-1} \Lambda_o Q^T = I_n, \qquad (5.70)$$

is a counterpart of (4.29). Under the slow adaptation constraint (5.36) one gets

$$
\begin{aligned}
\sum_{i=0}^{\infty} H_{\text{LMS}}^2(i) &= \mu^2 Q \left[\sum_{i=0}^{\infty} (I_n - \mu \Lambda_o)^{2i} \right] \Lambda_o^2 Q^T \\
&= \mu Q \, (2 I_n - \mu \Lambda_o)^{-1} \Lambda_o Q^T \cong \frac{\mu}{2} Q \Lambda_o Q^T = \frac{\mu}{2} \Phi_o,
\end{aligned}
$$

leading to

$$\frac{1}{n} \, \text{tr} \left\{ \sum_{i=0}^{\infty} H_{\text{LMS}}^2(i) \right\} \cong \frac{1}{l_\infty^{\text{LMS}}}, \qquad (5.71)$$

which is a counterpart of (4.30). Finally, we note that results identical with (5.70) and (5.71) can be derived for normalized LMS algorithms.

5.4.2 Frequency response associated with LMS estimators

The frequency response associated with the LMS estimator can be defined as the Fourier transform of the impulse response $H_{\text{LMS}}(i)$:

$$H_{\text{LMS}}(\omega) = \sum_{i=0}^{\infty} H_{\text{LMS}}(i) e^{-j\omega i}. \qquad (5.72)$$

Unlike the WLS case, the frequency response of the LMS estimator is represented by a transfer *matrix*; this is an obvious consequence of the fact that tracking properties of LMS filters may be strongly direction dependent. Using (5.69) one arrives at

$$H_{\text{LMS}}(\omega) =$$

$$Q \, \text{diag} \left\{ \frac{\mu \lambda_1 (\Phi_o)}{1 - (1 - \mu \lambda_1(\Phi_o)) \, e^{-j\omega}} \,, \, \ldots \,, \, \frac{\mu \lambda_n (\Phi_o)}{1 - (1 - \mu \lambda_n(\Phi_o)) \, e^{-j\omega}} \right\} Q^T. \qquad (5.73)$$

It is straightforward to check that the transfer matrix (5.73) is symmetric, i.e.

$$H_{\text{LMS}}^T(\omega) = H_{\text{LMS}}(\omega).$$

Associated frequency characteristics can be used to specify the frequency distribution of the bias error,

$$\overline{\mathcal{T}}_b = \mathop{\text{E}}_{\Theta(t)} \left[\| \bar{\theta}(t) - \theta(t) \|^2 \right] \cong \frac{1}{\pi} \int_0^\pi \text{tr} \{ E_{\text{LMS}}(\omega) S_\theta(\omega) \} \, d\omega \qquad (5.74)$$

where

$$E_{\text{LMS}}(\omega) = [I_n - H_{\text{LMS}}(\omega)]^\dagger \, [I_n - H_{\text{LMS}}(\omega)] \qquad (5.75)$$

denotes the matrix generalization of the scalar parameter tracking characteristic defined for WLS estimators (4.39); † denotes complex conjugate transpose.

5.5   Comparison of the EWLS and LMS estimators

5.5.1   *Initial convergence*

Consider a time-invariant system subject to a stationary excitation governed by

$$y(t) = \varphi^T(t)\theta + v(t).$$

By the initial convergence period of an adaptive filter we mean the time needed for the estimation algorithm to reach its steady-state behavior. Since the WLS algorithms are (asymptotically) unbiased for $t \geq n$, it holds that

$$E[\widetilde{\theta}(t)] = E[\theta - \widehat{\theta}_{\mathrm{EWLS}}(t)] \cong 0, \qquad \forall t \geq n. \tag{5.76}$$

This means that the mean estimation error of the EWLS estimator is reduced from its initial value of $\widetilde{\theta}_{\mathrm{EWLS}}(0)$ to zero in a *finite number of steps irrespective of initial conditions* $\widehat{\theta}_{\mathrm{EWLS}}(0)$ (usually set to zero).

The excess mean square prediction and mean square parameter tracking errors are given by

$$\mathcal{P}[\widehat{\theta}_{\mathrm{EWLS}}(t)] \cong \frac{n\sigma_v^2}{l_t}, \tag{5.77}$$

and

$$\mathcal{T}[\widehat{\theta}_{\mathrm{EWLS}}(t)] \cong \frac{\mathrm{tr}\{\Phi_o^{-1}\}\sigma_v^2}{l_t}, \tag{5.78}$$

respectively. Both quantities tend to their steady-state values at the rate at which

$$l_t = \frac{(1 - \lambda^t)(1 + \lambda)}{(1 + \lambda^t)(1 - \lambda)}$$

approaches

$$l_\infty = \frac{1 + \lambda}{1 - \lambda}.$$

Note that

$$l_t = \frac{1 - \lambda^t}{1 + \lambda^t}l_\infty.$$

If the value of the forgetting constant λ is sufficiently close to 1 then

$$\ln \lambda^t = t \ln \lambda \cong -t(1 - \lambda) \cong -\frac{2t}{l_\infty},$$

or equivalently,

$$\lambda^t \cong e^{-2t/l_\infty}.$$

Using these approximation one obtains

$$\frac{l_t}{l_\infty} \cong \begin{cases} 0.75 & \text{for} \quad t = l_\infty \\ 0.90 & \text{for} \quad t = 1.5l_\infty \\ 0.96 & \text{for} \quad t = 2l_\infty \end{cases}$$

which means that when the number of steps approaches twice the equivalent window width, the process of initial convergence is practically finished.

The initial convergence behavior of LMS filters differs from that observed for WLS algorithms.

Using averaging theory one can show that

$$F(t,0) = \prod_{i=1}^{t}(I_n - \mu\varphi(i)\varphi^T(i)) \xrightarrow[\mu\to 0]{} (I_n - \mu\Phi_o)^t, \qquad (5.79)$$

and

$$E[\widetilde{\theta}_{\text{LMS}}(t)] \cong (I_n - \mu\Phi_o)^t\widetilde{\theta}_{\text{LMS}}(0) = Q(I_n - \mu\Lambda_o)^t Q^T\widetilde{\theta}_{\text{LMS}}(0). \qquad (5.80)$$

Observe that

$$(I_n - \mu\Lambda_o)^t = \text{diag}\left\{(1 - \mu\lambda_1(\Phi_o))^t, \cdots, (1 - \mu\lambda_n(\Phi_o))^t\right\}. \qquad (5.81)$$

Hence the bias can be decomposed into n terms which decay exponentially to zero. Expressing the stepsize μ in terms of the equivalent memory span of the LMS algorithm, the ith diagonal element of (5.81) can be written as

$$1 - \mu\lambda_i(\Phi_o) = 1 - \frac{2\lambda_i(\Phi_o)}{l_\infty\lambda_{\text{ave}}(\Phi_o)}. \qquad (5.82)$$

There are two major practical problems associated with (5.80). First, the value of the bias at any time t depends on the initial bias

$$\widetilde{\theta}_{\text{LMS}}(0) = \theta - \widehat{\theta}_{\text{LMS}}(0),$$

which is beyond our control (because the true value of θ is not known). Second, the speed of convergence is different in different directions of the parameter space, set by the eigenvectors of Φ_o. Since

$$\frac{1}{n}\sum_{i=1}^{n}(1 - \mu\lambda_i(\Phi_o)) = 1 - \frac{2}{l_\infty},$$

and

$$\left(1 - \frac{2}{l_\infty}\right)^t \cong e^{-\frac{2t}{l_\infty}},$$

the average time constant of the initial mean convergence can be defined as

$$\tau_{\text{ave}} = \frac{l_\infty}{2}.$$

Note, however, that the time constants of the natural learning modes of the LMS algorithm may significantly differ since

$$\frac{1}{n} < \frac{1 + (n-1)/\delta(\Phi_o)}{n} \le \frac{\lambda_{\text{ave}}(\Phi_o)}{\lambda_i(\Phi_o)} \le \frac{1 + (n-1)\delta(\Phi_o)}{n} < \delta(\Phi_o). \qquad (5.83)$$

According to (5.83) the time constant of the slowest mode of convergence,

$$\tau_{\text{max}} = \frac{l_\infty}{2}\frac{\lambda_{\text{ave}}(\Phi_o)}{\lambda_{\text{min}}(\Phi_o)},$$

may be almost as much as $\delta(\Phi_o)$ times larger than τ_{ave} and the ratio between the slowest-mode and fastest-mode time constants $\tau_{\max}/\tau_{\min}$ may approach the value $n\delta(\Phi_o)$. The upper inner bound in (5.83) is achieved if Φ_o has one eigenvalue equal to $\lambda_{\min}$ and $(n-1)$ eigenvalues equal to $\lambda_{\max}$.

Remark

Identical conclusions can be reached if the tracking-oriented memory measure is used instead of l_∞. Actually, observe that

$$1 - \mu\lambda_i(\Phi_o) = 1 - \frac{2\lambda_i(\Phi_o)\lambda_{\text{ave}}(\Phi_o^{-1})}{l_\infty^\star}, \tag{5.84}$$

and

$$\frac{1}{n} < \frac{1 + (n-1)/\delta(\Phi_o)}{n} \leq \lambda_i(\Phi_o)\lambda_{\text{ave}}(\Phi_o^{-1}) \leq \frac{1 + (n-1)\delta(\Phi_o)}{n} < \delta(\Phi_o). \tag{5.85}$$

Once more the ratio $\tau_{\max}/\tau_{\min}$ may be as large as $n\delta(\Phi_o)$.

∎

The transient mean square prediction and mean square tracking errors of the LMS algorithm can be established by analyzing the difference equation (5.38).

Setting $\widetilde{\Pi}(t) = Q^T\Pi(t)Q$ and $\Phi(t) = \Phi_o$ one can rewrite (5.81) in the form

$$\widetilde{\Pi}(t+1) = (I_n - \mu\Lambda_o)\widetilde{\Pi}(t)(I_n - \mu\Lambda_o) + \mu^2\Lambda_o\sigma_v^2. \tag{5.86}$$

According to (5.86)

$$\begin{aligned}
\widetilde{\pi}_{ii}(t) &= (1 - \mu\lambda_i(\Phi_o))^2\widetilde{\pi}_{ii}(t-1) + \mu^2\lambda_i(\Phi_o)\sigma_v^2 \\
&= (1 - \mu\lambda_i(\Phi_o))^{2t}\widetilde{\pi}_{ii}(0) + \mu\frac{1 - (1 - \mu\lambda_i(\Phi_o))^{2t}}{2 - \mu\lambda_i(\Phi_o)}\sigma_v^2 \\
&\cong (1 - \mu\lambda_i(\Phi_o))^{2t}\widetilde{\pi}_{ii}(0) + \frac{\mu}{2}[1 - (1 - \mu\lambda_i(\Phi_o))^{2t}]\sigma_v^2,
\end{aligned}$$

where the approximation holds under the slow adaptation condition.

Note that

$$\mathcal{P}[\widehat{\theta}_{\text{LMS}}(t)] = \text{tr}[\Pi(t)\Phi_o] = \text{tr}[\widetilde{\Pi}(t)\Lambda_o].$$

Hence

$$\mathcal{P}[\widehat{\theta}_{\text{LMS}}(t)] = \sum_{i=1}^{n}\lambda_i(\Phi_o)\widetilde{\pi}_{ii}(t)$$

$$\cong \sum_{i=1}^{n}\lambda_i(\Phi_o)(1 - \mu\lambda_i(\Phi_o))^{2t}\widetilde{\pi}_{ii}(0) + \frac{\mu}{2}\sum_{i=1}^{n}\lambda_i(\Phi_o)[1 - (1 - \mu\lambda_i(\Phi_o))^{2t}]\sigma_v^2. \tag{5.87}$$

For a stable LMS filter the first term on the right-hand side of (5.87), which is due to the estimation bias, decays to zero from its initial value of

$$\widetilde{\pi}_{ii}(0) = \sum_{i=1}^{n}\lambda_i(\Phi_o)\widetilde{\pi}_{ii}(0) = \text{tr}[\widetilde{\Pi}(0)\Lambda_o] = \text{tr}[\Pi(0)\Phi_o] = \text{E}[\epsilon_{ex}^2(0)],$$

and the second term grows from zero to its steady state value equal to

$$\frac{\mu}{2}\sum_{i=1}^{n}\lambda_i(\Phi_o)\sigma_v^2 = \frac{\mu}{2}\,\mathrm{tr}\{\Phi_o\}\sigma_v^2.$$

(already determined in Section 5.3). In both cases the rate of decay or growth is different for different learning modes of the LMS algorithm, associated with the eigenvalues of Φ_o. The one-step rate of change for the ith mode is equal to

$$(1 - \mu\lambda_i(\Phi_o))^2 = \left(1 - \frac{2\lambda_i(\Phi_o)}{l_\infty\lambda_{\mathrm{ave}}(\Phi_o)}\right)^2$$

and suffers from the same eigenvalue-disparity sensitivity problem which was discussed earlier.

The analysis of the mean square parameter tracking error

$$\mathcal{T}[\widehat{\theta}_{\mathrm{LMS}}(t)] \cong \mathrm{tr}[\Pi(t)] = \mathrm{tr}[\widetilde{\Pi}(t)]$$

can be carried out along the same lines as analysis of the mean square prediction error $\mathcal{P}[\cdot]$ and leads to identical conclusions from a qualitative viewpoint.

Remark

We note that the secret of rapid initial convergence of the EWLS algorithm lies in the fact that the 'matrix stepsize' $P(t)$ in (4.8) which decides upon the rate of change of parameter estimates,

$$\widehat{\theta}(t) = \widehat{\theta}(t-1) + K(t)\epsilon(t) = \widehat{\theta}(t-1) + P(t)\varphi(t)\epsilon(t),$$

is time-varying. It is 'large' in the initial phase of adaptation, where the estimation accuracy is poor, and gradually decreases down to the asymptotic (average) value of $(1-\lambda)\Phi_o^{-1}$ as the algorithm reaches its steady-state behavior.

In contrast with this, the LMS algorithm is equipped with a constant stepsize which may significantly limit the rate of readjustment of parameter estimates in the initial phase of convergence, especially for small values of the stepsize μ. One of the advantages of the trace LMS algorithm is due to the fact that the normalized stepsize $\tilde{\mu}/r(t)$ in (5.61) follows the time-varying gain pattern of the EWLS algorithm, which guarantees faster initial convergence.

5.5.2   Tracking performance

The speed of convergence of gradient algorithms in their startup phase is sensitive to the distribution of eigenvalues of Φ_o and this created a rather bad reputation for LMS filters as devices which can be used to track system changes. The opinion that the tracking performance of LMS algorithms is inferior to that of WLS algorithms is as widespread as it is unsubstantiated. The point is that the poor initial convergence behavior does *not* imply poor steady-state tracking performance. And conversely, the fast initial convergence of the WLS filter does not guarantee its quick response to parameter changes in the steady state [127, Fig.3].

To compare the tracking characteristics of EWLS/LMS filters we will assume that process parameters vary according to the random walk model

$$\theta(t) = \theta(t-1) + w(t), \tag{5.88}$$

where $\{w(t)\}$ denotes white noise, independent of $\{v(t)\}$. Consider the 'general' tracking algorithm

$$\widehat{\theta}(t) = \widehat{\theta}(t-1) + \gamma A(t)\varphi(t)\epsilon(t), \tag{5.89}$$

where γ denotes a small adaptation gain and $A(t)$ is an asymptotically gain-invariant matrix. When

$$\gamma = \mu, \qquad A(t) = I_n, \tag{5.90}$$

equation (5.89) coincides with the LMS algorithm. Similarly, when

$$\gamma = 1 - \lambda, \qquad A(t) = \frac{1}{1-\lambda}R^{-1}(t), \tag{5.91}$$

equation (5.89) is identical with the EWLS algorithm.

Denote by

$$\widetilde{\theta}(t) = \theta(t+1) - \widehat{\theta}(t)$$

the one-step-ahead parameter prediction error. Combining (5.88), (5.89) and the system equation

$$y(t) = \varphi^T(t)\theta(t) + v(t),$$

one arrives at the following recursive error equation:

$$\widetilde{\theta}(t) = (I_n - \gamma A(t)\varphi(t)\varphi^T(t))\widetilde{\theta}(t-1) - \gamma A(t)\varphi(t)v(t) + w(t+1), \tag{5.92}$$

which was thoroughly analyzed in the seminal paper of Guo and Ljung [68]. Imposing some weak persistence of excitation, boundedness and mixing conditions on $\{\varphi(t), v(t), w(t)\}$ Guo and Ljung showed that the covariance matrix $E[\widetilde{\theta}(t)\widetilde{\theta}^T(t)]$ can be closely approximated with the matrix $\Pi(t)$ obeying the following deterministic linear difference equation:

$$\Pi(t) = (I_n - \gamma G(t))\Pi(t-1)(I_n - \gamma G(t))^T + \gamma^2\sigma_v^2(t)H(t) + W(t), \tag{5.93}$$

where

$$\begin{aligned} G(t) &= E[A(t)\varphi(t)\varphi^T(t)], \\ H(t) &= E[A(t)\varphi(t)\varphi^T(t)A(t)], \\ W(t) &= \text{cov}[w(t)]. \end{aligned} \tag{5.94}$$

The approximation holds in the sense that

$$\| E[\widetilde{\theta}(t)\widetilde{\theta}^T(t) - \Pi(t) \| \le \sigma(\gamma) \| \Pi(t) \|, \tag{5.95}$$

where

$$\sigma(\gamma) \xrightarrow[\mu \to 0]{} 0.$$

Since

$$\mathrm{E}[\widetilde{\theta}(t)\widetilde{\theta}^T(t)] = \mathrm{E}[\theta_\star(t)\theta_\star^T(t)] + W(t),$$

where $\theta_\star(t) = \theta(t) - \widehat{\theta}(t)$, the covariance matrix of the parameter tracking error can be approximated with

$$\Pi^\star(t) = \Pi(t) - W(t). \tag{5.96}$$

For time-invariant systems the one-step-ahead parameter prediction error $\widetilde{\theta}(t)$ is identical with the parameter tracking error $\theta_\star(t)$, i.e. $\Pi^\star(t) = \Pi(t)$.

Note that the difference equation (5.38), which was used in Section 5.3 to analyze the steady-state properties of the LMS estimator for time-invariant systems, is a special case of (5.94). It can be obtained from (5.94) by setting $\gamma = \mu$, $A(t) = I_n$, $\sigma_v^2(t) = \sigma_v^2$ and $W(t) = O$.

Assuming that the processes $\{v(t)\}$, $\{w(t)\}$ and $\{\varphi(t)\}$ are mutually independent and wide-sense stationary ($\sigma_v^2(t) = \sigma_v^2$, $W(t) = W$, $\mathrm{cov}[\varphi(t)] = \Phi_o$), which effectively reduces the analysis to FIR systems, we shall examine solutions of (5.93) for LMS and EWLS estimators.

LMS algorithm

Substituting (5.90) into (5.94) one obtains

$$G(t) = H(t) = \mathrm{E}[\varphi(t)\varphi^T(t)] = \Phi_o,$$

hence (5.93) can be rewritten as

$$\Pi(t) = (I_n - \mu\Phi_o)\Pi(t-1)(I_n - \mu\Phi_o) + \mu^2\Phi_o\sigma_v^2 + W. \tag{5.97}$$

The steady-state solution of (5.97) can be obtained from

$$\Pi_\infty = (I_n - \mu\Phi_o)\Pi_\infty(I_n - \mu\Phi_o) + \mu^2\Phi_o\sigma_v^2 + W \tag{5.98}$$

which, after neglecting the term proportional to $\mu^2\Pi_\infty$ (see the remark in Section 5.3), can be rewritten in the form

$$\Phi_o\Pi_\infty + \Pi_\infty\Phi_o \cong \mu\Phi_o\sigma_v^2 + \frac{1}{\mu}W. \tag{5.99}$$

According to (5.96)

$$\mathcal{P}_\infty^{\mathrm{LMS}} \cong \mathrm{tr}\{\Pi_\infty^\star\Phi_o\} = \mathrm{tr}\{\Pi_\infty\Phi_o\} - \mathrm{tr}\{W\Phi_o\}. \tag{5.100}$$

Evaluating the trace of both sides of (5.99) one obtains

$$\mathrm{tr}\{\Pi_\infty\Phi_o\} \cong \frac{\mu}{2}\sigma_v^2\,\mathrm{tr}\{\Phi_o\} + \frac{1}{2\mu}\,\mathrm{tr}\{W\}. \tag{5.101}$$

For any two positive definite matrices A and B,

$$\lambda_{\min}(A)\,\mathrm{tr}\{B\} \le \mathrm{tr}\{AB\} \le \lambda_{\max}(A)\,\mathrm{tr}\{B\}. \tag{5.102}$$

Hence, in the case where

$$\mu \ll \frac{1}{2\operatorname{tr}\{\Phi_o\}} < \frac{1}{2\lambda_{\max}\{\Phi_o\}} \, ,$$

or equivalently

$$l_\infty \gg 4n, \qquad (5.103)$$

which is a slightly modified variant of the slow adaptation condition (5.52), then

$$\operatorname{tr}\{W\Phi_o\} \ll \frac{1}{2\mu}\operatorname{tr}\{W\} \le \lambda_{\max}(\Phi_o)\operatorname{tr}\{W\}. \qquad (5.104)$$

Combining (5.101) and (5.104) one obtains

$$\mathcal{P}_\infty^{\mathrm{LMS}} \cong \operatorname{tr}\{\Pi_\infty \Phi_o\} = \frac{\mu}{2}\sigma_v^2 \operatorname{tr}\{\Phi_o\} + \frac{1}{2\mu}\operatorname{tr}\{W\}. \qquad (5.105)$$

To obtain the approximate value of the mean square parameter tracking error

$$\mathcal{T}_\infty^{\mathrm{LMS}} = \operatorname{tr}\{\tilde{\Pi}_\infty^\star\},$$

postmultiply both sides of (5.99) by Φ_o^{-1}, giving

$$\Phi_o\Pi_\infty\Phi_o^{-1} + \Pi_\infty \cong \mu\sigma_v^2 I_n + \frac{1}{\mu}W\Phi_o^{-1},$$

and note that

$$\operatorname{tr}\{\Phi_o\Pi_\infty\Phi_o^{-1}\} = \operatorname{tr}\{\Pi_\infty\Phi_o^{-1}\Phi_o\} = \operatorname{tr}\{\Pi_\infty\}.$$

Hence

$$\operatorname{tr}\{\Pi_\infty\} \cong \frac{\mu}{2}n\sigma_v^2 + \frac{1}{2\mu}\operatorname{tr}\{W\Phi_o^{-1}\}.$$

Observe that under (5.103)

$$\frac{1}{2\mu}\operatorname{tr}\{W\Phi_o^{-1}\} \ge \frac{1}{2\mu}\lambda_{\min}(\Phi_o^{-1})\operatorname{tr}\{W\} = \frac{\operatorname{tr}\{W\}}{2\mu\lambda_{\max}(\Phi_o)} \gg \operatorname{tr}\{W\},$$

which results in

$$\mathcal{T}_\infty^{\mathrm{LMS}} \cong \operatorname{tr}\{\Pi_\infty\} \cong \frac{\mu}{2}n\sigma_v^2 + \frac{1}{2\mu}\operatorname{tr}\{W\Phi_o^{-1}\}. \qquad (5.106)$$

EWLS algorithm

Substitute (5.91) into (5.94) and use the fact that under the slow adaptation constraint $(1 - \lambda \ll 1)$ and for large values of t it holds that

$$R(t) \cong \frac{1}{1-\lambda}\Phi_o, \quad R^{-1}(t) \cong (1-\lambda)\Phi_o^{-1},$$

leading to

$$\begin{aligned} A(t) &\cong \Phi_o^{-1}, \\ G(t) &\cong I_n, \\ H(t) &\cong \Phi_o^{-1}. \end{aligned}$$

Hence (5.93) can be rewritten in the form

$$\Pi(t) = \lambda^2 \Pi(t-1) + (1-\lambda)^2 \Phi_o^{-1} \sigma_v^2 + W, \tag{5.107}$$

yielding the steady-state solution

$$\Pi_\infty = \frac{(1-\lambda)^2}{1-\lambda^2} \Phi_o^{-1} \sigma_v^2 + \frac{1}{1-\lambda^2} W \cong \Pi_\infty^\star. \tag{5.108}$$

The last transition holds under the slow adaptation (i.e. long memory) constraint, since in the case where $(1-\lambda^2) \ll 1$, or equivalently

$$l_\infty \gg 4, \tag{5.109}$$

it holds that

$$\frac{1}{1-\lambda^2} W \gg W.$$

Using (5.108) one arrives at

$$\mathcal{P}_\infty^{EWLS} \cong \operatorname{tr}\{\Pi_\infty \Phi_o\} = \frac{(1-\lambda)^2}{1-\lambda^2} n\sigma_v^2 + \frac{1}{1-\lambda^2} \operatorname{tr}\{W\Phi_o\} \tag{5.110}$$

and

$$\mathcal{T}_\infty^{EWLS} \cong \operatorname{tr}\{\Pi_\infty\} = \frac{(1-\lambda)^2}{1-\lambda^2} \operatorname{tr}\{\Phi_o^{-1}\}\sigma_v^2 + \frac{1}{1-\lambda^2} \operatorname{tr}\{W\}. \tag{5.111}$$

Comparison of predictive abilities

To compare the predictive abilities of EWLS and LMS algorithms in a sensible way, we will express the corresponding excess mean square prediction errors in terms of the equivalent number of observations:

$$\mathcal{P}_\infty^{LMS} \cong \frac{n\sigma_v^2}{l_\infty} + \frac{\operatorname{tr}\{W\}\operatorname{tr}\{\Phi_o\}}{4n} l_\infty, \tag{5.112}$$

$$\mathcal{P}_\infty^{EWLS} \cong \frac{n\sigma_v^2}{l_\infty} + \frac{\operatorname{tr}\{W\Phi_o\}}{4} l_\infty. \tag{5.113}$$

The first terms on the right-hand side of (5.112) and (5.113) are identical and can be interpreted as variance components of the excess mean square prediction error. They are caused by fluctuations of parameter estimates around their mean values. The second terms in (5.112) and (5.113) can be interpreted as bias components of the error and are caused by the fact that the mean trajectory of parameter estimates is always a distorted version of the true parameter trajectory; estimation delay is the main source of distortion.

Since the variance terms are the same in both expressions, a consequence of 'equalizing' the estimation memory spans of the compared algorithms, we need to compare the bias terms only.

First of all, note that if $W = \sigma_w^2 I_n$, i.e. if parameter changes occur with the same probability in all directions in the parameter space, then

$$\frac{\operatorname{tr}\{W\}\operatorname{tr}\{\Phi_o\}}{4n} = \frac{\operatorname{tr}\{W\Phi_o\}}{4} = \frac{\sigma_w^2}{4} \operatorname{tr}\{\Phi_o\},$$

i.e. both bias terms are identical.

If parameter changes occur with greater probability in some directions of the parameter space than in the other directions, the performance of the compared algorithms depends on the structure of covariance matrices W and Φ_o.

According to (5.102)

$$\lambda_{\min}(\Phi_o)\,\mathrm{tr}\{W\} \leq \mathrm{tr}\{W\Phi_o\} \leq \lambda_{\max}(\Phi_o)\,\mathrm{tr}\{W\}, \tag{5.114}$$

which clearly demonstrates that the bias component of $\mathcal{P}_\infty^{\mathrm{EWLS}}$ may strongly depend on the structure of Φ_o. On the other hand, since

$$\frac{\mathrm{tr}\{W\}\mathrm{tr}\{\Phi_o\}}{n} = \lambda_{\mathrm{ave}}(\Phi_o)\,\mathrm{tr}\{W\}, \tag{5.115}$$

the bias term of $\mathcal{P}_\infty^{\mathrm{LMS}}$ is a function of the average eigenvalue of Φ_o, i.e. it is *less sensitive* to the covariance structure of regressors.

The same conclusion can be reached after analyzing the best achievable performance of both estimation algorithms. Since the variance components of the mean square prediction errors are inversely proportional to l_∞, and the bias components are proportional to l_∞, the equivalent memory length of both adaptive filters can be chosen so as to provide an optimal trade-off between the two error sources.

Minimization of $\mathcal{P}_\infty^{\mathrm{LMS}}$ and $\mathcal{P}_\infty^{\mathrm{EWLS}}$ with respect to l_∞ is straightforward. The corresponding formulas are

$$\left(l_\infty^{\mathrm{LMS}}\right)_{\mathrm{opt}} = \frac{2n\sigma_v}{\sqrt{\mathrm{tr}\{W\}\mathrm{tr}\{\Phi_o\}}}, \tag{5.116}$$

$$\left(l_\infty^{\mathrm{EWLS}}\right)_{\mathrm{opt}} = \frac{2\sqrt{n}\sigma_v}{\sqrt{\mathrm{tr}\{W\Phi_o\}}}, \tag{5.117}$$

leading to

$$\left(\mathcal{P}_\infty^{\mathrm{LMS}}\right)_{\min} = \sigma_v\sqrt{\mathrm{tr}\{W\}\mathrm{tr}\{\Phi_o\}}, \tag{5.118}$$

$$\left(\mathcal{P}_\infty^{\mathrm{EWLS}}\right)_{\min} = \sigma_v\sqrt{n\,\mathrm{tr}\{W\Phi_o\}}. \tag{5.119}$$

Even though for a particular choice of the covariance matrices W and Φ_o comparison is inconclusive – depending on the environmental conditions one of the algorithms yields the better performance – it is obvious that the LMS algorithm is *more robust* than the EWLS algorithm. Good robustness properties of gradient algorithms can be explained by the fact that they converge and track faster in those directions of the parameter space which are more important from the viewpoint of adaptive prediction.

Comparison of parameter tracking abilities

When minimization of parameter tracking errors, rather than prediction errors, is our main concern, the corresponding performance measures should be expressed in terms of the parameter-tracking-oriented equivalent memory span $l_\infty^\star$

$$\mathcal{T}_\infty^{\mathrm{LMS}} \cong \frac{\mathrm{tr}\{\Phi_o^{-1}\}\sigma_v^2}{l_\infty^\star} + \frac{n\,\mathrm{tr}\{W\Phi_o^{-1}\}}{4\,\mathrm{tr}\{\Phi_o^{-1}\}}l_\infty^\star, \tag{5.120}$$

$$\mathcal{T}_\infty^{\mathrm{EWLS}} \cong \frac{\mathrm{tr}\{\Phi_o^{-1}\}\sigma_v^2}{l_\infty^\star} + \frac{1}{4}\,\mathrm{tr}\{W\}l_\infty^\star. \tag{5.121}$$

Basically, the conclusions that stem from comparison of $\mathcal{T}_\infty^{\text{LMS}}$ and $\mathcal{T}_\infty^{\text{EWLS}}$ are opposite to those reached when comparing $\mathcal{P}_\infty^{\text{LMS}}$ and $\mathcal{P}_\infty^{\text{EWLS}}$. The bias component of the mean square parameter tracking error $\mathcal{T}_\infty^{\text{EWLS}}$ does not depend on the matrix Φ_o, which is a straightforward consequence of our earlier observation that the parameter tracking capabilities of the WLS estimators do not depend on the direction of changes in the parameter space. In contrast with this, the bias term in (5.120) is sensitive to the structure of Φ_o. Note that the bias component of $\mathcal{T}_\infty^{\text{LMS}}$ can be expressed as

$$\frac{\text{tr}\{W\Phi_o^{-1}\}}{4\lambda_{\text{ave}}\{\Phi_o^{-1}\}}l_\infty^\star,$$

and that

$$\lambda_{\min}(\Phi_o^{-1})\,\text{tr}\{W\} \le \text{tr}\{W\Phi_o^{-1}\} \le \lambda_{\max}(\Phi_o^{-1})\,\text{tr}\{W\}.$$

According to the results of analysis carried out above in all applications where minimization of the mean square parameter tracking error is the ultimate goal of modeling, the EWLS estimators should be preferred to the LMS estimators. The same conclusions can be reached by comparing the best achievable parameter tracking performance of both algorithms. Straightforward calculations yield

$$\left(l_\infty^{\star\ \text{LMS}}\right)_{\text{opt}} = \frac{2\text{tr}\{\Phi_o^{-1}\}\sigma_v}{\sqrt{n\,\text{tr}\{W\Phi_o^{-1}\}}}, \tag{5.122}$$

$$\left(l_\infty^{\star\ \text{EWLS}}\right)_{\text{opt}} = \frac{2\sqrt{\text{tr}\{\Phi_o^{-1}\}}\sigma_v}{\sqrt{\text{tr}\{W\}}}, \tag{5.123}$$

leading to

$$\left(\mathcal{T}_\infty^{\text{LMS}}\right)_{\text{min}} = \sigma_v\sqrt{n\,\text{tr}\{W\Phi_o^{-1}\}}, \tag{5.124}$$

$$\left(\mathcal{T}_\infty^{\text{EWLS}}\right)_{\text{min}} = \sigma_v\sqrt{\text{tr}\{W\}\text{tr}\{\Phi_o^{-1}\}}. \tag{5.125}$$

Note, however, that practically all applications mentioned in Section 1.5 are prediction oriented, so the conclusions reached above are of *minor* importance.

Remark

The optimal values of l_∞ and $l_\infty^\star$ can be 'translated back' to the optimal values of μ, λ and $\mu^\star$, $\lambda^\star$, respectively:

$$\mu_{\text{opt}} = \frac{\sqrt{\text{tr}\{W\}}}{\sigma_v\sqrt{\text{tr}\{\Phi_o\}}},$$

$$\mu_{\text{opt}}^\star = \frac{\sqrt{\text{tr}\{W\Phi_o^{-1}\}}}{\sigma_v\sqrt{n}},$$

$$\lambda_{\text{opt}} = 1 - \frac{\sqrt{\text{tr}\{W\Phi_o\}}}{\sigma_v\sqrt{n}},$$

$$\lambda_{\text{opt}}^\star = 1 - \frac{\sqrt{\text{tr}\{W\}}}{\sigma_v\sqrt{\text{tr}\{\Phi_o^{-1}\}}}. \tag{5.126}$$

5.6   Computer simulations

Figures 5.2 to 5.11 show identification results obtained using the LMS algorithm for the time-varying finite impulse response systems (FIR1 and FIR2) and autoregressive processes (AR1 and AR2) described in Section 2.9. Since for the LMS filter the tracking-oriented and prediction-oriented memory spans are different, the first sequence of plots (Figures 5.2 - 5.6) correspond to the case where $l_\infty^\star = 50$ (the parameter tracking oriented analysis) and the second sequence of plots (Figures 5.7 - 5.11) corresponds to the case where $l_\infty = 50$ (the prediction oriented analysis).

Figures 5.2 and 5.7 show the output of the linear filter associated with the LMS estimator. For both FIR systems the 'theoretical' plots stay in very good agreement with the results of averaging (over 100 simulation runs) the estimated parameter trajectories (Figures 5.4 and 5.9). Estimation results obtained for a single realization of the identified process are shown in Figures 5.3 and 5.8. The plots obtained for the system with abrupt parameter changes clearly show the effect of cross-coupling between the estimates of different system parameters (the jump change of one coefficient affects the estimate of the other one) which is typical of LMS filters. No such effect was observed for WLS filters. Note also the difference in the parameter tracking behavior between LMS filters with $l_\infty = 50$ and $l_\infty^\star = 50$.

Figures 5.5, 5.6 and 5.10, 5.11 show identification results obtained for the time-varying autoregressive processes AR1 and AR2. Since the tracking-oriented and prediction-oriented estimation memory spans of the LMS filter depend on the covariance structure of regressors, for a nonstationary AR process they are not constant; the stepsize μ was chosen so as to enforce $l_\infty = 50$ (or $l_\infty^\star = 50$) at the beginning of the analysis interval, i.e. for the starting values of process coefficients. The 'variable memory' feature of the LMS algorithm explains its faster response to parameter changes which can be observed for AR processes compared to their FIR counterparts.

Remark 1

All plots show the steady-state tracking behavior of the LMS algorithm. To reach the steady state, parameter estimation was initialized at instant $t = -500$, i.e. before the analysis interval $T = [0, 500]$ started. In the initial convergence period, the system parameters were constant $(a_1(t) = a_1(0), a_2(t) = a_2(0)$ for $t < 0)$.

Remark 2

The greater variance of the estimates in the AR case, compared to the FIR case, is mainly the consequence of a smaller signal-to-noise ratio.

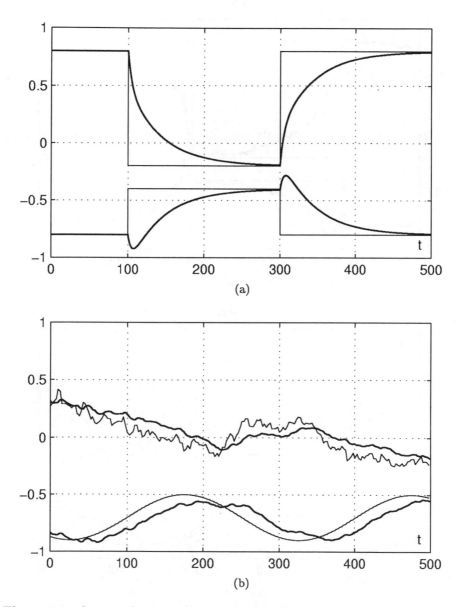

(a)

(b)

Figure 5.2 Output of a linear filter associated with the least mean squares (LMS) estimator for a system with jump parameter changes (a) and continuous parameter changes (b); the equivalent memory of the estimation algorithm was set to $l_\infty^* = 50$ ($\mu = 0.04$).

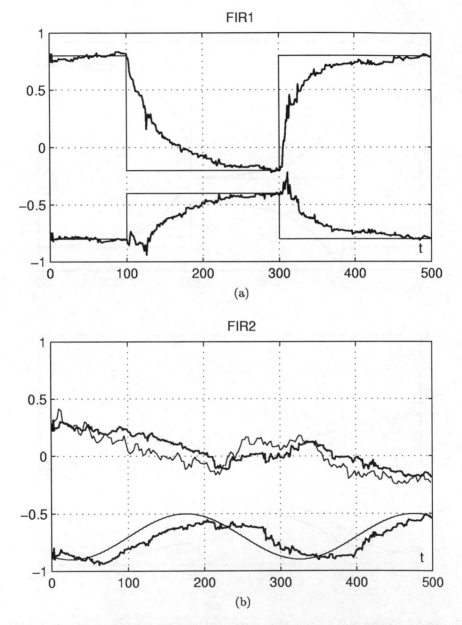

Figure 5.3 Parameter estimates yielded by the least mean squares (LMS) algorithm for a single realization of an FIR system with jump parameter changes (a) and continuous parameter changes (b); the equivalent memory of the estimation algorithm was set to $l_\infty^\star = 50$ ($\mu = 0.04$).

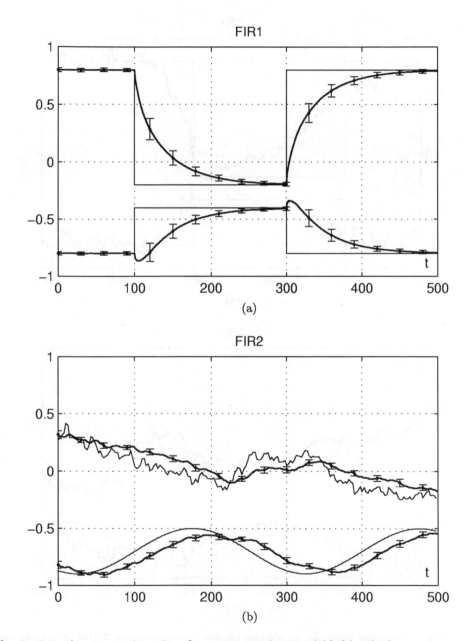

Figure 5.4 Average trajectories of parameter estimates yielded by the least mean squares (LMS) algorithm for an FIR system with jump parameter changes (a) and continuous parameter changes (b); vertical bars show standard deviation of the estimates.

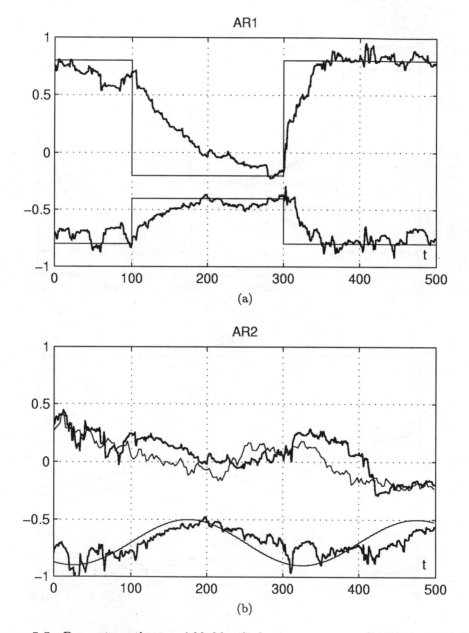

Figure 5.5 Parameter estimates yielded by the least mean squares (LMS) algorithm for a single realization of an AR system with jump parameter changes (a) and continuous parameter changes (b); the starting value of the equivalent memory of the estimation algorithm was set to $l_\infty^\star = 50$.

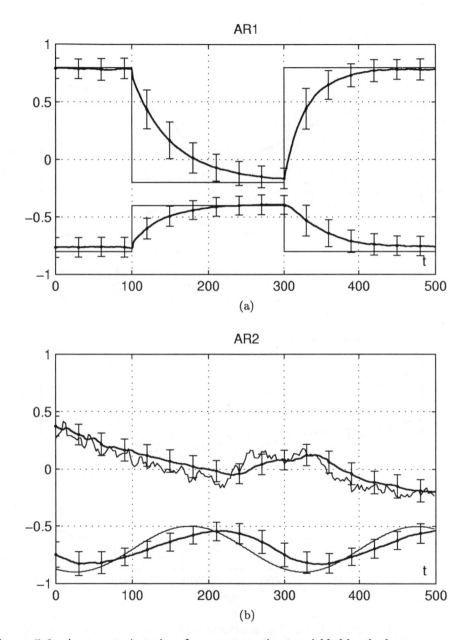

Figure 5.6 Average trajectories of parameter estimates yielded by the least mean squares (LMS) algorithm for an AR system with jump parameter changes (a) and continuous parameter changes (b); vertical bars show standard deviation of the estimates.

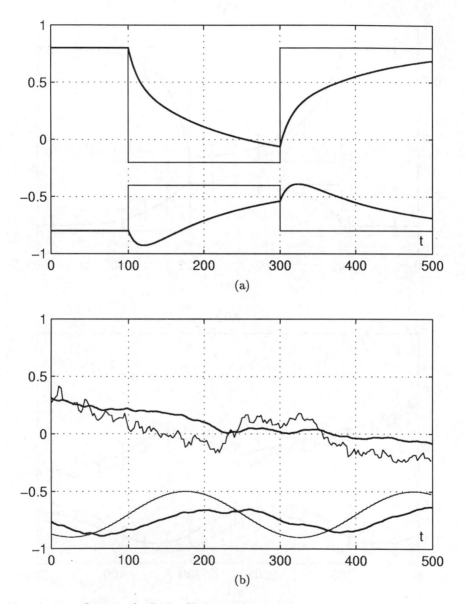

Figure 5.7 Output of a linear filter associated with the least mean squares (LMS) estimator for a system with jump parameter changes (a) and continuous parameter changes (b); the equivalent memory of the estimation algorithm was set to $l_\infty = 50$ ($\mu = 0.0144$).

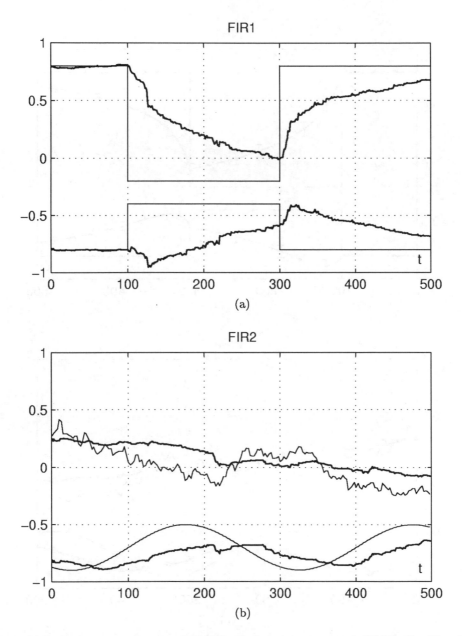

Figure 5.8   Parameter estimates yielded by the least mean squares (LMS) algorithm for a single realization of an FIR system with jump parameter changes (a) and continuous parameter changes (b); the equivalent memory of the estimation algorithm was set to $l_\infty = 50$ ($\mu = 0.0144$).

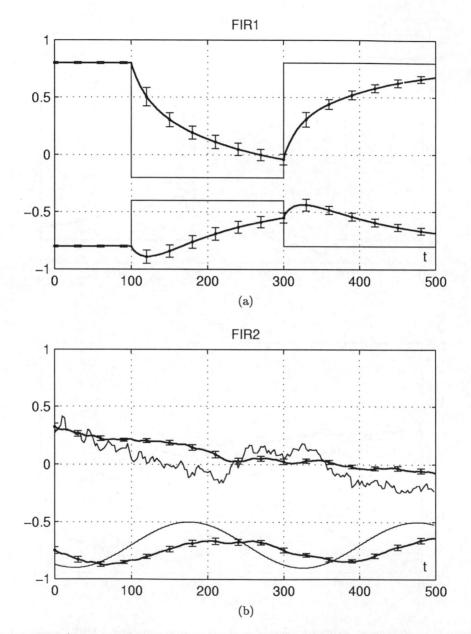

Figure 5.9 Average trajectories of parameter estimates yielded by the least mean squares (LMS) algorithm for an FIR system with jump parameter changes (a) and continuous parameter changes (b); vertical bars show standard deviation of the estimates.

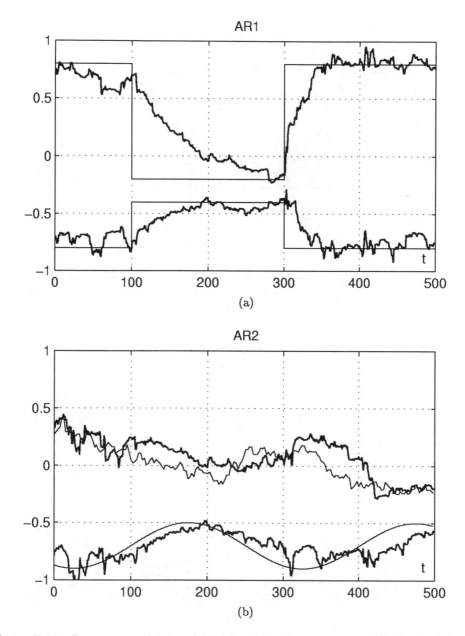

Figure 5.10 Parameter estimates yielded by the least mean squares (LMS) algorithm for a single realization of an AR system with jump parameter changes (a) and continuous parameter changes (b); the starting value of the equivalent memory of the estimation algorithm was set to $l_\infty = 50$.

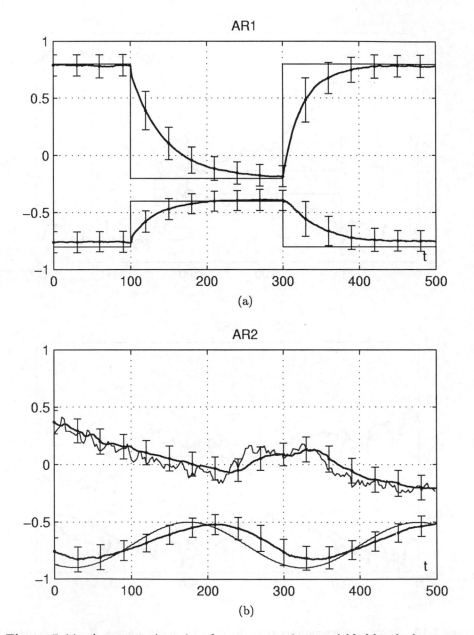

Figure 5.11 Average trajectories of parameter estimates yielded by the least mean squares (LMS) algorithm for an AR system with jump parameter changes (a) and continuous parameter changes (b); vertical bars show standard deviation of the estimates.

5.7   Extension to ARMAX processes

Extension of the LMS algorithm to ARMAX systems is straightforward. The corresponding algorithm is identical with the LMS algorithm for ARX systems except that the regression vector $\varphi(t)$ is replaced with its approximation $\widetilde{\varphi}(t)$:

$$
\begin{aligned}
\widehat{\theta}(t) &= \widehat{\theta}(t-1) + \mu\widetilde{\varphi}(t)\epsilon(t), \\
\epsilon(t) &= y(t) - \widetilde{\varphi}^T(t)\widehat{\theta}(t-1).
\end{aligned}
\tag{5.127}
$$

Analysis of a decreasing gain (i.e. growing memory) counterpart of (5.127),

$$
\widehat{\theta}(t) = \widehat{\theta}(t-1) + \mu(t)\widetilde{\varphi}(t)\epsilon(t)
\tag{5.128}
$$

where

$$
\sum_{t=1}^{\infty}\mu(t) = \infty, \quad \sum_{t=1}^{\infty}\mu^2(t) < \infty,
\tag{5.129}
$$

reveals that for a time-invariant ARMAX system, the parameter estimates converge to their true values provided that the identified system obeys the following strict positive realness condition:

$$
\theta \in D_{\text{SPR}}^{\star}, \quad D_{\text{SPR}}^{\star} = \{\theta : \text{Re}\left\{C(e^{j\omega},\theta) - \frac{1}{2}\right\} > 0, \quad \forall\omega\}.
\tag{5.130}
$$

Note that the SPR condition (5.130) is different from the analogous condition (3.56) established for the ELS algorithm.

Comments and extensions

Section 5.2

- A comprehensive analysis of different learning characteristics of the LMS-type stochastic gradient-descent algorithms was provided by Gardner [51]. The analysis, carried out for time-invariant systems, sheds a lot of light on the problem of initial convergence of LMS estimates and its dependence on the covariance structure and distribution of regressors.

- The $\mathcal{H}^{\infty}$, or minimax, robustness of LMS filters was shown by Hassibi et al. [74]. Although interesting from a theoretical viewpoint, such worst-case analysis (the LMS filter can be shown to deal most favorably with the least favorable disturbance patterns) seems to be of little practical significance; all worst-case results are very conservative.

- The class of 'multistep' LMS-type tracking algorithms is analyzed in an interesting paper of Benveniste [20]. The same paper shows the usefulness of the ODE (ordinary differential equation) technique for analyzing the steady-state tracking properties of gradient algorithms.

- Another useful extension of the LMS technique, the smoothed LMS approach, is discussed in the paper of Feuer and Berman [48].

Section 5.3

- Success has many fathers. The normalized LMS algorithm, commonly attributed to Nagumo and Noda [129] and Albert and Gardner [3], was in fact proposed by the Polish mathematician Kaczmarz [84] in 1937!

Section 5.4

- One of the first studies of nonstationary learning characteristics of LMS filters (for independent regressors) was provided by Widrow et al. [192]; see also Widrow and Walach [193]. A more thorough analysis of the tracking performance of the LMS algorithm (including the concept of the associated time and frequency characteristics) can be found in Gardner [52].

6

Basis Functions

The common feature of estimation schemes described in Chapter 3 (the segmentation approach), Chapter 4 (the weighted least squares approach) and Chapter 5 (the least mean squares approach) is the lack of an explicit mathematical model of the process parameter variation. Instead of adopting a particular description of changes, we assumed that process parameters were 'slowly time-varying' and hence can be followed using the local estimation approach.

In contrast with this, the method of basis functions (BF) is based on an explicit model of parameter variation; namely, it is assumed that the parameter trajectory can be approximated by a linear combination of known functions of time – the basis functions. The method can be used in two different ways. First, the entire time domain can be divided into constant-length or variable-length analysis frames, and system changes can be modeled independently in each frame. The resulting identification scheme can be regarded as an extension of the segmentation approach described in Chapter 3 and can be used in typical parameter matching applications. Whenever parameter tracking is required, which is typical of adaptive prediction and control applications, one can combine the functional series approximation with information weighting to obtain the finite-memory weighted basis function estimators. The weighted basis function estimators can be regarded as an extension of the WLS estimators described in Chapter 4.

6.1   Approach based on process segmentation

6.1.1   Estimation principles

Consider the analysis frame $T = [1, \ldots, N]$ of length N and let

$$\{f_j(t), j = 1, \ldots, k\}, \quad t \in T \tag{6.1}$$

be a set of linearly independent discrete-time functions (sequences) defined on T, called basis functions. We shall assume that each time-varying process coefficient can be represented by a linear combination of basis functions

$$\theta_i(t) = \sum_{j=1}^{k} c_{ij} f_j(t), \quad t \in T, \tag{6.2}$$

$$i = 1, \ldots, n.$$

There are two ways to select the basis functions. If some prior knowledge about the nature of process time variation is available, one may attempt to choose the basis functions so as to capture the 'dominant' trends in coefficient changes. This is the case with identification of mobile radio channels (Section 2.5.3).

Unfortunately, the situation where the right set of basis functions is known a priori is extremely rare. If no insights or hints are available, selection of the basis has to rely on some general approximation guidelines. The most frequently used general-purpose bases involve the powers of time (Taylor series approximation)

$$f_j(t) = t^{j-1}, \quad j = 1, 2, \ldots, \tag{6.3}$$

and cosines (Fourier series approximation)

$$f_j(t) = \cos \frac{\pi(j-1)t}{N-1}, \quad j = 1, 2, \ldots, \tag{6.4}$$

but in principle any set of linearly independent basis functions can be used.

Remark

Denote by $l^2(T)$ the space of all square summable sequences on T with the inner product and vector norm defined as

$$< f, g > = \sum_{t=1}^{N} f(t)g(t), \quad \|f\|^2 = < f, f >,$$

and denote by $F_k(T)$ the subspace of $l^2(T)$ spanned by the basis functions (6.1). It is straightforward to show that $F_N(T) = l^2(T)$, i.e. any function $g(\cdot) \in l^2(T)$ can be written down *exactly* as a linear combination of N basis functions

$$g(t) = \sum_{j=1}^{N} c_j f_j(t), \quad t \in T.$$

Since $F_1(T) \subset F_2(T) \subset \ldots \subset F_N(T) = l^2(T)$, by increasing the number of basis functions one increases the flexibility of the functional series model of parameter variations. There is, however, a natural limit to k set by the number of available data points. According to the principle of parsimony, the total number of estimated parameters should be much smaller than the number of available data points. This leads to the following recommendation: $kn \ll N$, or equivalently

$$k \ll \frac{N}{n}. \tag{6.5}$$

Note that in any case, even if the parsimony principle is ignored, the number of basis functions must obey

$$k \leq \frac{N}{n}$$

to guarantee that the total number of degrees of freedom does not exceed the number of data points.   ∎

To facilitate further analysis we will introduce a Kronecker product $A \otimes B$ $[im \times jn]$ of two matrices $A[i \times j]$ and $B[m \times n]$:

$$A \otimes B \triangleq \begin{bmatrix} a_{11}B & \cdots & a_{1j}B \\ \vdots & & \vdots \\ a_{i1}B & \cdots & a_{ij}B \end{bmatrix}.$$

We will show that under (6.2) the time-varying process

$$y(t) = \varphi^T(t)\theta(t) + v(t)$$

can be described by a time-invariant model. Denote by

$$\psi(t) = \varphi(t) \otimes \mathbf{f}(t) = [\varphi_1(t)\mathbf{f}^T(t), \dots, \varphi_n(t)\mathbf{f}^T(t)]^T,$$

where

$$\mathbf{f}(t) = [f_1(t), \dots, f_k(t)]^T$$

and $\varphi_i(t)$ is the ith component of $\varphi(t)$, the $[nk \times 1]$ generalized regression vector associated with the analyzed process. Furthermore, let

$$\gamma = [\gamma_1^T, \dots, \gamma_n^T]^T, \quad \gamma_i = [c_{i1}, \dots, c_{ik}]^T,$$

be the $[nk \times 1]$ vector of all coefficients used to describe the process time variation. Using the shorthand notation introduced above and assuming that the model (6.2) applies to the entire data segment $\Xi(T) = \{\xi(t), t \in T\}$, the process equation

$$y(t) = \varphi^T(t)\theta(t) + v(t)$$

can be rewritten as

$$y(t) = \psi^T(t)\gamma + v(t), \quad t \in T, \tag{6.6}$$

i.e. the time-varying process of order n can be represented by a linear time-invariant model of order nk.

The estimate of γ can be obtained using the method of least squares

$$\hat{\gamma}(T) = \arg\min_{\gamma} \sum_{t=1}^{N} [y(t) - \psi^T(t)\gamma]^2$$

$$= \left(\sum_{t=1}^{N} \psi(t)\psi^T(t) \right)^{-1} \left(\sum_{t=1}^{N} y(t)\psi(t) \right) = G^{-1}(T)h(T), \tag{6.7}$$

provided of course that the regression matrix in (6.7) is invertible. Based on (6.7) one can obtain an estimate of the parameter trajectory in the analysis interval T

$$\hat{\theta}_i(t) = \mathbf{f}^T(t)\hat{\gamma}_i(T), \quad i = 1, \dots, n, \quad t \in T, \tag{6.8}$$

or equivalently

$$\hat{\theta}(t) = Z(t)\hat{\gamma}(T), \quad t \in T, \tag{6.9}$$

where
$$Z(t) = I_n \otimes \mathbf{f}^T(t).$$

The estimate of the input noise variance can be obtained from

$$\widehat{p}(T) = \frac{1}{N} \sum_{t=1}^{N} [y(t) - \psi^T(t)\widehat{\gamma}(T)]^2. \tag{6.10}$$

Note that both the parameter estimate (6.9) and the variance estimate (6.10) can be evaluated recursively using the algorithms derived in Section 3.1, provided that the regression vector $\varphi(t)$ is replaced with the generalized regression vector $\psi(t)$, namely

$$\begin{aligned}
\widehat{\gamma}(t) &= \widehat{\gamma}(t-1) + L(t)\epsilon(t), \\
\epsilon(t) &= y(t) - \psi^T(t)\widehat{\gamma}(t-1), \\
L(t) &= \frac{Q(t-1)\psi(t)}{1 + \psi^T(t)Q(t-1)\psi(t)}, \\
Q(t) &= Q(t-1) - \frac{Q(t-1)\psi(t)\psi^T(t)Q(t-1)}{1 + \psi^T(t)Q(t-1)\psi(t)},
\end{aligned} \tag{6.11}$$

and

$$\widehat{\sigma}^2(t) = \frac{p(t)}{t},$$

where

$$\begin{aligned}
p(t) &= p(t-1) + [1 - L^T(t)\psi(t)]\epsilon^2(t) \\
&= p(t-1) + \frac{\epsilon^2(t)}{1 + \psi^T(t)Q(t-1)\psi(t)},
\end{aligned} \tag{6.12}$$

Remark

Defining the generalized regression vector $\psi(t)$ in a slightly modified form

$$\psi'(t) = \mathbf{f}(t) \otimes \varphi(t) = [f_1(t)\varphi^T(t), \ldots, f_k(t)\varphi^T(t)]^T,$$

and redefining the parameter vector as

$$\gamma' = [(\gamma_1')^T, \ldots, (\gamma_k')^T]^T, \quad \gamma_i' = [c_{1i}, \ldots, c_{ni}]^T,$$

one can rewrite the process equation in a form analogous to (6.6):

$$y(t) = (\psi'(t))^T \gamma' + v(t), \quad t \in T. \tag{6.13}$$

The least squares estimate of γ' is given by

$$\widehat{\gamma}'(T) = \left(\sum_{t=1}^{N} \psi'(t)(\psi'(t))^T \right)^{-1} \left(\sum_{t=1}^{N} y(t)\psi'(t) \right) = (G'(T))^{-1} h'(T). \tag{6.14}$$

Since $\gamma' = \text{perm}(\gamma)$, where $\text{perm}(\cdot)$ denotes a permutation operation, it holds that $\widehat{\gamma}'(T) = \text{perm}(\widehat{\gamma}(T))$, i.e. both estimates are identical up to the ordering of vector coordinates. Even though the ordering of parameter and regression vector elements does not affect the estimation results, it may be important from a computational viewpoint [71].

■

6.1.2 Invariance under the change of coordinates

We will show that the results of the BF identification depend on our choice of the subspace $F_k(T)$ but do *not* depend on the particular choice of the basis set of $F_k(T)$. Actually, let $\{f_1^\star(t), \ldots, f_k^\star(t), t \in T\}$ denote another basis set of $F_k(T)$, different from $\{f_1(t), \ldots, f_k(t), t \in T\}$ and let $\widehat{\theta}_i^\star(t), t \in T$,

$$\widehat{\theta}_i^\star(t) = f_\star^T(t)\widehat{\gamma}_i^\star, \quad i = 1, \ldots, n,$$
$$f_\star(t) = [f_1^\star(t), \ldots, f_k^\star(t)]^T,$$

denote the corresponding BF estimates of process parameter trajectories.

Since $f(t)$ and $f_\star(t)$ are basis vectors of the same subspace, there exists a nonsingular transformation matrix A such that

$$f_\star(t) = Af(t).$$

Straightforward calculations yield

$$\widehat{\gamma}_\star(T) = \left(\sum_{t=1}^N \psi_\star(t)\psi_\star^T(t)\right)^{-1} \left(\sum_{t=1}^N y(t)\psi_\star(t)\right)$$

$$= \left[H\left(\sum_{t=1}^N \psi(t)\psi^T(t)\right)H^T\right]^{-1} \left[H\left(\sum_{t=1}^N y(t)\psi(t)\right)\right]$$

$$(H^T)^{-1}\left(\sum_{t=1}^N \psi(t)\psi^T(t)\right)^{-1} \left(\sum_{t=1}^N y(t)\psi(t)\right) = (H^T)^{-1}\widehat{\gamma}(T),$$

where $\psi_\star(t) = \varphi(t) \otimes f_\star(t)$ and $H = I_n \otimes A$. Hence

$$\widehat{\theta}_\star(t) = Z_\star(t)\widehat{\gamma}_\star(T) = Z(t)H^T\widehat{\gamma}_\star(T) = Z(t)\widehat{\gamma}(T) = \widehat{\theta}(t),$$

where $Z_\star(t) = I_n \otimes f_\star^T(t)$.

Most of the results derived in this chapter will be specified in terms of orthonormal basis vectors

$$f_o(t) = [f_1^o(t), \ldots, f_k^o(t)]^T,$$

i.e. vectors that obey

$$\sum_{t=1}^N f_o(t)f_o^T(t) = I_k. \tag{6.15}$$

The orthonormal basis can be obtained from the nonorthonormal basis by applying the Gram–Schmidt procedure:

$$f_1^o = \frac{f_1}{\|f_1\|},$$

$$f_2^o = \frac{f_2 - <f_2, f_1^o>f_1^o}{\|f_2 - <f_2, f_1^o>f_1^o\|},$$

$$\vdots$$

$$f_k^o = \frac{f_k - \sum_{i=1}^{k-1} <f_k, f_i^o>f_i^o}{\|f_k - \sum_{i=1}^{k-1} <f_k, f_i^o>f_i^o\|}. \tag{6.16}$$

Orthogonalizing the basis (6.3) one obtains a set of discrete, normalized, shifted
Legendre polynomials ($k = 3$)

$$f_1^o(t) = \frac{1}{\sqrt{N}},$$

$$f_2^o(t) = \sqrt{\frac{3(N-1)}{N(N+1)}} \left[1 - 2\frac{t-1}{N-1} \right],$$

$$f_3^o(t) = \sqrt{\frac{5(N-1)(N-2)}{N(N+1)(N+2)}} \left[1 - 6\frac{t-1}{N-1} + 6\frac{(t-1)(t-2)}{(N-1)(N-2)} \right]. \quad (6.17)$$

For this reason the basis (6.3) will be further referred to as a Legendre basis.

Applying the Gram–Schmidt procedure to (6.4) one obtains ($k = 3$)

$$f_1^o(t) = \frac{1}{\sqrt{N}},$$

$$f_2^o(t) = \sqrt{\frac{2}{N+1}} \cos\frac{\pi(t-1)}{N-1},$$

$$f_3^o(t) = \sqrt{\frac{2N}{(N+2)(N-1)}} \left[\cos\frac{2\pi(t-1)}{N-1} - \frac{1}{N} \right]. \quad (6.18)$$

The basis set (6.4) will be called a Fourier basis.

The plots of the first three orthonormal functions of the Legendre and Fourier bases
are shown in Figure 6.1.

Remark 1

Consider the problem of a functional series approximation of a measurable signal
$\{y(t), t \in T\}$. In the case considered, note how it is the signal $y(t)$ itself, not its
parameters, that is described by the functional series. If N data points are available,
the least squares solution to the problem is given by

$$\widehat{\gamma}(T) = \arg\min_\gamma \sum_{t=1}^{N} (y(t) - f^T(t)\gamma)^2$$

$$= \left(\sum_{t=1}^{N} f(t)f^T(t) \right)^{-1} \left(\sum_{t=1}^{N} y(t)f(t) \right),$$

which resembles (6.7). As in the case considered before, the LS approximation

$$\widehat{y}(t) = f^T(t)\widehat{\gamma}(T), \quad t \in T,$$

is invariant with respect to the choice of the basis set of $F_k(T)$. A clear advantage of
choosing the orthonormal basis $\{f_i^o(t), i = 1, \ldots, k\}$ is that, owing to (6.15), there is
no need to invert the regression matrix, i.e. one obtains

$$\widehat{y}(t) = f_o^T(t)\widehat{\gamma}_o(T)$$

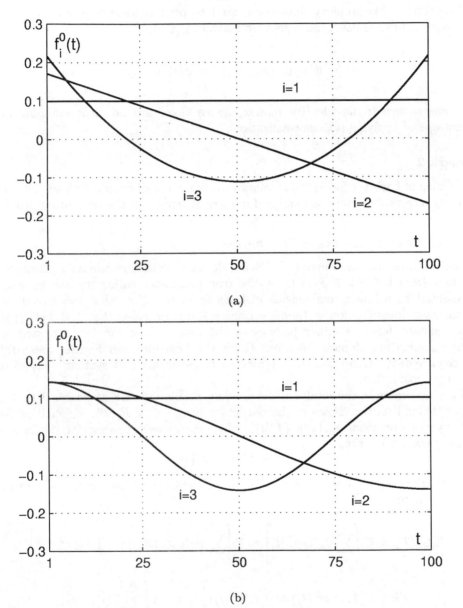

Figure 6.1 The plots of the first three orthonormal functions of the Legendre (a) and Fourier (b) bases ($N = 100$).

where

$$\hat{\gamma}_o(T) = \sum_{t=1}^{N} y(t) f_o(t).$$

Unfortunately, this property does not extend to process identification. Even if the orthonormal basis is used, the regression matrix in (6.7),

$$\sum_{t=1}^{N} \psi_o(t)\psi_o^T(t), \quad \psi_o(t) = \varphi(t) \otimes f_o(t),$$

does *not* reduce to an identity matrix, hence there are no clear computational advantages of applying orthonormalization.

Remark 2

The choice of basis set for $F_k(T)$ is immaterial from a mathematical viewpoint but it certainly affects the computational and numerical aspects of the estimation problem.

6.1.3 Static characteristics of BF estimators

We will consider the case where $\{\varphi(t)\}$ is a wide-sense stationary sequence (assumption C1) and $\{\theta(t), t \in T\} \in F_k(T)$, i.e. the true parameter trajectory can be *exactly* represented by a linear combination of basis functions. Note that this covers, as a special case, identification of time-invariant systems provided that $1 \in F_k(T)$; that is, a constant function either belongs to the basis set or can be expressed as a linear combination of basis functions (both the Legendre and Fourier bases fulfill this requirement). Condition (C1) is met for time-varying FIR systems subject to a stationary excitation.

Since, according to the results in the previous section, the BF estimates $\hat{\theta}(t)$, $t \in T$ do not depend on the choice of the particular basis set of $F_k(T)$, we shall perform analysis for an orthonormal basis $\{f_j^o(t)\}$, which corresponds to adopting the following system parameterization:

$$y(t) = \psi_o^T(t)\gamma_o + v(t)$$

where $\psi_o(t) = \varphi(t) \otimes f_o(t)$. Under (C1) the expectation of the regression matrix in (6.7) is given by

$$E[G_o(T)] = E\left[\sum_{t=1}^{N} \psi_o(t)\psi_o^T(t)\right] = E\left[\sum_{t=1}^{N} (\varphi(t) \otimes f_o(t))\left(\varphi^T(t) \otimes f_o^T(t)\right)\right]$$

$$= E\left[\sum_{t=1}^{N} (\varphi(t)\varphi^T(t)) \otimes (f_o(t)f_o^T(t))\right] = \Phi_o \otimes \left[\sum_{t=1}^{N} f_o(t)f_o^T(t)\right]$$

$$= \Phi_o \otimes I_k = G_o. \tag{6.19}$$

The evaluation is based on (6.15) and the following identity which holds for Kronecker products [33]:

$$(A \otimes B)(C \otimes D) = AC \otimes BD. \tag{6.20}$$

We will perform analysis for a simplified BF estimator

$$\widehat{\widehat{\gamma}}_o(T) = G_o^{-1} \sum_{t=1}^{N} y(t)\psi_o(t), \tag{6.21}$$

$$\widehat{\widehat{\theta}}(t) = Z_o(t)\widehat{\widehat{\gamma}}_o(T), \quad t \in T, \tag{6.22}$$

$$Z_o(t) = I_n \otimes f_o^T(t),$$

obtained using the approximation

$$G^{-1}(T) \cong (\mathrm{E}[G(T)])^{-1} = G_o^{-1}, \tag{6.23}$$

valid for sufficiently long analysis intervals (one can show that the matrix $G(T)$ tends to G_o, in a probabilistic sense, when the length N of the analysis interval tends to infinity [139]).

Combining (6.19) and (6.21) one obtains

$$\mathrm{E}[\widehat{\gamma}_o(T)] \cong \mathrm{E}[\widehat{\widehat{\gamma}}_o(T)] = G_o^{-1}\mathrm{E}\left[\sum_{t=1}^{N}\psi_o(t)\psi_o^T(t)\gamma_o + \sum_{t=1}^{N} v(t)\psi_o(t)\right] = \gamma_o,$$

hence

$$\mathrm{E}[\widehat{\theta}(t)] \cong \mathrm{E}[\widehat{\widehat{\theta}}(t)] = Z_o(t)\gamma_o = \theta(t), \quad \forall t \in T, \tag{6.24}$$

which means that the BF estimator considered here is unbiased. To arrive at an expression for the covariance matrix of the estimation error, note that

$$\mathrm{cov}[\widehat{\theta}(t)] = Z_o(t)\,\mathrm{cov}[\widehat{\gamma}_o(T)]Z_o^T(t).$$

Furthermore

$$\mathrm{cov}[\widehat{\gamma}_o(T)] \cong \mathrm{cov}[\widehat{\widehat{\gamma}}_o(T)] = G_o^{-1}\mathrm{E}\left[\sum_{t=1}^{N}\sum_{s=1}^{N}\psi_o(t)\psi_o^T(s)v(t)v(s)\right]G_o^{-1}$$

$$= \sigma_v^2 G_o^{-1} = \sigma_v^2(\Phi_o \otimes I_k)^{-1} = \sigma_v^2(\Phi_o^{-1} \otimes I_k)$$

Hence (cf. (6.20))

$$\mathrm{cov}[\widehat{\theta}(t)] \cong \sigma_v^2\left(I_n \otimes f_o^T(t)\right)\left(\Phi_o^{-1} \otimes I_k\right)\left(I_n \otimes f_o(t)\right)$$

$$= \sigma_v^2\left(\Phi_o^{-1} \otimes f_o^T(t)\right)(I_n \otimes f_o(t)) = \sigma_v^2\left(\Phi_o^{-1} \otimes f_o^T(t)f_o(t)\right)$$

$$= \sigma_v^2 f_o^T(t)f_o(t)\Phi_o^{-1}, \tag{6.25}$$

or alternatively

$$\mathrm{cov}[\widehat{\theta}_{\mathrm{BF}}(t)] \cong \frac{\sigma_v^2\Phi_o^{-1}}{l_{t|N}^{\mathrm{BF}}}, \tag{6.26}$$

where

$$l_{t|N}^{\mathrm{BF}} = \frac{1}{\mathbf{f}_o^T(t)\mathbf{f}_o(t)} \, , \tag{6.27}$$

is the quantity which can be interpreted as an instantaneous (local) equivalent number of observations corresponding to a particular choice of the time instant t within the analysis interval $T = \{1, \ldots, N\}$. The interpretation stems from the fact that under time-invariant conditions the covariance matrix of the BF estimator is exactly the same as the covariance matrix of the LS estimator incorporating $l_{t|N}$ data points. Since the average value of the error covariance matrix (evaluated across the entire analysis interval) can be expressed in the form

$$\frac{1}{N} \sum_{t=1}^{N} \mathrm{cov}[\widehat{\theta}_{\mathrm{BF}}(t)] \cong \frac{\sigma_v^2 \Phi_o^{-1}}{\bar{l}_N^{\mathrm{BF}}} \, , \tag{6.28}$$

where

$$\bar{l}_N^{\mathrm{BF}} = \frac{N}{\sum_{t=1}^{N} \mathbf{f}_o^T(t)\mathbf{f}_o(t)} = \frac{N}{k} \, . \tag{6.29}$$

The quantity $\bar{l}_N$ can be regarded as an average equivalent memory of a BF estimator. It is interesting to note that $\bar{l}_N$ depends only on the size of the analysis interval and the number of basis functions. It does not depend on the choice of basis functions, or, more precisely, on the choice of the subspace $F_k(T)$.

The plots depicted in Figure 6.2 show dependence of the normalized variance multipliers

$$\frac{l_{t|N}}{\bar{l}_N} = \frac{N}{k} l_{t|N}$$

on t (i.e. on the location within the analysis window) for Legendre bases of different dimensions. According to these plots, the best accuracy is achieved in the middle of the analysis interval and it gradually decreases towards both ends. This is hardly surprising, because in the vicinity of the left (right) segment end the estimation is based on the future (past) samples only, whereas in the middle it incorporates both past and future samples. For large values of k the difference between the worst accuracy and the best accuracy may be significant.

6.1.4 Dynamic characteristics of BF estimators

The unbiasedness result (6.24) is of little practical importance. The assumption that the parameter trajectory belongs to $F_k(T)$ is rather naive. It would be more realistic to assume that (6.2) is an *approximate* expression for $\theta_i(t)$, i.e. that $\{\theta(t), t \in T\} \notin F_k(T)$.

We will investigate the problem for an arbitrarily time-varying FIR system subject to a stationary excitation. Observe that for the system governed by

$$y(t) = \varphi^T(t)\theta(t) + v(t),$$

one obtains

$$\bar{\theta}(t) = \mathrm{E}[\widehat{\theta}(t)] \cong \mathrm{E}[\widehat{\bar{\theta}}(t)] = \mathrm{E}\left[Z_o(t) G_o^{-1} \sum_{i=1}^{N} \psi_o(i)\varphi^T(i)\theta(i) \right].$$

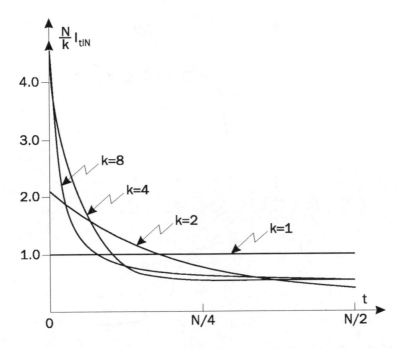

Figure 6.2 Dependence of the normalized variance multipliers for Legendre-based BF estimators of different orders on the location within the analysis window; since all plots are symmetric with respect to $t = N/2$ they are shown only for $t \in [1, \ldots, N/2]$.

Using the identity (6.20) one arrives at

$$\mathrm{E}\left[\psi_o(i)\varphi^T(i)\right] = \mathrm{E}\left[(\varphi(i) \otimes f_o(i))\varphi^T(i)\right] = Z_o^T(i)\Phi_o = G_o Z_o^T(i).$$

Hence

$$\bar{\theta}(t) \cong Z_o(t) \sum_{i=1}^{N} Z_o^T(i)\theta(i) = f_o^T(t) \sum_{i=1}^{N} f_o(i)\theta(i). \qquad (6.30)$$

We note that (6.30) has an interesting geometric interpretation: the term on the right-hand side can be interpreted as an orthogonal projection of $\{\theta(t), t \in T\}$ on the subspace $F_k(T)$ (Figure 6.3).

Remark

Since

$$f_o^T(t)f_o(i) = f^T(t)\left[\sum_{s=1}^{N} f(s)f^T(s)\right]^{-1} f(i),$$

expression (6.30) can be easily rewritten in terms of *any* basis set of $F_k(T)$. ∎

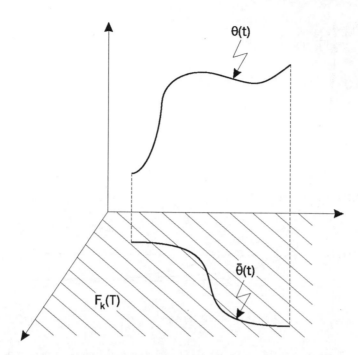

Figure 6.3 Geometric interpretation of the average trajectory of parameter estimates obtained using the method of basis functions.

6.1.5 *Impulse response associated with BF estimators*

Rewrite (6.30) in the form

$$\bar{\theta}_{\text{BF}}(t) \cong \sum_{i=t-1}^{t-N} h_{\text{BF}}(t,i)\theta(t-i),$$
(6.31)

where

$$h_{\text{BF}}(t,i) = \begin{cases} \mathbf{f}_o^T(t)\mathbf{f}_o(t-i) & t-N \le i \le t \\ 0 & \text{elsewhere} \end{cases}$$
(6.32)

According to (6.31) the process $\{\bar{\theta}(t)\}$ can be approximately regarded as a result of passing the process $\{\theta(t)\}$ through the linear noncausal time-varying filter of the impulse response $h_{\text{BF}}(t,i)$. The sequence $\{h_{\text{BF}}(t,i), i \in [t-N,t)\}$ will be further called the impulse response associated with the BF estimator.

The properties of $h_{\text{BF}}(t,i)$ closely resemble those of $h_{\text{WLS}}(i)$ – an impulse response associated with WLS estimators.

First of all, observe that

$$\sum_i h_{\text{BF}}^2(t,i) = \sum_{i=1}^{N} \left(\mathbf{f}_o^T(t)\mathbf{f}_o(i)\right)^2 = \mathbf{f}_o^T(t) \left[\sum_{i=1}^{N} \mathbf{f}_o(i)\mathbf{f}_o^T(i)\right]\mathbf{f}_o(t) = \mathbf{f}_o^T(t)\mathbf{f}_o(t),$$

hence

$$\sum_i h_{\text{BF}}^2(t,i) = \frac{1}{l_{t|N}^{\text{BF}}}, \quad \forall t \in T,$$
(6.33)

which is an obvious analog of (4.30).

The second property,

$$\sum_i h_{\mathrm{BF}}(t,i) = 1, \quad \forall t \in T, \tag{6.34}$$

holds for bases that fulfill the condition $1 \in F_k(T)$. Actually, choosing the first orthonormal basis function of $F_k(T)$ in the form $f_1^o(t) = 1/\sqrt{N}$ and exploiting the orthogonality condition $< f_1^o, f_i^o >= 0, \forall i \neq 1$ one obtains

$$\sum_{i=1}^{N} f_o^T(t) f_o(i) = \sum_{i=1}^{N} (f_1^o(i))^2 = 1,$$

which is nothing but (6.34). Note that (6.34) is a counterpart of (4.29).

6.1.6   Frequency response associated with BF estimators

The instantaneous (frozen) frequency response associated with the BF estimator can be defined as a discrete Fourier transform of the instantaneous impulse response $h_{\mathrm{BF}}(t,i)$ (regarded as a function of i):

$$H_{\mathrm{BF}}(t,\omega) = \sum_i h_{\mathrm{BF}}(t,i) e^{-j\omega i} = A_{\mathrm{BF}}(t,\omega) e^{j\phi_{\mathrm{BF}}(t,\omega)}, \tag{6.35}$$

where $A_{\mathrm{BF}}(t,\omega) = |H_{\mathrm{BF}}(t,\omega)|$ and $\phi_{\mathrm{BF}}(t,\omega) = \arg H_{\mathrm{BF}}(t,\omega)$ are the amplitude and phase responses, respectively. Note that

$$H_{\mathrm{BF}}(t,\omega) = f_o^T(t) \sum_{i=t-N}^{t-1} f_o(t-i) e^{-j\omega i} = f_o^T(t) \sum_{k=1}^{N} f_o(k) e^{-j\omega(t-k)}$$

$$= e^{-j\omega t} f_o^T(t) F_o^\star(\omega), \tag{6.36}$$

where

$$F_o(\omega) = \sum_{t=1}^{N} f_o(t) e^{-j\omega t}$$

and $F_o^\star(\omega)$ denotes the complex conjugate of $F_o(\omega)$.

We will show that the instantaneous frequency characteristics, if handled with due caution, may be a valuable tool for analyzing the parameter matching properties of BF estimators. As we did with WLS estimators, we will examine the frequency distribution of the averaged mean square parameter matching error due to the estimation bias,

$$\overline{M}_b(t) = \mathop{\mathrm{E}}_{\Theta(t)} [\|\bar{\theta}(t) - \theta(t)\|^2],$$

in the case where $\{\theta(t)\}$ is a zero-mean wide-sense stationary process with a spectral density matrix $S_\theta(\omega)$. According to (6.31)

$$\bar{\theta}(t) - \theta(t) \cong \Delta \delta(t),$$

$$\Delta = [\theta(1)| \ldots |\theta(N)], \quad \delta(t) = b(t) - 1_t,$$

where $b(t) = [h_{\mathrm{BF}}(t, t-1), \ldots, h_{\mathrm{BF}}(t, t-N)]^T$ and 1_t is the N-dimensional vector with the tth element equal to one and the remaining elements equal to zero. Observe that

$$\overline{\mathcal{M}}_b(t) \cong \delta^T(t) \mathrm{E}[\Delta^T \Delta] \delta(t)$$

and

$$\mathrm{E}[\Delta^T \Delta]_{ij} = \mathrm{E}[\,\theta^T(i)\theta(j)\,] = \mathrm{tr}\,\{R_\theta(i-j)\}\,.$$

Since

$$R_\theta(\tau) = \frac{1}{2\pi} \int_{-\pi}^{\pi} S_\theta(\omega) e^{j\omega\tau} \, d\omega,$$

one can rewite $\overline{\mathcal{M}}_b(t)$ in the form

$$\overline{\mathcal{M}}_b(t) \cong \frac{1}{2\pi} \int_{-\pi}^{\pi} \delta^T(t) z(\omega) z^\dagger(\omega) \delta(t) \mathrm{tr}\{S_\theta(\omega)\} \, d\omega, \qquad (6.37)$$

where $z(\omega) = [e^{-j\omega(N-1)}, \ldots, e^{-j\omega}, 1]^T$ and $z^\dagger(\omega)$ denotes the conjugate transpose of $z(\omega)$.

Note that

$$\delta^T(t) z(\omega) = e^{-j\omega(N-t)} \left[1 - H_{\mathrm{BF}}^\star(t, \omega)\right].$$

Hence

$$\delta^T(t) z(\omega) z^\dagger(\omega) \delta(t) = E_{\mathrm{BF}}(t, \omega), \qquad (6.38)$$

where

$$E_{\mathrm{BF}}(t, \omega) = |1 - H_{\mathrm{BF}}(t, \omega)|^2 \qquad (6.39)$$

denotes the instantaneous parameter matching characteristics of a BF estimator. Combining (6.37) with (6.38) and taking advantage of the fact that both $S_\theta(\omega)$ and $E_{\mathrm{BF}}(t, \omega)$ are even functions of ω, one obtains

$$\overline{\mathcal{M}}_b(t) \cong \frac{1}{\pi} \int_0^{\pi} E_{\mathrm{BF}}(t, \omega) \mathrm{tr}\{S_\theta(\omega)\} \, d\omega. \qquad (6.40)$$

It is interesting to note that although the interpretation of the instantaneous frequency characteristics is somewhat heuristic, the derivation of (6.40) is free from heuristic elements. In particular, it was *not* necessary to assume that the instantaneous frequency response $H_{\mathrm{BF}}(t, \omega)$ is a slowly varying function of t – the assumption underlying almost all considerations referring to the concept of instantaneous (frozen) frequency characteristics. The approximate equality sign in (6.40) follows *only* from the fact that there is an approximate equality sign in (6.31).

The average value of the bias error evaluated across the entire analysis interval,

$$\overline{\mathcal{M}}_b = \frac{1}{N} \sum_{t=1}^{N} \overline{\mathcal{M}}_b(t),$$

can be obtained from

$$\overline{\mathcal{M}}_b \cong \frac{1}{\pi} \int_0^{\pi} \overline{E}_{\mathrm{BF}}(\omega) \mathrm{tr}\{S_\theta(\omega)\} \, d\omega, \qquad (6.41)$$

where

$$\overline{E}_{\mathrm{BF}}(\omega) = \frac{1}{N} \sum_{t=1}^{N} E_{\mathrm{BF}}(t,\omega). \tag{6.42}$$

Note that (6.41) is a counterpart of (4.52).

It is easy to prove the following identity:

$$\overline{E}_{\mathrm{BF}}(\omega) = 1 - \frac{1}{N} \| F_o(\omega) \|^2 = 1 - \frac{1}{N} \sum_{i=1}^{k} |F_i^o(\omega)|^2, \tag{6.43}$$

where

$$F_i^o(\omega) = \sum_{t=1}^{N} f_i^o(t) e^{-j\omega t}.$$

Actually, note that

$$E_{\mathrm{BF}}(t,\omega) = 1 - H_{\mathrm{BF}}(t,\omega) - H_{\mathrm{BF}}^\star(t,\omega) + |H_{\mathrm{BF}}(t,\omega)|^2. \tag{6.44}$$

Using (6.36) one obtains

$$\frac{1}{N} \sum_{t=1}^{N} H_{\mathrm{BF}}(t,\omega) = \frac{1}{N} \sum_{t=1}^{N} \mathbf{f}_o^T(t) e^{-j\omega t} F_o^\star(\omega) = \frac{1}{N} F_o^T(\omega) F_o^\star(\omega)$$

$$= \frac{1}{N} \| F_o(\omega) \|^2 \tag{6.45}$$

Similarly

$$\frac{1}{N} \sum_{t=1}^{N} H_{\mathrm{BF}}^\star(t,\omega) = \frac{1}{N} \| F_o(\omega) \|^2, \tag{6.46}$$

and due to orthogonality of the basis set,

$$\frac{1}{N} \sum_{t=1}^{N} |H_{\mathrm{BF}}(t,\omega)|^2 = \frac{1}{N} \sum_{t=1}^{N} F_o^T(\omega) \mathbf{f}_o(t) \mathbf{f}_o^T(t) F_o^\star(\omega) = \frac{1}{N} \| F_o(\omega) \|^2 . \tag{6.47}$$

Combining (6.41) and (6.42) with (6.44) to (6.47) one arrives at (6.43).

6.1.7 Properties of the associated frequency characteristics

Integral approximations

If the analysis interval T is sufficiently long, the properties of the associated frequency characteristics can be studied by analyzing the corresponding integral approximations. Suppose that

$$f_j(t) = \tilde{f}_j\left(\frac{t}{N}\right), \quad t \in T,$$

$$j = 1, \ldots, k$$

where $\widetilde{f}_1(\cdot), \ldots, \widetilde{f}_k(\cdot) \in L_2[0,1]$ are the continuous-time linearly independent basis-generating functions – analog 'prototypes' of $f_1(\cdot), \ldots, f_k(\cdot)$.

Furthermore, denote by $\widetilde{f}_1^o(s), \ldots, \widetilde{f}_k^o(s), s \in [0,1]$ the orthonormal basis set, obtained as a result of applying the Gram–Schmidt orthogonalization procedure to $\widetilde{f}_j(s)$ successively for $j = 1, \ldots, k$, and let

$$\widetilde{\mathbf{f}}_o(s) = [\,\widetilde{f}_1^o(s), \ldots, \widetilde{f}_k^o(s)\,]^T.$$

One can show that

$$\mathbf{f}_o(t) \xrightarrow[N \mapsto \infty]{} \frac{1}{\sqrt{N}}\widetilde{\mathbf{f}}_o\left(\frac{t}{N}\right), \quad t \in T.$$

The proof is straightforward after noting that, due to orthonormality of $\{\widetilde{f}_j^o(s)\}$,

$$\frac{1}{N}\sum_{t=1}^{N}\widetilde{\mathbf{f}}_o\left(\frac{t}{N}\right)\widetilde{\mathbf{f}}_o^T\left(\frac{t}{N}\right) \simeq \frac{1}{N}\int_0^N \widetilde{\mathbf{f}}_o\left(\frac{t}{N}\right)\widetilde{\mathbf{f}}_o^T\left(\frac{t}{N}\right)dt$$

$$= \int_0^1 \widetilde{\mathbf{f}}_o(s)\widetilde{\mathbf{f}}_o^T(s)\,ds = I_k.$$

If the analysis interval is sufficiently long, the following integral approximation of $F_o(\omega)$ can be used in the range of small values of ω (for the Legendre and Fourier bases the limitation is $\omega < \pi k/N, k/N \ll 1$)

$$F_o(\omega) = \sum_{t=1}^{N}\mathbf{f}_o(t)e^{-j\omega t} \simeq \frac{1}{\sqrt{N}}\sum_{t=1}^{N}\widetilde{\mathbf{f}}_o\left(\frac{t}{N}\right)e^{-j\omega t}$$

$$\simeq \frac{1}{\sqrt{N}}\int_0^N \widetilde{\mathbf{f}}_o\left(\frac{t}{N}\right)e^{-j\omega t}dt = \sqrt{N}\int_0^1 \widetilde{\mathbf{f}}_o(s)e^{-j\omega N s}ds = \sqrt{N}\widetilde{F}_o(\omega N),$$

where

$$\widetilde{F}_o(\omega) = \int_0^1 \widetilde{\mathbf{f}}_o(s)e^{-j\omega s}ds = [\,\widetilde{F}_1^o(\omega), \ldots, \widetilde{F}_k^o(\omega)\,]^T,$$

and

$$\widetilde{F}_i^o(\omega) = \int_0^1 \widetilde{f}_i^o(s)e^{-j\omega s}ds.$$

Consequently

$$H_{\mathrm{BF}}(t, \omega) = e^{-j\omega t}\mathbf{f}_o^T(t)F_o^\star(\omega) \cong e^{-j\omega t}\widetilde{\mathbf{f}}_o^T\left(\frac{t}{N}\right)\widetilde{F}_o^\star(\omega N)$$

$$= e^{-j\omega t}\sum_{i=1}^{k}\widetilde{f}_i^o\left(\frac{t}{N}\right)\widetilde{F}_i^o(-\omega N), \tag{6.48}$$

and

$$\overline{E}_{\mathrm{BF}}(\omega) = 1 - \frac{1}{N}\|F_o(\omega)\|^2 \cong 1 - \|\widetilde{F}_o(\omega)\|^2 = 1 - \sum_{i=1}^{k}|\widetilde{F}_i^o(\omega N)|^2. \tag{6.49}$$

Example 6.1

The first three 'prototype' functions comprising the Legendre basis are

$$
\begin{aligned}
\tilde{f}_1^o(s) &= 1, \\
\tilde{f}_2^o(s) &= \sqrt{3}(1-2s), \\
\tilde{f}_3^o(s) &= \sqrt{5}(1-6s+6s^2),
\end{aligned}
$$

and the corresponding transforms are given by

$$
\tilde{F}_1^o(\eta) = e^{-j\eta}\frac{\sin\eta}{\eta},
$$

$$
\tilde{F}_2^o(\eta) = \sqrt{3}je^{-j\eta}\left[\frac{\cos\eta}{\eta}-\frac{\sin\eta}{\eta^2}\right],
$$

$$
\tilde{F}_3^o(\eta) = \sqrt{5}e^{-j\eta}\left[\frac{\sin\eta}{\eta}+3\frac{\cos\eta}{\eta^2}-3\frac{\sin\eta}{\eta^3}\right],
$$

where $\eta = \omega/2$.

Hence, for the Legendre basis of order $k=3$, one obtains the following integral approximation:

$$
H_{BF}(t,\omega) \cong e^{-j\omega(t-\frac{N}{2})}\sum_{i=1}^{3}g_i(t,\omega), \tag{6.50}
$$

where

$$
g_1(t,\omega) = \frac{\sin\frac{\omega N}{2}}{\frac{\omega N}{2}},
$$

$$
g_2(t,\omega) = 3j\left(\frac{2t}{N}-1\right)\left[\frac{\sin\frac{\omega N}{2}}{(\frac{\omega N}{2})^2}-\frac{\cos\frac{\omega N}{2}}{\frac{\omega N}{2}}\right],
$$

$$
g_3(t,\omega) = 5\left(1-\frac{6t}{N}+\frac{6t^2}{N^2}\right)\left[\frac{\sin\frac{\omega N}{2}}{\frac{\omega N}{2}}+3\frac{\cos\frac{\omega N}{2}}{(\frac{\omega N}{2})^2}-3\frac{\sin\frac{\omega N}{2}}{(\frac{\omega N}{2})^3}\right],
$$

Example 6.2

The first three functions comprising the Fourier basis are

$$
\begin{aligned}
\tilde{f}_1^o(s) &= 1, \\
\tilde{f}_2^o(s) &= \sqrt{2}\cos(\pi s), \\
\tilde{f}_3^o(s) &= \sqrt{2}\cos(2\pi s),
\end{aligned}
$$

and the corresponding transforms are given by

$$
\tilde{F}_1^o(\eta) = e^{-j\eta}\frac{\sin\eta}{\eta},
$$

$$
\tilde{F}_2^o(\eta) = \sqrt{2}je^{-j\eta}\frac{\eta\cos\eta}{(\frac{\pi}{2})^2-\eta^2},
$$

$$
\tilde{F}_3^o(\eta) = \sqrt{2}e^{-j\eta}\frac{\eta\sin\eta}{\pi^2-\eta^2},
$$

Hence, for the Fourier basis of order $k = 3$, the integral approximation of $H_{BF}(t, \omega)$ takes the form (6.50) with

$$g_1(t, \omega) = \frac{\sin \frac{\omega N}{2}}{\frac{\omega N}{2}},$$

$$g_2(t, \omega) = -2j \cos \frac{\pi t}{N} \frac{\frac{\omega N}{2} \cos \frac{\omega N}{2}}{(\frac{\pi}{2})^2 - (\frac{\omega N}{2})^2},$$

$$g_3(t, \omega) = -2 \cos \frac{2\pi t}{N} \frac{\frac{\omega N}{2} \sin \frac{\omega N}{2}}{(\pi)^2 - (\frac{\omega N}{2})^2}.$$

Estimation bandwidth

It is obvious that for a given lowpass density $S_\theta(\omega)$ the bias error $\overline{\mathcal{M}}_b$ depends on the 'average bandwidth' of the corresponding associated filter. The average estimation bandwidth $\overline{\omega}_\beta$ of a BF estimator can be defined analogously to the estimation bandwidth of a WLS estimator

$$\overline{\omega}_\beta : \begin{cases} \sup\limits_{\omega \in [0, \overline{\omega}_\beta]} \overline{E}(\omega) \leq \beta \\ \text{or} \\ \int_0^{\overline{\omega}_\beta} \overline{E}(\omega) d\omega \leq \beta \end{cases} \tag{6.51}$$

Regardless of the definition and for a sensible choice of β, the greater $\overline{\omega}_\beta$ then the smaller the average mean square matching error one can expect. Note that in the range of small values of ω, where the integral approximations (6.48) and (6.49) hold, the associated frequency characteristics are functions of ωN, which means that $\overline{\omega}_\beta$ is inversely proportional to the length of the analysis interval N. On the other hand, since the functions

$$|F_i^o(\omega)|^2, \quad i = 1, \ldots, k,$$

are nonnegative, $\overline{\omega}_\beta$ is an increasing function of k (Figure 6.4). Combining the two observations made above, one gets approximately

$$\overline{\omega}_\beta \sim \frac{k}{N} = \frac{1}{\overline{l}_N} \tag{6.52}$$

where $\sim$ denotes proportionality.

Comparing (6.26) with (6.52) one can easily see that the choice of the equivalent memory in BF process identification must be the result of a compromise: small values of $\overline{l}_N$ yield estimators with large bandwidth (i.e. capable of tracking parameter changes in a wide frequency range) but also with a large variability, and vice versa.

6.1.8 Comparing the matching properties of different BF estimators

Suppose we want to compare the parameter matching properties of BF estimators corresponding to different basis sets, i.e. sets differing in number and/or type of basis functions. We will assume that system coefficients are 'slowly varying', i.e. they can be modeled as lowpass signals. Following the general guidelines set before, we should

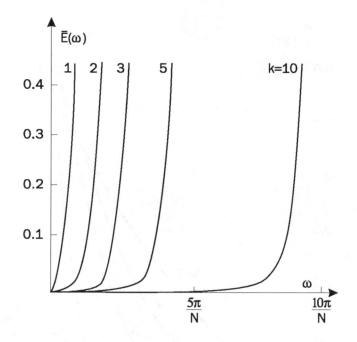

Figure 6.4 Average matching characteristics of some Legendre-based BF estimators.

compare estimators characterized by the same values of $\bar{l}_N$, i.e. yielding the same average estimation accuracy under time-invariant conditions.

The direct comparison of different frequency characteristics is possible provided they are expressed as functions of a normalized frequency

$$\Omega = \omega \bar{l}_N = \frac{\omega N}{k}.$$

As an example consider the normalized error characteristics of Fourier-based BF estimators of different orders. Using the technique of integral approximations one can easily show that in the range of small values of Ω ($\Omega < \pi$) it holds

$$\overline{E}_k(\Omega) \cong 1 - \sum_{i=1}^{k} G_k^i(\Omega),$$

where

$$G_k^1(\Omega) = \left[\frac{\sin \frac{\Omega k}{2}}{\frac{\Omega k}{2}} \right]^2$$

and

$$G_k^i(\Omega) = \left[\frac{\frac{\Omega k}{2}}{\left(\frac{\pi i}{2}\right)^2 - \left(\frac{\Omega k}{2}\right)^2} \right]^2 \left[1 - (-1)^i \cos \Omega k\right]$$

$$= \left[\frac{2\Omega k}{(\pi i)^2 - (\Omega k)^2} \right]^2 \left[1 - (-1)^i \cos \Omega k\right], \quad i > 1.$$

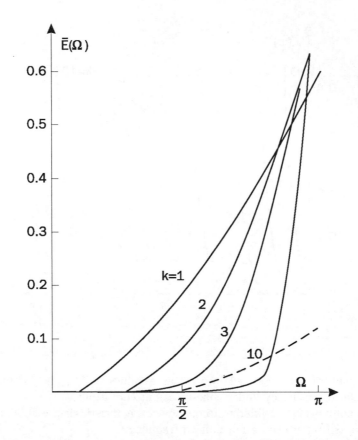

Figure 6.5 Normalized average matching characteristics of Legendre-based BF estimators of different orders (solid line) and the corresponding characteristic of a sliding window WLS estimator (dashed line).

The plots of $\overline{E}_k(\Omega)$ for different values of k are shown in Figure 6.5.

First, it is clear that even though the normalized estimation bandwidth of the compared BF estimators grows with the number of basis functions, the achievable bandwidth increments may be pretty small for large values of k. Second, since for large values of k the transition between the bandpass and bandstop areas becomes increasingly sharp, the performance of high-order BF estimators may be very *sensitive* to the choice of the average estimation memory.

It is interesting to compare the parameter matching characteristics of BF estimators with those of WLS estimators. The plot of a normalized matching error curve for the sliding window least squares estimator is superimposed on BF error plots in Figure 6.5. It is clear from comparison of all plots that the parameter matching properties of an SWLS estimator are much better than the average matching properties of BF estimators – irrespective of the number of basis functions. This can be easily explained by increased matching errors at both ends of the analysis interval. For higher-order BF estimators, both components of the mean square matching error are strongly affected by location within the analysis interval. As observed in Section 6.1.3, the variance component tends to increase towards both ends of the interval. A different kind of estimation 'misbehavior' can be expected by examining the instantaneous frequency characteristics of BF estimators. First, the relative bumpiness of the amplitude characteristics at both ends of the analysis interval (Figure 6.6) allows one to expect large dynamic bias errors due to overshoot effects for t close to 1 or N. Second, nonlinearity of the phase characteristics (one can show that the phase delay is equal to zero *only* in the middle of the analysis interval) causes some extra increase in matching errors due to the estimation delay ($t > N/2$) and estimation advance ($t < N/2$) effects. Quite clearly, the compared estimation algorithms serve different purposes: the SWLS algorithm yields a *point* estimate $\widehat{\theta}(t)$, while the BF algorithm provides an *interval* estimate of the parameter trajectory $\{\widehat{\theta}(t), t \in T\}$. It is obvious from the analysis carried out above that the estimates of parameter trajectory based on process segmentation are less accurate than those obtained by repetitive use of WLS algorithms (sometimes known as running WLS estimates). Hence, if segmentation is not required by the application at hand, it should generally be avoided.

6.2   Weighted basis function estimation

It would be pretty naive to assume that the expansion coefficients c_{ij} in (6.2) are constant in the entire time domain. This assumption would mean, in fact, that the analyzed process could be regarded as reducible, i.e. (perfectly) predictably time-varying.

When the segmentation approach is used, one fits different BF models to different data frames, which is a practical way of coping with possible fluctuations in c_{ij}'s. Another approach, which serves the same purpose, results when the method of basis functions is combined with information weighting. Weighting forces the estimation to be more focused on the most recent data samples and therefore allows one to track slow variations in the expansion coefficients. We shall call this approach the weighted basis function (WBF) estimation.

6.2.1   Estimation principles

Denote by
$$T_t = \{1, \ldots, t\}$$
the expanding analysis window and denote by $\{w(\cdot)\}$ the weighting sequence obeying (4.1) and (4.2). WBF estimators can be defined in two different ways.

Running-basis estimators

Let $F_k^+(t)$ be a subspace spanned by $\{f_1(s), \ldots, f_k(s), s \in T_t\}$ with the inner product and a vector norm defined as

$$< f, g >_w = \sum_{i=0}^{t-1} w(i) f(t-i) g(t-i), \quad \| f \|_w^2 = < f, f >_w .$$

We will assume that
$$\{f_1(s), \ldots, f_k(s), s \in T_\infty\}$$

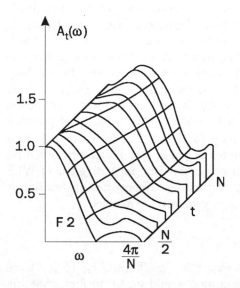

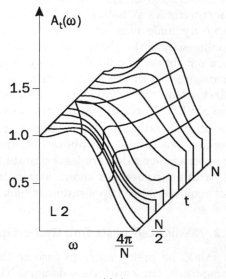

(a)

(b)

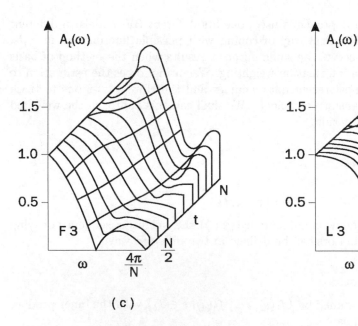

(c)

(d)

Figure 6.6 Dependence of the instantaneous amplitude characteristics of BF estimators on the location within the analysis window; since all plots are symmetric with respect to $t = N/2$ they are shown only for $t \in [1, \dots, N/2]$: (a) second-order Fourier, (b) second-order Legendre, (c) third-order Fourier, (d) third-order Legendre.

is a set of basis functions chosen so that for every $t \geq k$ the sequences $\{f_1(s), \ldots, f_k(s), s \in T_t\}$ are linearly independent with weight $\{w(\cdot)\}$; that is, for every $t > k$ the condition

$$\| \sum_{j=1}^{k} a_j f_j(s) \|_w = 0, \quad \forall s \in T_t$$

holds if and only if $a_1 = \ldots = a_k = 0$. Note that both the Legendre and Fourier bases fulfill this requirement.

The running-basis WBF estimators can be obtained by combining the previously used model of parameter variation,

$$\theta_i(s) = \sum_{j=1}^{k} c_{ij}^+ f_j(s), \quad s \in T_t, \tag{6.53}$$

or equivalently

$$y(s) = \psi^T(s)\gamma + v(s), \quad s \in T_t,$$

with the elements of data weighting. Using the method of weighted least squares one arrives at

$$\hat{\gamma}(t) = \arg \min_{\gamma} \sum_{i=0}^{t-1} w(i)[y(t-i) - \psi^T(t-i)\gamma]^2$$

$$= \left(\sum_{i=0}^{t-1} w(i)\psi(t-i)\psi^T(t-i) \right)^{-1} \left(\sum_{i=0}^{t-1} w(i)y(t-i)\psi(t-i) \right)$$

$$= G_+^{-1}(t)h_+(t), \tag{6.54}$$

and

$$\hat{\theta}_+(t) = Z(t)\hat{\gamma}(t). \tag{6.55}$$

Formally γ is written down as a constant vector but its components, the expansion coefficients c_{ij}, may be (slowly) time-varying. We already know that the method of weighted least squares is capable of tracking such slow time variations.

The term 'running basis' refers to the fact that as the time runs from say t_1 to t_2, the basis functions are simply extended to cover the additional time interval $(t_1, t_2 >$.

Fixed-basis estimators

Denote by $F_k^-(t)$ a subspace spanned by $\{f_1(t-s+1), \ldots, f_k(t-s+1), s \in T_t\}$ with the inner product defined as

$$(f, g)_w = \sum_{i=0}^{t-1} w(i)f(i+1)g(i+1).$$

and consider the following 'reverse-time' description of the parameter trajectory:

$$\theta_i(s) = \sum_{j=1}^{k} c_{ij}^- f_j(t-s+1), \quad s \in T_t, \tag{6.56}$$

leading to

$$y(s) = \psi_t^T(s)\beta + v(s), \quad s \in T_t, \tag{6.57}$$

where

$$\psi_t(s) = \varphi(s) \otimes f(t - s + 1),$$

and

$$\beta = [\beta_1^T, \ldots, \beta_n^T]^T, \quad \beta_i = [c_{i1}^-, \ldots, c_{ik}^-]^T.$$

Note that as long as local approximation is concerned, the 'reverse-time' model (6.56) is no less justified than (6.53); as a matter of fact, it better fits the truncated functional series expansion argument referred to before.

Applying the method of weighted least squares to estimation of β, one arrives at

$$\widehat{\beta}(t) = \arg\min_{\beta} \sum_{i=0}^{t-1} w(i)[y(t-i) - \psi_t^T(t-i)\beta]^2$$

$$= \left(\sum_{i=0}^{t-1} w(i)\psi_t(t-i)\psi_t^T(t-i)\right)^{-1} \left(\sum_{i=0}^{t-1} w(i)y(t-i)\psi_t(t-i)\right)$$

$$= G_-^{-1}(t)h_-(t), \tag{6.58}$$

and

$$\widehat{\theta}_-(t) = Z(1)\widehat{\beta}(t). \tag{6.59}$$

Since the basis set

$$\{f_1(t-s+1), \ldots, f_k(t-s+1), \quad s \in T_t\} \tag{6.60}$$

underlying our identification scheme is redefined for each new value of t, we will call (6.59) the fixed-basis WBF estimator.

Remark 1

If the basis set is *direction invariant* in the sense that for every t there exists a nonsingular $k \times k$ matrix $A(t)$ such that

$$f(t-s+1) = A(t)f(s), \quad \forall s \in T_t, \tag{6.61}$$

the running-basis and fixed-basis WBF estimators coincide:

$$\widehat{\theta}_+(t) = \widehat{\theta}_-(t). \tag{6.62}$$

The proof of (6.62) is basically the same as the proof of the invariance result given in Section 6.1.2. For basis sets that are not direction invariant, the subspaces $F_k^-(t)$ and $F_k^+(t)$ differ, which means that the corresponding running-basis and fixed-basis estimators are not identical.

Since the functions $(t - s + 1)^{i-1}$, $i = 1, \ldots, k$, can be expressed as linear combinations of s^{i-1}, $i = 1, \ldots, k$, it is easy to construct the matrix $A(t)$ for the Legendre basis. We note that the direction invariance condition is not fulfilled for the Fourier basis (6.4) but holds for the modified (mixed sine/cosine) Fourier basis:

$$f_1(t) = 1,$$

$$f_{2i}(t) = \sin \omega_i t, \quad f_{2i+1}(t) = \cos \omega_i t, \quad i = 1, 2, \ldots \tag{6.63}$$

Even when leading to identical results, the running-basis and fixed-basis estimation schemes offer different computational and/or numerical advantages [136].

Remark 2

If the basis set is reduced to a single ($k = 1$) constant function $f_1(t) = 1$, both WBF estimators are identical with the WLS estimator.

6.2.2 Recursive WBF estimators

From practical viewpoint it is important to find weighting and basis sequences which make the WBF estimators recursively computable. To guarantee recursive computability one can use:

1. Any window which allows for recursive computation of weighted sums
2. Any set of recursively computable basis functions

The first condition is met, for example, by rectangular and exponential windows. To fulfill the second condition we will require that there exists a constant $k \times k$ matrix B such that

$$f(t + 1) = Bf(t), \quad \forall t \in T_\infty. \tag{6.64}$$

We note that under (6.64) the components of $f(t)$ must have the form $t^{i-1} \cos(\omega_j t + \phi_j)$, $i, j = 1, 2, \ldots$, or must be the exponentially weighted linear combinations of such functions.

Example 6.3

For the Legendre basis of order k, the matrix B obeying (6.64) has the form

$$B = \begin{bmatrix} 1 & & & \\ \begin{pmatrix} 1 \\ 1 \end{pmatrix} & \ddots & & 0 \\ \vdots & & & \\ \begin{pmatrix} k-1 \\ k-1 \end{pmatrix} & \cdots & \begin{pmatrix} k-1 \\ 1 \end{pmatrix} & 1 \end{bmatrix}$$

and for the modified Fourier basis (6.63) of order $2k + 1$ one obtains

$$B = \begin{bmatrix} 1 & 0 & \cdots & 0 \\ 0 & \begin{matrix} \cos \omega_1 & \sin \omega_1 \\ -\sin \omega_1 & \cos \omega_1 \end{matrix} & & 0 \\ \vdots & & \ddots & \\ 0 & 0 & \cdots & \begin{matrix} \cos \omega_k & \sin \omega_k \\ -\sin \omega_k & \cos \omega_k \end{matrix} \end{bmatrix}$$

∎

When the recursive computability conditions are fulfilled, it is easy to design algorithms for recursive updating of the matrices $G_+(t)$, $G_-(t)$ and the vectors $h_+(t)$, $h_-(t)$ in (6.54) and (6.58).

If exponential weighting is used $(w(i) = \lambda^i,\ 0 < \lambda < 1)$ then the corresponding recursions are

$$
\begin{aligned}
G_+(t) &= \lambda G_+(t-1) + \psi(t)\psi^T(t), \\
h_+(t) &= \lambda h_+(t-1) + y(t)\psi(t),
\end{aligned}
\tag{6.65}
$$

and

$$
\begin{aligned}
G_-(t) &= \lambda C G_-(t-1)C^T + \psi_t(t)\psi_t^T(t), \\
h_-(t) &= \lambda C h_-(t-1) + y(t)\psi_t(t),
\end{aligned}
\tag{6.66}
$$

where $C = I_k \otimes B$.

Based on (6.65) and (6.66) it is possible to derive the recursive algorithms for direct updating of $\widehat{\gamma}(t)$ and $\widehat{\beta}(t)$.

Running-basis estimators

Let $Q_+(t) = G_+^{-1}(t)$. Using the matrix inversion lemma (3.10) one arrives at the following recursive algorithm for computation of $\widehat{\gamma}(t) = G_+^{-1}(t)h_+(t)$:

$$
\begin{aligned}
\widehat{\gamma}(t) &= \widehat{\gamma}(t-1) + L_+(t)\epsilon(t), \\
\epsilon(t) &= y(t) - \psi^T(t)\widehat{\gamma}(t-1), \\
L_+(t) &= \frac{Q_+(t-1)\psi(t)}{\lambda + \psi^T(t)Q_+(t-1)\psi(t)}, \\
Q_+(t) &= \frac{1}{\lambda}\left[Q_+(t-1) - \frac{Q_+(t-1)\psi(t)\psi^T(t)Q_+(t-1)}{\lambda + \psi^T(t)Q_+(t-1)\psi(t)}\right],
\end{aligned}
\tag{6.67}
$$

which is an obvious analog of (4.8).

If the basis sequences are not bounded for $t \mapsto \infty$ (which is the case when the Legendre basis is used) the elements of $G_+(t)$ grow with t, which may cause numerical problems. The simplest way out of difficulty, suggested by Xianya and Evans [197] for second-order Legendre-based estimators, is to 'shift back' the basis sequences by t_o every t_o time steps. This can be achieved by appropriate scaling of the corresponding matrices and vectors:

$$
Q_+(kt_o) := D^{-T}Q_+(kt_o)D^{-1},
$$

$$
f(kt_o + i) := f(i), \quad i = 1,\ldots,t_o,
$$

where $D = I_k \otimes B^{t_o}$ and $D^{-T} = (D^T)^{-1} = (D^{-1})^T$.

Since resetting is equivalent to replacing the original basis vectors $f(t)$ with $B^{-t_o}f(t) = f(t - t_o)$, the change does not affect the estimate $\widehat{\gamma}(t)$. Unfortunately, even with periodic resetting, the algorithm may be numerically unreliable for large values of t. This is because, for the Legendre basis, all eigenvalues of D can be shown to lie on the unit circle [197] so the resetting procedure can itself become numerically unstable. In such cases the complete startup of the algorithm may be necessary every Kt_o steps, where K is a suitably chosen integer.

Resetting is not necessary if one works with bounded basis sequences such as sinusoidal and/or cosinusoidal functions.

Fixed-basis estimators

Denote $Q_-(t) = G_-^{-1}(t)$. Using the matrix inversion lemma (3.10) one obtains

$$Q_-(t) =$$

$$\frac{1}{\lambda} \left[C^{-T} Q_-(t-1) C^{-1} - \frac{C^{-T} Q_-(t-1) C^{-1} \psi_t(t) \psi_t^T(t) C^{-T} Q_-(t-1) C^{-1}}{\lambda + \psi_t^T(t) C^{-T} Q_-(t-1) C^{-1} \psi_t(t)} \right].$$

Observe that

$$C^{-1} \psi_t(t) = \left(I_k \otimes B^{-1} \right) \left(\varphi(t) \otimes f(1) \right) = \varphi(t) \otimes \left(B^{-1} f(1) \right)$$

$$= \varphi(t) \otimes f(0) = \zeta(t).$$

Combining (6.66) with the formulas derived above it is possible to express $\widehat{\beta}(t) = G_-^{-1}(t) \, h_-(t)$ in the following recursively computable form:

$$\widehat{\beta}(t) = C^{-T} \left[\widehat{\beta}(t-1) + L_-(t) \epsilon(t) \right],$$

$$\epsilon(t) = y(t) - \zeta^T(t) \widehat{\beta}(t-1),$$

$$L_-(t) = \frac{Q_-(t-1) \zeta(t)}{\lambda + \zeta^T(t) Q_-(t-1) \zeta(t)},$$

$$Q_-(t) = \frac{1}{\lambda} C^{-T} \left[Q_-(t-1) - \frac{Q_-(t-1) \zeta(t) \zeta^T(t) Q_-(t-1)}{\lambda + \zeta^T(t) Q_-(t-1) \zeta(t)} \right] C^{-1}. \quad (6.68)$$

One can show that in the case considered the elements of $G_-(t)$ are bounded in the mean square sense, provided the input and output sequences are bounded, even if the basis functions do not obey the uniform boundedness condition. Hence resetting is not necessary if the above algorithm is used.

Remark

Note that for the Legendre basis (6.3) we have $f(0) = [1, 0, \ldots, 0]^T$, which further simplifies the fixed-basis algorithm.

6.2.3 Static characteristics of WBF estimators

To evaluate the equivalent memory of WBF estimators, we will assume that $\{\varphi(t)\}$ is a wide-sense stationary sequence and that $\{\theta(s), s \in T_t\} \in F_k^+(t)$ (for running-basis estimators) or $\{\theta(s), s \in T_t\} \in F_k^-(t)$ (for fixed-basis estimators).

Running-basis estimators

Denote by

$$\{f_1^+(s), \ldots, f_k^+(s), \quad s \in T_t\}$$

any basis set of $F_k^+(t)$ which is orthonormal with weight $\{w(\cdot)\}$ (*w*-orthonormal), namely

$$\sum_{i=0}^{t-1} w(i) \mathrm{f}_+(t-i) \mathrm{f}_+^T(t-i) = I_k \tag{6.69}$$

where

$$\mathrm{f}_+(s) = [\, f_1^+(s), \ldots, f_k^+(s) \,]^T.$$

The orthonormal basis can be obtained by applying the Gram–Schmidt procedure (6.18), provided that the inner product $< \cdot, \cdot >$ and vector norm $\| \cdot \|$ are replaced with $< \cdot, \cdot >_w$ and $\| \cdot \|_w$, respectively.

Expressing the process equation in terms of the *w*-orthonormal basis set, one obtains

$$y(t) = \varphi^T(t)\theta(t) + v(t) = \psi_+^T(t)\gamma_+ + v(t), \tag{6.70}$$

where $\psi_+(t) = \varphi(t) \otimes \mathrm{f}_+(t)$.

Under (C1) the expectation of the regression matrix in (6.54) is given by

$$\mathrm{E}[G_+(t)] = \mathrm{E}\left[\sum_{i=0}^{t-1} w(i)\psi_+(t-i)\psi_+^T(t-i)\right]$$

$$= \mathrm{E}\left[\sum_{i=0}^{t-1} w(i)\left(\varphi(t)\varphi^T(t)\right) \otimes \left(\mathrm{f}_+(t-i)\mathrm{f}_+^T(t-i)\right)\right]$$

$$= \Phi_o \otimes \left[\sum_{i=0}^{t-1} w(i)\mathrm{f}_+(t-i)\mathrm{f}_+^T(t-i)\right] = \Phi_o \otimes I_k = G_o.$$

It is easy to show that

$$\mathrm{E}[\widehat{\gamma}_+(t)] \cong G_o^{-1}\mathrm{E}\left[\sum_{i=0}^{t-1} w(i)y(t-i)\psi_+(t-i)\right] = \gamma_+$$

and that

$$\mathrm{cov}[\widehat{\gamma}_+(t)] \cong G_o^{-1}\mathrm{E}\left[\sum_{i=0}^{t-1}\sum_{j=0}^{t-1} w(i)w(j)\psi_+(t-i)\psi_+(t-j)v(t-i)v(t-j)\right] G_o^{-1}$$

$$= \sigma_v^2 \left(\Phi_o^{-1} \otimes I_k\right)\left(\Phi_o \otimes \left[\sum_{i=0}^{t-1} w^2(i)\mathrm{f}_+(t-i)\mathrm{f}_+^T(t-i)\right]\right)\left(\Phi_o^{-1} \otimes I_k\right)$$

$$= \sigma_v^2 \Phi_o^{-1} \otimes \left[\sum_{i=0}^{t-1} w^2(i)\mathrm{f}_+(t-i)\mathrm{f}_+^T(t-i)\right].$$

Note that $\theta(t) = Z_+(t)\gamma_+$ where $Z_+(t) = I_n \otimes \mathsf{f}_+^T(t)$. Consequently

$$\mathrm{E}[\widehat{\theta}_+(t)] = \mathrm{E}[Z_+(t)\widehat{\gamma}_+(t)] \cong \theta(t) \tag{6.71}$$

and

$$\mathrm{cov}[\widehat{\theta}_+(t)] = Z_+(t)\mathrm{cov}[\widehat{\gamma}_+(t)]Z_+^T(t)$$

$$\cong \sigma_v^2 \mathsf{f}_+^T(t) \left[\sum_{i=0}^{t-1} w^2(i)\mathsf{f}_+(t-i)\mathsf{f}_+^T(t-i) \right] \mathsf{f}_+(t)\Phi_o^{-1}, \tag{6.72}$$

or equivalently

$$\mathrm{cov}[\widehat{\theta}_+ \,\mathrm{WBF}(t)] \cong \frac{\sigma_v^2 \Phi_o^{-1}}{l_t^{+\,\mathrm{WBF}}} \tag{6.73}$$

where

$$l_t^{+\,\mathrm{WBF}} = \frac{1}{\mathsf{f}_+^T(t) \left[\sum_{i=0}^{t-1} w^2(i)\mathsf{f}_+(t-i)\mathsf{f}_+^T(t-i) \right] \mathsf{f}_+(t)}$$

$$= \frac{1}{\sum_{i=0}^{t-1} \left[w(i)\mathsf{f}_+^T(t)\mathsf{f}_+(t-i) \right]^2}\,, \tag{6.74}$$

denotes the equivalent memory span of the running-basis WBF estimator.

Remark

The finite-memory condition

$$l_t^+ \le c < \infty, \quad \forall t$$

does not guarantee that l_t^+ stabilizes at a constant value as t tends to infinity, e.g. when the Fourier basis is adopted l_t^+ changes in an asymptotically periodic manner.

Fixed-basis estimators

Denote by

$$\{f_1^-(s), \ldots, f_k^-(s), \quad s \in T_t\}$$

any w-orthonormal basis set of $F_k^-(t)$, i.e. any set satisfying

$$\sum_{i=0}^{t-1} w(i)\mathsf{f}_-(i+1)\mathsf{f}_-^T(i+1) = I_k \tag{6.75}$$

where

$$\mathsf{f}_-(s) = [\, f_1^-(s), \ldots, f_k^-(s) \,]^T.$$

Using the same technique as before, one can show that

$$\mathrm{E}[\widehat{\theta}_-(t)] \cong \theta(t) \tag{6.76}$$

and

$$\mathrm{cov}[\widehat{\theta}_- \,\mathrm{WBF}(t)] \cong \frac{\sigma_v^2 \Phi_o^{-1}}{l_t^{-\,\mathrm{WBF}}} \tag{6.77}$$

where

$$l_t^{-\text{WBF}} = \frac{1}{f_-^T(1)\left[\sum_{i=0}^{t-1} w^2(i)f_-(i+1)f_-^T(i+1)\right]f_-(1)}$$

$$= \frac{1}{\sum_{i=0}^{t-1}\left[w(i)f_-^T(1)f_-(i+1)\right]^2} . \tag{6.78}$$

If the finite-memory condition

$$l_t^- \le c < \infty, \quad \forall t$$

is fulfilled, the equivalent number of observations l_t^- reaches its steady-state value as t tends to infinity:

$$l_\infty^- = \frac{1}{\sum_{i=0}^{\infty}\left[w(i)f_-^T(1)f_-(i+1)\right]^2} \tag{6.79}$$

where $f_-(s)$ denotes any w-orthonormal basis vector of $F_k^-(\infty)$.

Remark

When the basis is direction invariant, i.e. it obeys condition (6.61), the running-basis and fixed-basis WBF estimators coincide, hence

$$l_t^+ = l_t^-, \quad \forall t.$$

Example 6.4

Consider the fixed-basis WBF estimator which combines a Legendre basis of order $k = 2$ ($f_1(t) = 1$, $f_2(t) = t$) with an exponential weighting ($w(t) = \lambda^t$, $0 < \lambda < 1$). Using the Gram–Schmidt procedure one obtains

$$f_1^-(s) = \sqrt{1-\lambda},$$

$$f_2^-(s) = (1-\lambda)\sqrt{\frac{1-\lambda}{\lambda}}(s-1) - \sqrt{\lambda(1-\lambda)}, \quad s \in T_\infty$$

and

$$l_\infty^- = \frac{(1+\lambda)^3}{(1-\lambda)(5\lambda^2 + 4\lambda + 1)} \cong \frac{4}{5(1-\lambda)},$$

where the approximation holds for λ close to 1. When the order of the basis is increased to $k = 3$ ($f_1(t) = 1$, $f_2(t) = t$, $f_3(t) = t^2$) the approximate value of l_∞^- can be obtained from

$$l_\infty^- \cong \frac{1}{4(1-\lambda)} .$$

According to this formula, for $\lambda = 0.99$ the equivalent memory of the third-order estimator is equal to 25 samples, rather than 200 samples as one might expect based on the experience with EWLS estimators. This shows that when estimation is carried out using the exponentially weighted BF algorithms, the forgetting constant should be chosen with caution; the rules of thumb which work for the EWLS scheme (i.e. the first-order EWBF) cannot be mechanically extended to higher-order EWBF estimators.

6.2.4   Impulse response associated with WBF estimators

Consider an arbitrarily time-varying FIR system subject to a wide-sense stationary excitation and suppose that the infinitely long observation history is available at instant t (which guarantees that WBF estimators reach their steady-state behavior).

Running-basis estimators

Using the same technique as in the previous section, one can show that

$$\bar{\theta}_+(t) = \mathrm{E}[\hat{\theta}_+(t)] \cong \mathrm{E}\left[Z_+(t)G_o^{-1}\sum_{i=0}^{\infty} w(i)\psi_+(t-i)\varphi(t-i)\theta(t-i) \right]$$

$$= \sum_{i=0}^{\infty} w(i)\mathrm{f}_+^T(t)\mathrm{f}_+(t-i)\theta(t-i), \tag{6.80}$$

where $\mathrm{f}_+(\cdot)$ are the w-orthonormal basis vectors obeying

$$\sum_{i=0}^{\infty} w(i)\mathrm{f}_+(t-i)\mathrm{f}_+^T(t-i) = I_k.$$

Alternatively

$$\bar{\theta}_{+\,\mathrm{WBF}}(t) \cong \sum_{i=0}^{\infty} h_{\mathrm{WBF}}^+(t,i)\theta(t-i), \tag{6.81}$$

where

$$h_{\mathrm{WBF}}^+(t,i) = \begin{cases} w(i)\mathrm{f}_+^T(t)\mathrm{f}_+(t-i) & i \ge 0 \\ 0 & i < 0 \end{cases} \tag{6.82}$$

denotes the impulse response associated with the WBF estimator. If the adopted basis is not direction invariant, the associated filter is time-varying.

Remark 1

Note that

$$\sum_{i=0}^{\infty} [h_{\mathrm{WBF}}^+(t,i)]^2 = \frac{1}{l_t^{+\,\mathrm{WBF}}}, \quad \forall t,$$

and (if $1 \in F_k^+(t), \forall t$)

$$\sum_{i=0}^{\infty} h_{\mathrm{WBF}}^+(t,i) = 1, \quad \forall t.$$

The first result is a steady-state variant of (6.74). The proof of the second result is a simple modification of the proof for (6.34) in Section 6.1.5.

Fixed-basis estimators

For the fixed-basis estimators one obtains

$$\bar{\theta}_-(t) = \mathrm{E}[\hat{\theta}_-(t)] \cong \sum_{i=0}^{\infty} w(i)\mathrm{f}_-^T(1)\mathrm{f}_-(i+1)\theta(t-i), \tag{6.83}$$

where $f_-(\cdot)$ are the w-orthonormal basis vectors obeying

$$\sum_{i=0}^{\infty} w(i)f_-(i+1)f_-^T(i+1) = I_k.$$

Hence

$$\bar{\theta}_{-\,\mathrm{WBF}}(t) \cong \sum_{i=0}^{\infty} h_{\mathrm{WBF}}^-(i)\theta(t-i), \qquad (6.84)$$

where

$$h_{\mathrm{WBF}}^-(i) = \left\{ \begin{array}{ll} w(i)f_-^T(1)f_-(i+1) & i \geq 0 \\ 0 & i < 0 \end{array} \right. \qquad (6.85)$$

Note that the associated impulse response of a fixed-basis WBF estimator is time invariant.

Remark 2

Similar to (6.79) observe that

$$\sum_{i=0}^{\infty} [h_{\mathrm{WBF}}^-(i)]^2 = \frac{1}{l_{\infty}^{-\mathrm{WBF}}}$$

and (if $1 \in F_k^-(\infty)$)

$$\sum_{i=0}^{\infty} h_{\mathrm{WBF}}^-(i) = 1.$$

6.2.5 Frequency response associated with WBF estimators

The frequency response associated with fixed-basis WBF estimator $\hat{\theta}_-(t)$ can be defined as a Fourier transform of its impulse response:

$$H_{\mathrm{WBF}}^-(\omega) = \sum_{i=0}^{\infty} h_{\mathrm{WBF}}^-(i)e^{-j\omega i} = A_{\mathrm{WBF}}^-(\omega)e^{j\phi_{\mathrm{WBF}}^-(\omega)}. \qquad (6.86)$$

As in the case of WLS estimators, the bias component of the mean square parameter tracking error can be conveniently analyzed in the frequency domain by studying properties of the parameter tracking characteristics

$$E_{\mathrm{WBF}}^-(\omega) = |1 - H_{\mathrm{WBF}}^-(\omega)|^2. \qquad (6.87)$$

Denote by $\tilde{w}(\cdot)$ and $\tilde{f}_1(\cdot), \ldots, \tilde{f}_k(\cdot)$ the continuous-time 'prototypes' of the weighting and basis functions, respectively. Setting

$$w^\eta(i) = \tilde{w}(\eta i),$$

$$f_j^\eta(i) = \tilde{f}_j(\eta i), \quad j = 1, \ldots, k,$$

one obtains scalable window/basis sequences which can be expanded $(0 < \eta < 1)$ or contracted $(\eta > 1)$ without changing their shape. Using the integral approximation technique one can show [141] that

$$(l_\infty^-)^\eta \cong \frac{l_\infty^-}{\eta}$$

and

$$(H_{\mathrm{WBF}}^-(\omega))^\eta \cong H_{\mathrm{WBF}}^- \left(\frac{\omega}{\eta}\right),$$

which means that, for small values of ω, the frequency characteristics can be approximately expressed as functions of normalized frequency

$$\Omega = l_\infty^- \omega.$$

Example 6.5

Once more consider the second-order fixed-basis exponentially weighted Legendre BF estimator $(f_1(s) = 1, \, f_2(s) = s)$. Using $\tilde{f}_1(t) = 1, \, \tilde{f}_2(t) = t, \, t \in [0, \infty)$ as a prototype continuous-time basis, one obtains

$$
\begin{aligned}
\tilde{h}_{\mathrm{WBF}}^-(t) &= e^{-\gamma t}[2\gamma - t\gamma^2], \\
\tilde{l}_\infty^- &= \frac{4}{5\gamma}, \\
\tilde{H}_{\mathrm{WBF}}^-(\omega) &= \frac{1 + j\frac{2\omega}{\gamma}}{\left(1 + j\frac{\omega}{\gamma}\right)^2},
\end{aligned}
$$

where $\gamma = -\ln \lambda > 0$. Consequently

$$\tilde{H}_{\mathrm{WBF}}^-(\tilde{\Omega}) = \frac{1 + j\frac{5}{2}\tilde{\Omega}}{\left(1 + j\frac{5}{4}\tilde{\Omega}\right)^2},$$

where $\tilde{\Omega} = \tilde{l}_\infty^- \omega$.

The exact calculations yield the following expressions:

$$
\begin{aligned}
h_{\mathrm{WBF}}^-(i) &= \lambda^i \left[(1 - \lambda^2) - i(1 - \lambda)^2\right], \\
l_\infty^- &= \frac{(1 + \lambda)^3}{(1 - \lambda)(5\lambda^2 + 4\lambda + 1)}, \\
H_{\mathrm{WBF}}^-(\omega) &= \frac{(1 - \lambda^2) - 2\lambda(1 - \lambda)e^{-j\omega}}{(1 - \lambda e^{-j\omega})^2}.
\end{aligned}
$$

First, note that for λ close to 1, $\gamma = -\ln \lambda \cong 1 - \lambda$, hence the values of $h_{\mathrm{WBF}}^-(i)$ and l_∞^- are practically identical with their integral approximations. For $\lambda = 0.98$ we have $l_\infty^- = 39.5$ and $\tilde{l}_\infty^- = 39.6$, i.e. the relative error is smaller than 0.3%.

Second, for $|\omega| < 0.14$ one has $\sin\omega \cong \omega$ and $\cos\omega \cong 1$ with error smaller than 1%, yielding the approximation $e^{-j\omega} \cong 1 - j\omega$. Using this approximation one gets

$$H^-_{\mathrm{WBF}}(\omega) \cong \frac{1 + \frac{2\lambda}{1-\lambda}j\omega}{\left(1 + \frac{\lambda}{1-\lambda}j\omega\right)^2} \;,$$

which practically coincides with $\tilde{H}^-_{\mathrm{WBF}}(\omega)$ for $\lambda \cong 1$.

■

Figure 6.7 shows normalized parameter tracking characteristics of exponentially weighted Legendre-based BF estimators of different orders (note that the first-order estimator is identical with the EWLS estimator). Comparing the plots of $E^-_{\mathrm{WBF}}(\Omega)$ amounts to comparing tracking properties of WBF estimators characterized by the same value of l^-_∞, that is, estimators yielding the same accuracy under time-invariant conditions. This makes comparison meaningful.

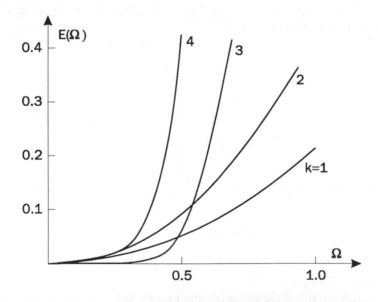

Figure 6.7 Normalized parameter tracking characteristics of exponentially weighted Legendre-based BF estimators of different orders.

Two important conclusions can be drawn. First, for the fixed value of l^-_∞ the estimation bandwidth ω_β, i.e. the frequency range in which parameter variations can be tracked 'successfully', does not significantly depend on the number of basis functions. As a matter of fact, unless the threshold β is assigned a very small value, the estimation bandwidth tends to *decrease* with k.

Second, the slope of normalized tracking characteristics in the bandstop region increases with k, making the higher-order estimators more and more sensitive to the choice of the equivalent memory. This means that for higher-order estimators operated under 'nonstandard' conditions (e.g. when parameters vary in a stochastic manner), even a small deviation of l_∞^- from its 'optimal' value may result in a dramatic deterioration of the algorithm's tracking performance. Such estimators may therefore be quite difficult to handle in practice.

The sensitivity problem is illustrated by Figures 6.8 and 6.9. Figure 6.8 shows the dependence of the mean square parameter tracking error on the equivalent memory for a second-order ARX plant with smoothly time-varying coefficients identified using exponentially weighted Legendre-based BF estimators of different orders. Not surprisingly, the higher-order estimators provide a consistently better performance than the first-order estimator.

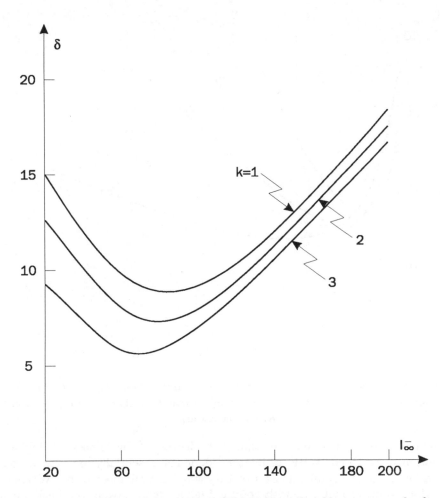

Figure 6.8 Dependence of the mean square parameter tracking errors on l_∞^- for the Legendre-based exponentially weighted WBF estimators of different orders – smooth parameter variation.

Figure 6.9 was obtained for the same plant subject to different time variations: one of the coefficients varied smoothly with time while the remaining two were slowly drifting according to the random walk model. The results support our earlier observations. First, the performance deteriorates with growing k. Second, and much more importantly, for the higher-order estimators the tracking error depends critically on our choice of the equivalent number of observations. The details of the simulations summarized in Figures 6.8 and 6.9 can be found in [136].

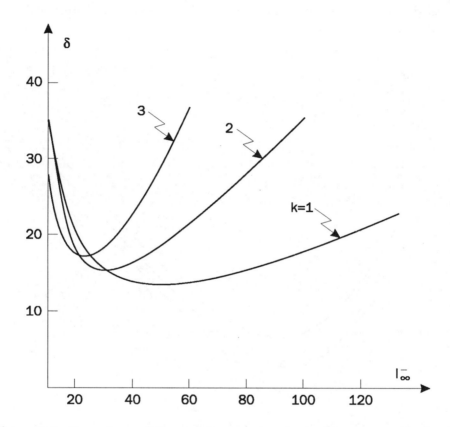

Figure 6.9 Dependence of the mean square parameter tracking errors on l_∞^- for the Legendre-based exponentially weighted WBF estimators of different orders – stochastic parameter variation.

It is clear from this analysis that the degree of usefulness of higher-order exponentially weighted WBF estimators depends essentially on the amount of information about the system nonstationarity which is available a priori. Whenever the prior information provides us with good guidelines for choosing the basis functions, the higher-order estimators may give better results than the EWLS estimator. If, however, all we know about the parameter variations is that they can be viewed as lowpass signals, application of higher-order exponentially weighted BF estimators may be hardly justifiable.

6.3   Computer simulations

Figures 6.10 to 6.19 show identification results obtained using two EWBF algorithms (second-order and third-order Legendre) for the time-varying finite impulse response systems (FIR1 and FIR2) and autoregressive processes (AR1 and AR2) described in Section 2.9.

Figures 6.10 and 6.15 show the output of linear filters associated with the EWBF estimators. Note that for both FIR systems the 'theoretical' plots stay in very good agreement with the results of averaging (over 100 simulation runs) the estimated parameter trajectories (Figures 6.12 and 6.17). Estimation results obtained for a single realization of the identified process are shown in Figures 6.11 and 6.16. The overshoot effects for a system with abrupt parameter changes illustrate the sensitivity problems typical of BF estimators: the larger the number of basis functions, the more sensitive the identification algorithm to a 'nonstandard' parameter variation.

Figures 6.13, 6.14 and 6.18, 6.19 show identification results obtained for the time-varying autoregressive processes AR1 and AR2. Even though all theoretical results were derived for FIR systems and cannot be extended to autoregressive processes, the plots shown in Figures 6.13, 6.18 (single realizations) and Figures 6.14, 6.19 (ensemble averages) very much resemble those obtained for the finite impulse response systems FIR1 and FIR2.

Remark 1

All plots show the steady-state tracking behavior of EWBF algorithms. To reach the steady state, parameter estimation was initialized at instant $t = -500$, i.e. before the analysis interval $T = [0, 500]$ started. In the initial convergence period the system parameters were constant $(a_1(t) = a_1(0), a_2(t) = a_2(0)$ for $t < 0)$.

Remark 2

The greater variance of the estimates in the AR case, compared to the FIR case, is the consequence of a smaller signal-to-noise ratio. The tracking capabilities of the WLS algorithm are approximately the same in both cases.

6.4   The method of basis functions: good news or bad news?

Whenever basis functions can be chosen in a 'conscious' way, e.g. based on the prior knowledge of system time variation and/or on preliminary identification experiments, the basis function approach offers a superb parameter tracking/matching performance, allowing one to identify nonstationary processes with dynamics that vary at an arbitrary rate. Adaptive equalization of rapidly fading radio channels, based on the deterministic model of channel variation (2.36), is one successful application of this technique.

First of all, we recall that the particular form of the basis (complex exponentials) incorporated in (2.36) stems directly from the analysis of a typical mobile radio channel under reasonable assumptions (few strong reflectors, linearly changing path delays). The sinusoidal variation of channel coefficients is forced by a Doppler effect occurring when a high-frequency modulated signal is picked up by a moving antenna [58]. Since in the case considered the channel is a wide-sense cyclostationary stochastic process,

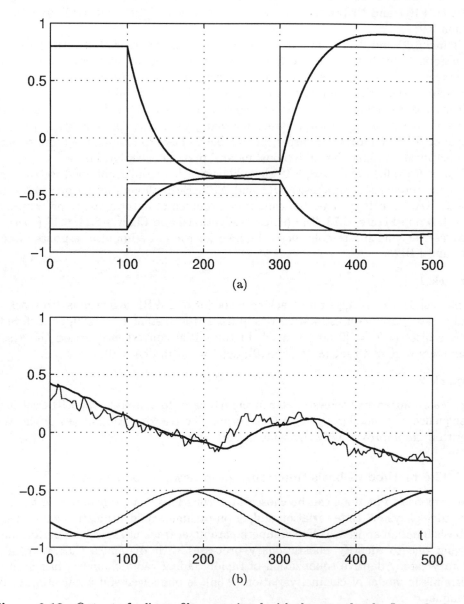

Figure 6.10 Output of a linear filter associated with the second-order Legendre-based exponentially weighted basis functions (EWBF) estimator for a system with jump parameter changes (a) and continuous parameter changes (b); the equivalent memory of the estimation algorithm was set to $l_\infty = 50$ ($\lambda = 0.984$).

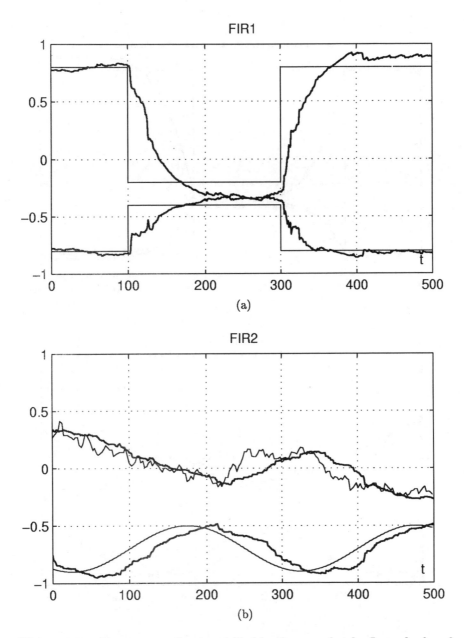

Figure 6.11 Parameter estimates yielded by the second-order Legendre-based exponentially weighted basis functions (EWBF) algorithm for a single realization of an FIR system with jump parameter changes (a) and continuous parameter changes (b); the equivalent memory of the estimation algorithm was set to $l_\infty = 50$ ($\lambda = 0.984$).

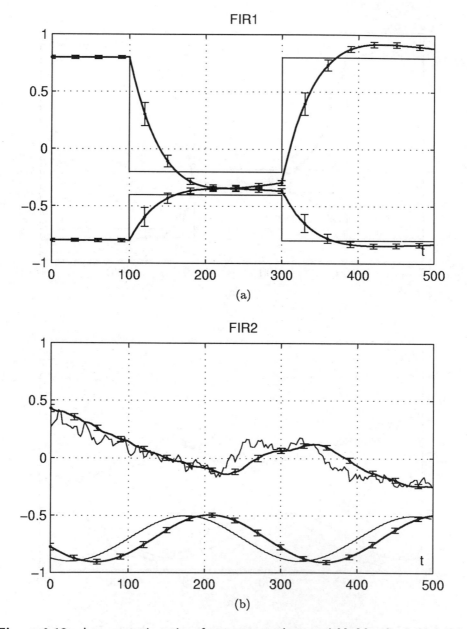

Figure 6.12 Average trajectories of parameter estimates yielded by the second-order Legendre-based exponentially weighted basis functions (EWBF) algorithm for an FIR system with jump parameter changes (a) and continuous parameter changes (b); vertical bars show standard deviation of the estimates.

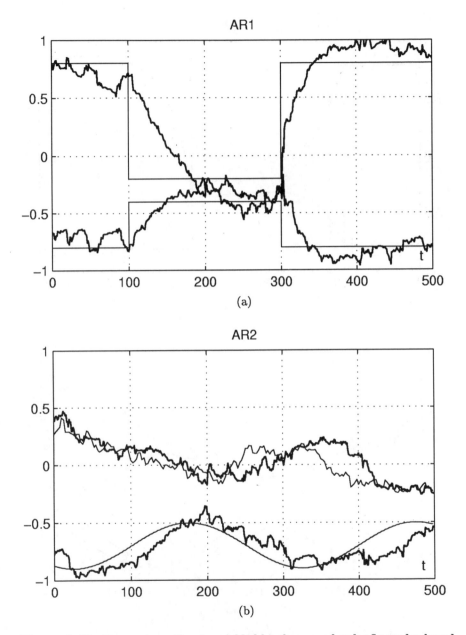

Figure 6.13 Parameter estimates yielded by the second-order Legendre-based exponentially weighted basis functions (EWBF) algorithm for a single realization of an AR system with jump parameter changes (a) and continuous parameter changes (b); the equivalent memory of the estimation algorithm was set to $l_\infty = 50$ ($\lambda = 0.984$).

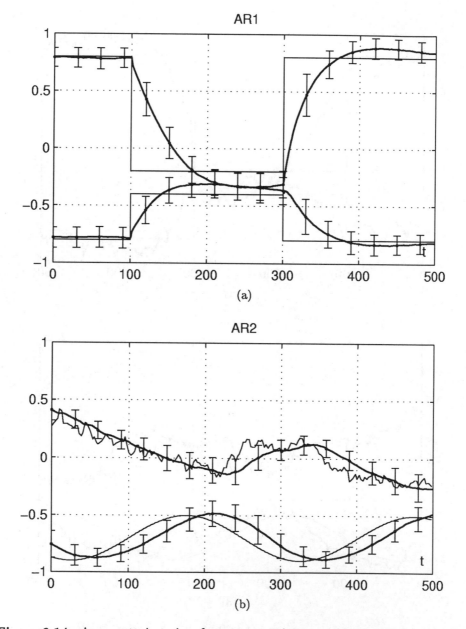

Figure 6.14 Average trajectories of parameter estimates yielded by the second-order Legendre-based exponentially weighted basis functions (EWBF) algorithm for an AR system with jump parameter changes (a) and continuous parameter changes (b); vertical bars show standard deviation of the estimates.

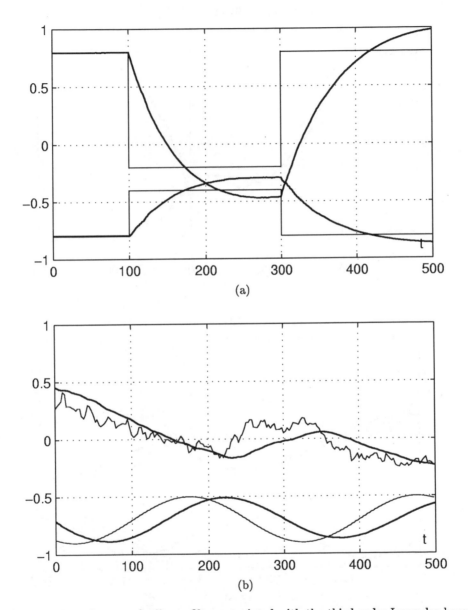

(a)

(b)

Figure 6.15 Output of a linear filter associated with the third-order Legendre-based exponentially weighted basis functions (EWBF) estimator for a system with jump parameter changes (a) and continuous parameter changes (b); the equivalent memory of the estimation algorithm was set to $l_\infty = 50$ ($\lambda = 0.995$).

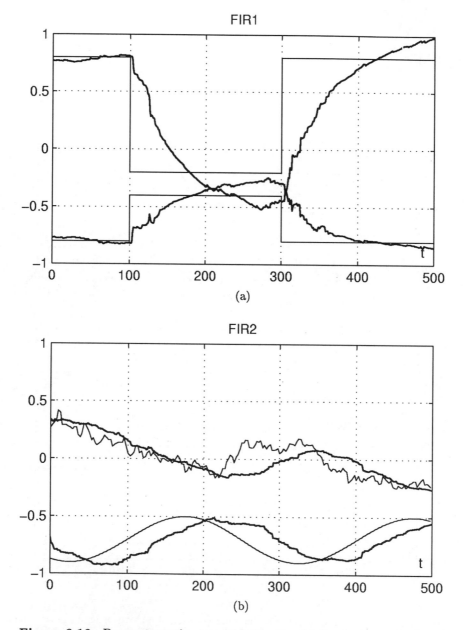

Figure 6.16   Parameter estimates yielded by the third-order Legendre-based
exponentially weighted basis functions (EWBF) algorithm for a single realization of an FIR
system with jump parameter changes (a) and continuous parameter changes (b); the
equivalent memory of the estimation algorithm was set to $l_\infty = 50$ ($\lambda = 0.995$).

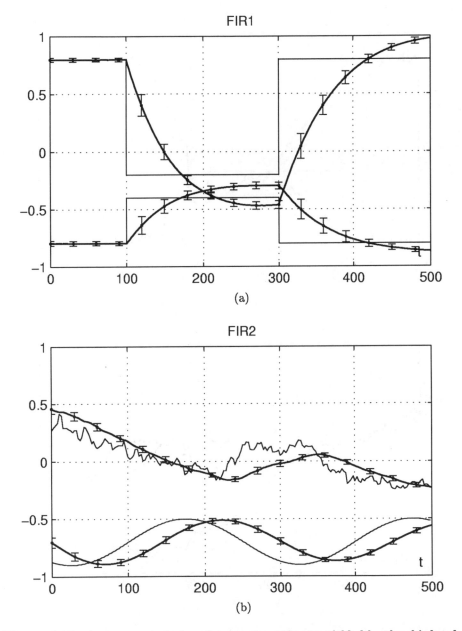

Figure 6.17 Average trajectories of parameter estimates yielded by the third-order Legendre-based exponentially weighted basis functions (EWBF) algorithm for an FIR system with jump parameter changes (a) and continuous parameter changes (b); vertical bars show standard deviation of the estimates.

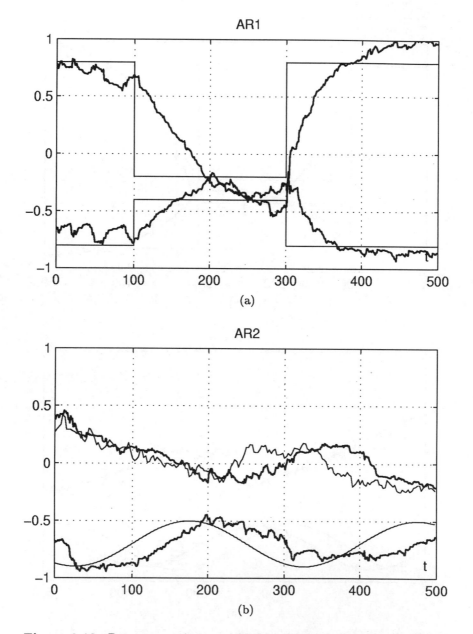

Figure 6.18 Parameter estimates yielded by the third-order Legendre-based exponentially weighted basis functions (EWBF) algorithm for a single realization of an AR system with jump parameter changes (a) and continuous parameter changes (b); the equivalent memory of the estimation algorithm was set to $l_\infty = 50$ ($\lambda = 0.995$).

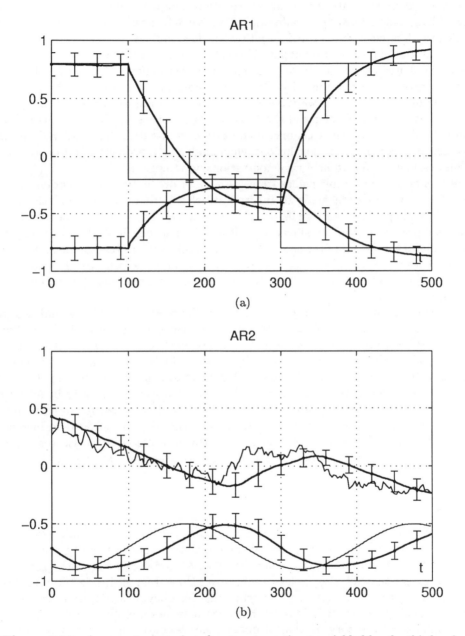

Figure 6.19 Average trajectories of parameter estimates yielded by the third-order Legendre-based exponentially weighted basis functions (EWBF) algorithm for an AR system with jump parameter changes (a) and continuous parameter changes (b); vertical bars show standard deviation of the estimates.

the frequencies of the exponential basis sequences can be relatively easily calculated from the so-called cyclic cumulants of the output signal [186]. Owing to a detailed knowledge about the nature of channel variation, adaptive equalizers based on the model (2.36) outperform the traditional adaptive schemes. In particular, they are significantly less sensitive to rapid variations or deep fadings.

The situation changes dramatically if the basis functions are chosen in a 'blind' manner, i.e. based on the 'universal approximation' arguments. The point is that such general approximations are usually not parsimonious in the sense that the increased model complexity does not reward one with the expected improvement in the tracking or matching behavior. Ziegler and Cioffi [206] report noise sensitivity problems (even at high SNRs) arising when channel equalization is based on 'general' BF models. Gardner [53] reaches similar conclusions when analyzing a problem of autocorrelation function estimation for an arbitrary nonstationary process.

The last negative example comes from the area of predictive speech coding. In the late 1970s attempts were made to incorporate the method of basis functions into predictive coding of speech signals [71]. In the classic LPC scheme each segment of speech is represented by a time-invariant AR model. The length of LPC speech frames is limited by the duration of the shortest speech sounds, such as transients, and usually does not exceed 20 ms. It was suggested that the time-invariant speech models can be replaced by the time-varying models. There are at least two possible advantages of time-varying LPC.

First, since the inclusion of time variations in the model allows for signal analysis over longer data windows, one may attempt to increase efficiency of the coding scheme by reducing the total number of parameters that have to be transmitted to recover speech signals of a given duration. Since time-varying LPC involves a larger number of coefficients than traditional LPC, the encoded nonstationary speech frames should be at least k times longer than the stationary frames for any coding gain to be achieved. Second, the time-varying representation is certainly better suited for describing the continuously changing behavior of speech signals. Therefore it can lead to increased accuracy in signal representation, e.g. it can provide a smoothed trajectory of the formants of a vocal tract. Practical experience with the time-varying LPC schemes never fulfilled this promise.

Remark

One of the problems with the time-varying LPC is due to the fact that (unlike the regular LPC) the time-varying 'poles' of the estimated model are not guaranteed to remain within the unit circle in the complex plane. Although in a time-varying context the requirement that all filter poles should at all times remain inside the unit circle is not a prerequisite to stability, filters which do not obey this condition are usually of no practical value as their impulse response may become excessively large. Not surprisingly, the trajectory of nonstationary poles tends to leave the unit circle at locations close to both ends of the analysis interval, i.e. at points where the accuracy of BF models significantly decreases (especially for large values of k).

The model stability problem was further analyzed in [135]. It was shown that the stability constraints are much easier to meet if the BF estimates are computed for the whole analysis interval but used in a smaller subinterval. Even though this clipping technique significantly reduces the probability of obtaining unstable pole trajectories and only slightly decreases the coding efficiency (the overlapping analysis frames must be used), it still does not guarantee stability of the synthesis filter. Stability can be enforced by using the lattice BF identification algorithms of Grenier [64].

■

Summing up, when rapid variations are present, only the structured (i.e. at least partially known) nonstationarities can be handled satisfactorily using the basis function approach. The 'general-purpose' bases, such as Legendre or Fourier, can be successfully used to describe the slow parameter variations only. If there is no prior knowledge on process parameter variation, the BF approach does not compare favorably with local estimation approaches. Even though it is possible to gain some increase in the estimation bandwidth, the improvement is achieved at a cost; the parameter tracking or matching algorithm becomes more sensitive to the choice of design parameters such as the equivalent estimation memory.

Comments and extensions

Section 6.1

- Application of functional series expansions to identification of time-varying systems was pioneered by Subba Rao [182], Mendel [126] and Liporace [110]. Computational aspects of BF estimation are considered in Hall et al. [71], Gersh and Yonemoto [56] and Grenier [64]. Almost no results on BF estimation of ARMAX systems are available. The only noticeable exceptions are the papers of Grenier [64] (lattice algorithms using the basis function expansion) and Grillenzoni [65] (general discussion of the problem).

- Powers of time and cosinusoidal functions are the most frequently used general-purpose bases. Another basis, the set of prolate spheroidal functions, was suggested by Grenier [64].

- For the Legendre-based BF estimators, the adaptive segmentation techniques described in Section 3.2 can be easily extended to the BF approach. The aggregate/split and global/local tests based on Akaike's information criterion carry over to the BF case almost without modification. The only change which must be introduced is due to a different model complexity evaluation: to account for the increased number of degrees of freedom, the quantity n in (3.28) and (3.30) should be replaced with nk.

- The time and frequency characteristics of BF estimators were established by Niedźwiecki [139].

- Our analysis of parameter matching properties for BF estimators was restricted to FIR systems. We note, however, that on the qualitative level the results obtained remain in full agreement with observations made by Hall et al. [71]. Examining the response of BF estimators to step parameter changes, Hall et al. come to the conclusion that the estimated trajectories of time-varying poles of an AR system resemble the step response of a noncausal lowpass filter. Even more interestingly, the response of the time-varying LPC system is reported to be approximately homogeneous and approximately additive; homogeneous in that the pole angle trajectory for a given center frequency change is proportional to to the size of the step change, and additive in that the response to two different jumps in one interval is approximately the same as the sum of the responses to each jump taken separately in the same interval. This shows that the concept of the associated time/frequency characteristics, introduced for FIR systems subject to a stationary excitation, may still be very useful for understanding properties of the BF estimators operated under less restrictive conditions.

Section 6.2

- The method of exponentially weighted basis functions (for the Legendre basis) was proposed by Xianya and Evans [197] and further explored by Li [105], [106]. The same technique, but in combination with the basis set consisting of complex exponentials, was used by Tsatsanis and Giannakis [186].

- The distinction between fixed-basis and running-basis estimation was introduced by Niedźwiecki [136]. The same paper provided the first analysis of the parameter tracking properties of WBF estimators.

- The low-complexity LMS-type algorithm for tracking slow variation in basis expansion coefficients was proposed by Tsatsanis and Giannakis [186] for adaptive channel equalization purposes. To circumvent the problem of slow initial convergence by the LMS filter, the initial estimates of channel coefficients were obtained using the method of least squares. This mixed-mode approach seems to be a very good way of combining the principal advantages of both estimation schemes: the fast initial convergence of LS algorithms and the low complexity of gradient algorithms (achieved without compromising tracking capabilities).

7

Kalman Filtering

7.1 Estimation principles

The method of basis functions was based on an explicit *deterministic* model of parameter variation. The statistical filtering approach assumes knowledge of a *stochastic* model of parameter changes.

Suppose that the evolution of process parameters can be described by the following stochastic equation:

$$\theta(t) = F\theta(t-1) + Gw(t), \tag{7.1}$$

$$\text{cov}[w(t)] = W,$$

where F and G are known $n \times n$ matrices and $\{w(t)\}$ denotes the parameter driving noise, independent of the measurement noise $\{v(t)\}$. Regarding the parameter vector $\theta(t)$ as a 'state' of a dynamic system with output governed by

$$y(t) = \varphi^T(t)\theta(t) + v(t), \tag{7.2}$$

one can formulate the problem of parameter tracking as a problem of filtering in the state space. F in (7.1) can be interpreted as a matrix of state transition coefficients and the regression vector $\varphi(t)$ in (7.2) can be interpreted as a vector of time-varying, but known, output (measurement) coefficients.

Under Gaussian assumptions

$$w(t) \sim \mathcal{N}(0, W),$$
$$v(t) \sim \mathcal{N}(0, \sigma_v^2),$$
$$\theta(0) \sim \mathcal{N}(\theta_o, \Sigma_o),$$

and given that $\{w(t)\}$, $\{v(t)\}$ and $\theta(0)$ are mutually orthogonal, the minimum mean square estimate of $\theta(t)$ based on the available data $\Xi(t)$,

$$\widehat{\theta}(t|t) = \mathrm{E}[\theta(t)|\Xi(t)],$$

can be computed recursively using the celebrated Kalman filtering algorithm [5]:

$$\begin{aligned}
\widehat{\theta}(t|t) &= \widehat{\theta}(t|t-1) + K(t)\epsilon(t), \\
\widehat{\theta}(t|t-1) &= F\widehat{\theta}(t-1|t-1), \\
\epsilon(t) &= y(t) - \varphi^T(t)\widehat{\theta}(t|t-1),
\end{aligned}$$

$$K(t) = \frac{\Sigma(t|t-1)\varphi(t)}{\sigma_v^2 + \varphi^T(t)\Sigma(t|t-1)\varphi(t)},$$

$$\Sigma(t|t-1) = F\Sigma(t-1|t-1)F^T + GWG^T,$$

$$\Sigma(t|t) = \Sigma(t|t-1) - \frac{\Sigma(t|t-1)\varphi(t)\varphi^T(t)\Sigma(t|t-1)}{\sigma_v^2 + \varphi^T(t)\Sigma(t|t-1)\varphi(t)}, \qquad (7.3)$$

with initial conditions

$$\Sigma(0|0) = \Sigma_o, \qquad \widehat{\theta}(0|0) = \theta_o.$$

Here the $n \times n$ matrix $\Sigma(t|t)$ can be interpreted as a covariance matrix of the filtered (a posteriori) state estimate $\widehat{\theta}(t|t)$,

$$\Sigma(t|t) = \text{cov}[\widehat{\theta}(t|t)] = \text{cov}[\theta(t)|\Xi(t)],$$

and the $n \times n$ matrix $\Sigma(t|t-1)$ can be interpreted as a covariance matrix of a predicted (a priori) state estimate $\widehat{\theta}(t|t-1) = E[\theta(t)|\Xi(t-1)]$,

$$\Sigma(t|t-1) = \text{cov}[\widehat{\theta}(t|t-1)] = \text{cov}[\theta(t)|\Xi(t-1)].$$

Observe that the Kalman filtering algorithm (7.3) can be rewritten in the form

$$\widehat{\theta}(t|t) = \widehat{\theta}(t|t-1) + K(t)\epsilon(t),$$

$$\widehat{\theta}(t|t-1) = F\widehat{\theta}(t-1|t-1),$$

$$\epsilon(t) = y(t) - \varphi^T(t)\widehat{\theta}(t|t-1),$$

$$K(t) = \frac{\widetilde{\Sigma}(t|t-1)\varphi(t)}{1 + \varphi^T(t)\widetilde{\Sigma}(t|t-1)\varphi(t)},$$

$$\widetilde{\Sigma}(t|t-1) = F\widetilde{\Sigma}(t-1|t-1)F^T + G\widetilde{W}G^T,$$

$$\widetilde{\Sigma}(t|t) = \widetilde{\Sigma}(t|t-1) - \frac{\widetilde{\Sigma}(t|t-1)\varphi(t)\varphi^T(t)\widetilde{\Sigma}(t|t-1)}{1 + \varphi^T(t)\widetilde{\Sigma}(t|t-1)\varphi(t)}, \qquad (7.4)$$

where

$$\widetilde{\Sigma}(t|t) = \frac{\Sigma(t|t)}{\sigma_v^2}, \quad \widetilde{\Sigma}(t|t-1) = \frac{\Sigma(t|t-1)}{\sigma_v^2}, \quad \widetilde{W} = \frac{W}{\sigma_v^2}$$

are the *normalized* covariance matrices and the initial conditions are

$$\widetilde{\Sigma}(0|0) = \frac{\Sigma_o}{\sigma_v^2}, \quad \widehat{\theta}(0|0) = \theta_o.$$

Remark 1

The fact that the output vector in (7.2) is not only time dependent but also *data* dependent is clearly a nonstandard feature of the state space description given by (7.1) and (7.2). In particular, note that for AR(X) processes the vector $\varphi(t)$ depends on system outputs. It is known, however, that the theory of Kalman filtering can be extended to the case where system coefficients are functions of *past* input and output variables, which is true in the case considered.

Remark 2

The Kalman filtering approach allows one to incorporate into the process of system identification the prior knowledge about the estimated coefficients. If no such knowledge is available, one can adopt 'noninformative' priors by setting

$$\Sigma_o^{-1} = O. \tag{7.5}$$

Note that condition (7.5) is approximately met when

$$\Sigma(0|0) = \delta I_n \quad \text{or} \quad \tilde{\Sigma}(0|0) = \delta I_n, \tag{7.6}$$

where δ is a 'sufficiently large' positive constant.

7.2　Estimation based on the random walk model

To use the Kalman filtering algorithm one should know the state transition matrix F and the covariance matrix W of the parameter driving noise. If our prior knowledge about the process nonstationarity is vague, which is typical in practice, it may be difficult to specify both matrices. Therefore, for practical reasons, a simplified random walk model of parameter variation is adopted instead of (7.1):

$$\theta(t) = \theta(t-1) + w(t), \tag{7.7}$$

$$\text{cov}[w(t)] = \sigma_w^2 I_n.$$

According to (7.7) the process parameters are assumed to change with independent, random increments and independently of each other. The variance σ_w^2 can be interpreted as the mean square rate of parameter change; the larger the value of σ_w^2, the faster the process parameters vary with time. Note that the random walk model (7.7) can be obtained from (7.1) by setting

$$F = I_n, \quad G = I_n, \quad W = \sigma_w^2 I_n.$$

Introducing shorthand notation

$$\widehat{\theta}(t) = \widehat{\theta}(t|t)$$

one can rewrite the general Kalman filtering algorithm (7.4) in a simpler form:

$$
\begin{aligned}
\widehat{\theta}(t) &= \widehat{\theta}(t-1) + K(t)\epsilon(t), \\
\epsilon(t) &= y(t) - \varphi^T(t)\widehat{\theta}(t-1), \\
K(t) &= \frac{\tilde{\Sigma}(t|t-1)\varphi(t)}{1 + \varphi^T(t)\tilde{\Sigma}(t|t-1)\varphi(t)}, \\
\tilde{\Sigma}(t|t-1) &= \tilde{\Sigma}(t-1|t-1) + \kappa^2 I_n, \\
\tilde{\Sigma}(t|t) &= \tilde{\Sigma}(t|t-1) - \frac{\tilde{\Sigma}(t|t-1)\varphi(t)\varphi^T(t)\tilde{\Sigma}(t|t-1)}{1 + \varphi^T(t)\tilde{\Sigma}(t|t-1)\varphi(t)},
\end{aligned} \tag{7.8}
$$

where $\kappa^2 = \sigma_w^2/\sigma_v^2$. We will call $\widehat{\theta}(t)$, governed by (7.8), the random walk Kalman filter (RWKF) estimator. In contrast with (7.4), the RWKF algorithm depends on only one user-dependent constant, the variance quotient κ^2. In Section 7.5 we will show that by changing κ, $0 \leq \kappa < \infty$, one can affect the equivalent estimation memory of the RWKF algorithm. Hence κ in the RWKF framework plays a similar role as the forgetting constant λ in the exponentially weighted least squares approach, or the stepsize μ in the least mean squares approach.

Denoting $P(t-1) = \widetilde{\Sigma}(t|t-1)$ one can express (7.8) in the following equivalent form:

$$
\begin{aligned}
\widehat{\theta}(t) &= \widehat{\theta}(t-1) + K(t)\epsilon(t), \\
\epsilon(t) &= y(t) - \varphi^T(t)\widehat{\theta}(t-1), \\
K(t) &= \frac{P(t-1)\varphi(t)}{1 + \varphi^T(t)P(t-1)\varphi(t)}, \\
P(t) &= P(t-1) - \frac{P(t-1)\varphi(t)\varphi^T(t)P(t-1)}{1 + \varphi^T(t)P(t-1)\varphi(t)} + \kappa^2 I_n,
\end{aligned}
\tag{7.9}
$$

which resembles the RLS algorithm (3.13). Note that both algorithms become identical when κ is set to zero, which amounts to assuming that the process coefficients are time invariant.

One might object that the random walk model is too naive to describe parameter variations in any real system. Surprisingly, as it may seem, the RWKF algorithms usually work very well in real-world environments [29]. It is not difficult to explain this supposed contradiction. The point is that due to its finite-memory property (for $\kappa > 0$) the Kalman filter based tracker does not rely heavily on information coming from the remote past and hence it does not require an adequate 'global' model of process coefficient variation. Vaguely speaking, any way of informing the identification routine that system coefficients should not be regarded as time invariant will result in a reasonable tracking performance, and adopting the random walk model is just a particular way of communicating this.

Zhang and Haykin [205] pointed to an interesting analogy between RWKF estimators and WLS estimators. To show the WLS connection we will express $\widehat{\theta}(t)$ in the form

$$
\widehat{\theta}(t) = \widetilde{\Sigma}(t|t)\widetilde{\sigma}(t|t),
\tag{7.10}
$$

where $\widetilde{\sigma}(t|t)$ is a recursively computable vector. It is straigtforward to check that

$$
K(t) = \widetilde{\Sigma}(t|t)\varphi(t).
$$

Since the matrix

$$
\widetilde{\Sigma}(t|t) = \left[\widetilde{\Sigma}^{-1}(t|t-1) + \varphi(t)\varphi^T(t)\right]^{-1}
\tag{7.11}
$$

is positive definite (note that the matrix $\widetilde{\Sigma}(t|t-1)$ is trivially bounded from below by $\kappa^2 I_n$; the relationship (7.11) follows directly from the matrix inversion lemma) the parameter update in (7.8) can be written as

$$
\widehat{\theta}(t) = \widetilde{\Sigma}(t|t)\left[\widetilde{\Sigma}^{-1}(t|t)\widehat{\theta}(t-1) + \varphi(t)\left(y(t) - \varphi^T(t)\widehat{\theta}(t-1)\right)\right].
$$

Therefore

$$
\widetilde{\sigma}(t|t) = \widetilde{\Sigma}^{-1}(t|t)\widehat{\theta}(t-1) + \varphi(t)\left(y(t) - \varphi^T(t)\widehat{\theta}(t-1)\right).
\tag{7.12}
$$

Using (7.11) one gets

$$\tilde{\Sigma}(t|t) = \left[[\tilde{\Sigma}(t|t-1) + \kappa^2 I_n]^{-1} + \varphi(t)\varphi^T(t)\right]^{-1},$$

hence

$$\tilde{\Sigma}^{-1}(t|t) = D(t-1)\tilde{\Sigma}^{-1}(t-1|t-1) + \varphi(t)\varphi^T(t) \tag{7.13}$$

where

$$D(t) = \left[I_n + \kappa^2 \tilde{\Sigma}^{-1}(t|t)\right]^{-1}. \tag{7.14}$$

According to (7.10)

$$\tilde{\Sigma}^{-1}(t-1|t-1)\hat{\theta}(t-1) = \tilde{\sigma}(t-1|t-1),$$

and after substituting (7.13) into (7.12) one gets

$$\tilde{\sigma}(t|t) = D(t-1)\tilde{\sigma}(t-1|t-1) + y(t)\varphi(t). \tag{7.15}$$

Now observe that $\tilde{\Sigma}^{-1}(t|t)$ and $\tilde{\sigma}(t|t)$ can be written down semiexplicitly as

$$\tilde{\Sigma}^{-1}(t|t) \quad = \quad \sum_{i=0}^{t-1} W_t(i)\varphi(t-i)\varphi^T(t-i) + W_t(t)\tilde{\Sigma}^{-1}(0|0),$$

$$\tilde{\sigma}(t|t) \quad = \quad \sum_{i=0}^{t-1} W_t(i)y(t-i)\varphi(t-i) + W_t(t)\tilde{\Sigma}^{-1}(0|0)\hat{\theta}(0), \tag{7.16}$$

where

$$W_t(0) \quad = \quad I_n,$$

$$W_t(i) \quad = \quad \prod_{j=1}^{i} D(t-j), \quad 0 < i \le t. \tag{7.17}$$

The term 'semiexplicitly' refers to the fact that the weighting matrices in (7.16) are *data dependent.*

According to (7.10) and (7.16), for noninformative priors

$$\tilde{\Sigma}^{-1}(0|0) = O,$$

one can rewrite the RWKF estimator in the form

$$\hat{\theta}(t) = \left(\sum_{i=0}^{t-1} W_t(i)\varphi(t-i)\varphi^T(t-i)\right)^{-1} \left(\sum_{i=0}^{t-1} W_t(i)y(t-i)\varphi(t-i)\right), \tag{7.18}$$

which closely resembles (4.5).

It can easily be shown that

$$\|W_t(i)\| \quad \ge \quad \|W_t(i+1)\|, \quad \forall\, t, \quad \forall\, 0 \le i \le t,$$

where $\|A\| = \lambda_{\max}(A)$ denotes a spectral norm of a nonnegative definite matrix A.

According to (7.18), from the qualitative viewpoint, the RWKF algorithm works rather like the WLS algorithm – the influence of past measurements on the parameter estimates diminishes with the age of the samples. The main difference lies in the weighting strategy; in the RWKF approach the scalar user-dependent weighting coefficients $w(i)$ are replaced with data-dependent weighting matrices $W_t(i)$.

Remark 1

Setting the variance quotient equal to κ^2 and applying the RWKF algorithm to a set of scaled measurements

$$y'(t) = \eta y(t), \qquad \varphi'(t) = \eta \varphi(t),$$

one obtains identical results as when running the RWKF algorithm with a variance ratio $(\kappa/\eta)^2$, on the original measurements. This means that the RWKF algorithm is *not* scale invariant, which is an obvious practical drawback.

Remark 2

By setting $W = \sigma_w^2 I_n$ one implicitly assumes that all process coefficients vary with the same speed. If some specific prior knowledge is available about the rate of change of different parameters, one can set

$$W = \mathrm{diag}\{\sigma_1^2, \ldots, \sigma_n^2\},$$

where σ_i^2 is the mean square rate of change of $\theta_i(t)$. In particular, if σ_i^2 is set to zero, the corresponding parameter is regarded as time invariant. When this more general model of process time variation is adopted, the matrix $\kappa^2 I_n$ in the covariance update of (7.8) or (7.9) should be replaced with $\mathrm{diag}\{\kappa_1^2, \ldots, \kappa_n^2\}$, where $\kappa_i^2 = \sigma_i^2/\sigma_v^2$.

7.3   Estimation based on the integrated random walk models

Note that the random walk model can be rewritten in the form

$$\nabla \theta(t) = w(t),$$

where

$$\nabla \theta(t) = (1 - q^{-1})\theta(t) = \theta(t) - \theta(t-1).$$

The integrated random walk (IRW) model of order k is defined in the form

$$\nabla^k \theta(t) = (1 - q^{-1})^k \theta(t) = \sum_{i=0}^{k} f_i \theta(t-i) = w(t), \qquad (7.19)$$

$$\mathrm{cov}[w(t)] = \sigma_w^2 I_n,$$

where

$$f_i = (-1)^i \binom{k}{i},$$

i.e. it is assumed that the kth order difference $\theta(t)$ is a white noise process.

Note that (7.19) can be rewritten as

$$\theta(t) = \sum_{i=1}^{k} (-f_i)\theta(t-i) + w(t). \tag{7.20}$$

For $k = 2$ one obtains

$$\theta(t) = 2\theta(t-1) - \theta(t-2) + w(t),$$

and for $k = 3$

$$\theta(t) = 3\theta(t-1) - 3\theta(t-2) + \theta(t-3) + w(t).$$

Since $\nabla^k \theta_i(t) = 0$ implies

$$\theta_i(t) = \sum_{j=1}^{k} c_{ij} t^{j-1}, \tag{7.21}$$

the integrated random walk model can be regarded as a local or 'perturbed' power series model of parameter variation. Generally, the larger the order of the IRW model, the smoother the corresponding parameter trajectories.

To obtain the parameter tracking algorithm based on (7.20) we will put the integrated random walk model in a state space form. By introducing the augmented parameter (state) vector θ_a and augmented regression vector φ_a,

$$\theta_a(t) = \begin{bmatrix} \theta(t) \\ \theta(t-1) \\ \vdots \\ \theta(t-k+1) \end{bmatrix}_{kn \times 1}, \quad \varphi_a(t) = \begin{bmatrix} \varphi(t) \\ 0 \\ \vdots \\ 0 \end{bmatrix}_{kn \times 1},$$

one can rewrite process equations (7.20) and (7.2) in the form

$$\begin{aligned} \theta_a(t) &= F_a \theta_a(t-1) + G_a w(t), \\ y(t) &= \varphi_a^T(t)\theta_a(t) + v(t), \end{aligned} \tag{7.22}$$

where

$$F_a = \begin{bmatrix} -f_1 I_n & -f_2 I_n & \cdots & -f_{k-1}I_n & -f_k I_n \\ I_n & O & \cdots & O & O \\ & & \ddots & & \\ O & O & \cdots & O & I_n \end{bmatrix}_{kn \times kn}, \quad G_a = \begin{bmatrix} I_n \\ O \\ \vdots \\ O \end{bmatrix}_{kn \times 1}.$$

Example 7.1

For the first three integrated random walk models one obtains

$$F_a = I_n, \quad G_a = I_n, \quad \text{for } k = 1,$$

$$F_a = \begin{bmatrix} 2I_n & -I_n \\ I_n & O \end{bmatrix}, \quad G_a = \begin{bmatrix} I_n \\ O \end{bmatrix}, \quad \text{for } k = 2,$$

$$F_a = \begin{bmatrix} 3I_n & -3I_n & I_n \\ I_n & O & O \\ O & I_n & O \end{bmatrix}, \quad G_a = \begin{bmatrix} I_n \\ O \\ O \end{bmatrix}, \quad \text{for } k = 3.$$

∎

Under Gaussian assumptions the minimum variance estimate of $\theta_a(t)$ can be obtained using the Kalman filtering algorithm (7.4), provided that the quantities F, G and $\varphi(t)$ are replaced with F_a, G_a and $\varphi_a(t)$, respectively:

$$\widehat{\theta}_a(t|t) = \widehat{\theta}_a(t|t-1) + K_a(t)\epsilon(t),$$

$$\widehat{\theta}_a(t|t-1) = F_a\widehat{\theta}_a(t-1|t-1),$$

$$\epsilon(t) = y(t) - \varphi_a^T(t)\widehat{\theta}_a(t|t-1),$$

$$K_a(t) = \frac{\widetilde{\Sigma}_a(t|t-1)\varphi_a(t)}{1 + \varphi_a^T(t)\widetilde{\Sigma}_a(t|t-1)\varphi_a(t)},$$

$$\widetilde{\Sigma}_a(t|t-1) = F_a\widetilde{\Sigma}_a(t-1|t-1)F_a^T + \kappa^2 G_a G_a^T,$$

$$\widetilde{\Sigma}_a(t|t) = \widetilde{\Sigma}_a(t|t-1) - \frac{\widetilde{\Sigma}_a(t|t-1)\varphi_a(t)\varphi_a^T(t)\widetilde{\Sigma}_a(t|t-1)}{1 + \varphi_a^T(t)\widetilde{\Sigma}_a(t|t-1)\varphi_a(t)}. \quad (7.23)$$

The estimate of the parameter vector $\theta(t)$ can be obtained from

$$\widehat{\theta}(t) = D_a\widehat{\theta}_a(t|t), \quad (7.24)$$

$$D_a = \begin{bmatrix} I_n & O & \cdots & O \end{bmatrix}_{n \times (kn)}.$$

As in the case of the RWKF filter, the tracking properties of the IRWKF algorithm (7.23) depend on the value of the scalar coefficient κ.

Remark 1

When $\kappa = 0$ the IRWKF algorithm (7.23) yields identical results as the Legendre-based BF estimator of order k (7.21); when $k = 1$ the results are identical with those of the random walk model.

7.4 Stability and convergence of the RWKF algorithm

Early stability proofs derived for Kalman filtering algorithms were based on the deterministic persistence of excitation condition (B3.1), i.e. it was assumed that the regression vector satisfies

$$\sum_{t=k+1}^{k+s} \varphi(t)\varphi^T(t) \geq cI_n, \quad \forall k, \quad (7.25)$$

for some deterministic positive constants c and s. Since the observability Gramian of the system described by (7.7) and (7.2) is

$$\mathcal{O}(k, s) = \sum_{t=k+1}^{k+s} \varphi(t)\varphi^T(t),$$

this condition can be interpreted as a uniform complete observability requirement for the associated time-varying linear system. Given that (B3.1) holds, the exponential stability of the RWKF algorithm follows from the general Kalman filter theory [82]. Unfortunately, as already remarked in Section 2.8.3, condition (7.25) is useless if the system works under nondeterministic excitation.

A more realistic study of stochastic stability and convergence was presented by Guo [66]. According to Guo, to prove the mean square and the almost sure boundedness of parameter tracking errors, namely

$$\limsup_{t \mapsto \infty} \ \mathrm{E}\left[\|\widehat{\theta}(t) - \theta(t)\|^2\right] < \infty,$$

$$\limsup_{N \mapsto \infty} \ \frac{1}{N}\sum_{t=1}^{N} \|\widehat{\theta}(t) - \theta(t)\| < \infty, \quad \text{a.s.}$$

it is sufficient to assume that regression vectors obey the stochastic persistence of excitation condition (B3.3) and that the sequences $\{v(t)\}$ and $\{w(t)\}$ have bounded higher-order moments. Remember that the above results were derived for very general measurement noise and parameter variation patterns; in particular, it was *not* assumed that $\{v(t)\}$ and $\{w(t)\}$ are white noise sequences.

If the persistence of excitation condition is not fulfilled, the RWKF algorithm may diverge (Section 8.1.3).

7.5   Estimation memory of the RWKF algorithm

When process parameters change according to the state space model (7.1) and the quantities F, G, W and σ_v^2 are known a priori, Kalman filtering is an optimal parameter tracking algorithm, in the sense that it provides estimates with the smallest achievable mean square errors. One should be careful, though, not to overemphasize this feature of the Kalman filtering approach. The point is that the lack of prior knowledge usually forces one to use instrumental state space models such as the random walk model (7.7) or its generalized versions (7.20). Even though such instrumental models can often be regarded as a good description of local parameter variation, it would be extremely naive to assume they provide an adequate description of process nonstationarity in the entire time domain. The same applies to the selection of κ. Since the true value of σ_w^2 is seldom known a priori (if it exists at all; it is a very crude assumption that the process coefficients vary with the same speed and independently of each other), κ should be regarded as an instrumental variable, a sort of user-dependent 'knob' allowing one to tune the Kalman filtering algorithm to the degree of nonstationarity of the identified process.

Of course, under these circumstances the Kalman filter parameter tracking algorithms can no longer be claimed optimal. They provide yet another way for recursive estimation of time-varying process coefficients, neither more nor less appropriate than the approaches presented earlier.

By changing the variance quotient κ^2 one can control the estimation memory of the RWKF algorithm; small values of κ, which correspond to slow hypothetical parameter variation, result in tracking behavior typical of long-memory adaptive filters, and vice versa. To put this statement into a more rigorous framework, we will check what happens when a KF algorithm is used to identify a time-invariant process ($W = O$)

$$y(t) = \varphi^T(t)\theta + v(t),$$

subject to a wide-sense stationary excitation. To evaluate the equivalent estimation memory of of the KF-based tracker, we should derive expression for excess mean square prediction error (prediction-oriented measure) and the mean square parameter estimation error (tracking-oriented measure) under such time-invariant conditions.

First of all, we will use the averaging technique to put the RWKF algorithm in a general form (5.89) considered by Guo and Ljung [68]. For small values of κ ($\kappa \ll 1$) it can be shown [19] that

$$\varphi^T(t)P(t-1)\varphi(t) \ll 1, \tag{7.26}$$

leading to the following approximate relationship (cf. (7.9)):

$$P(t) \cong P(t-1) - P(t-1)\varphi(t)\varphi^T(t)P(t-1) + \kappa^2 I_n. \tag{7.27}$$

Moreover, since for small values of κ variations in the covariance matrix $P(t)$ are much slower than in the regression vector $\varphi(t)$, one can set

$$P(t) \cong P(t-1) \cong \ldots \cong P(t-N), \tag{7.28}$$

where N denotes the length of a local analysis window. Combining (7.27) with (7.28) one arrives at

$$P(t)\left(\sum_{i=0}^{N-1} \varphi(t-i)\varphi^T(t-i)\right)P(t) \cong N\kappa^2 I_n.$$

Finally, since for reasonably large N ($N \gg 1$) it holds that

$$\frac{1}{N}\sum_{i=0}^{N-1} \varphi(t-i)\varphi^T(t-i) \cong E[\varphi(t)\varphi^T(t)] = \Phi_o,$$

one obtains

$$P(t)\Phi_o P(t) \cong \kappa^2 I_n,$$

leading to

$$P(t) \cong \kappa\Phi_o^{-1/2} \tag{7.29}$$

where $\Phi_o^{-1/2}$ denotes a *symmetric* square root of Φ_o^{-1} (with positive diagonal elements).

We will rewrite the RWKF algorithm in the 'standardized' form

$$\widehat{\theta}(t) = \widehat{\theta}(t-1) + \gamma A(t)\varphi(t)\epsilon(t),$$

where

$$\gamma = \kappa \tag{7.30}$$

denotes a small adaptation gain and

$$A(t) = \frac{P(t-1)}{\kappa \left(1 + \varphi^T(t)P(t-1)\varphi(t)\right)} \tag{7.31}$$

is an asymptotically gain-invariant matrix.

Observe that, according to (7.26) and (7.29), for large values of t and small values of κ it holds that

$$A(t) \cong \frac{1}{\kappa}P(t-1) \cong \Phi_o^{-1/2}. \tag{7.32}$$

Now we can use the main result of Guo and Ljung [68]. Substituting (7.30) and (7.32) into (5.94) one obtains

$$\begin{aligned}
G(t) &\cong \Phi_o^{1/2}, \\
H(t) &= I_n, \\
W(t) &= O,
\end{aligned}$$

which leads to the following approximate relationship describing evolution of the covariance matrix $\Pi(t)$ of the estimation error $\widetilde{\theta}(t) = \theta - \widehat{\theta}(t)$:

$$\Pi(t) = (I_n - \kappa\Phi_o^{1/2})\Pi(t-1)(I_n - \kappa\Phi_o^{1/2})^T + \kappa^2\sigma_v^2 I_n. \tag{7.33}$$

The steady-state solution of (7.33) can be obtained from

$$\Pi_\infty = (I_n - \kappa\Phi_o^{1/2})\Pi_\infty(I_n - \kappa\Phi_o^{1/2})^T + \kappa^2\sigma_v^2 I_n, \tag{7.34}$$

or after neglecting the term proportional to $\kappa^2\Pi_\infty$ on the right-hand side of (7.34), from

$$\Phi_o^{1/2}\Pi_\infty + \Pi_\infty\Phi_o^{1/2} \cong \kappa\sigma_v^2 I_n. \tag{7.35}$$

Solving (7.35) one obtains

$$\lim_{t \mapsto \infty} \mathrm{cov}[\widehat{\theta}_{\mathrm{RWKF}}(t)] = \Pi_\infty \cong \frac{\kappa\sigma_v^2\Phi_o^{-1/2}}{2}. \tag{7.36}$$

Prediction-oriented measure

Combining the excess mean square prediction error (5.45) with (7.36) one obtains

$$\mathcal{P}_\infty^{\mathrm{RWKF}} = \lim_{t \mapsto \infty} \mathcal{P}[\widehat{\theta}_{\mathrm{RWKF}}(t)] \cong \frac{\kappa\sigma_v^2 \operatorname{tr}\{\Phi_o^{1/2}\}}{2}. \tag{7.37}$$

Recall from (5.47) that

$$\mathcal{P}_\infty^{\mathrm{WLS}} \cong \frac{n}{l_\infty}\sigma_v^2.$$

Hence, solving

$$\mathcal{P}_\infty^{\mathrm{RWKF}} = \mathcal{P}_\infty^{\mathrm{WLS}},$$

for l_∞ one arrives at the following prediction-oriented measure of the estimation memory of the RWKF algorithm:

$$l_\infty^{\text{RWKF}} = \frac{2n}{\kappa\,\text{tr}\{\Phi_o^{1/2}\}} = \frac{2}{\kappa\lambda_{\text{ave}}(\Phi_o^{1/2})}\,, \qquad (7.38)$$

where

$$\lambda_{\text{ave}}(\Phi_o^{1/2}) = \frac{1}{n}\sum_{i=1}^n \lambda_i(\Phi_o^{1/2}) = \frac{1}{n}\sum_{i=1}^n \lambda_i^{1/2}(\Phi_o)\,.$$

As in the case of LMS filters, the equivalent memory of the RWKF estimator depends not only on the adaptation gain (κ) but also on the signal-related quantities ($\Phi_o^{1/2}$). For stationary FIR systems with uncorrelated regressors ($\Phi_o = \sigma_u^2 I_n$) one obtains $\Phi_o^{1/2} = \sigma_u I_n$ and hence

$$l_\infty = \frac{2}{\kappa\sigma_u}\,,$$

i.e. the equivalent memory can be expressed as a function of the standard deviation of the input signal. However, in a more general case, where Φ_o is not similar to an identity matrix, the equivalent memory of the RWKF tracker depends not only on the magnitude but also on the covariance structure of the regressors.

Remark 1

As in the case of LMS estimators, the scale-invariant RWKF filter can be obtained by replacing the variance quotient κ^2 in (7.8) or (7.9) with the appropriately scaled variance coefficient

$$\frac{\bar\kappa^2}{\epsilon + \|\varphi(t)\|^2} \qquad (7.39)$$

or

$$\frac{\tilde\kappa^2}{r(t)}, \quad r(t) = \lambda r(t-1) + \|\varphi(t)\|^2, \qquad (7.40)$$

$$0 < \lambda < 1,$$

resulting in the normalized random walk Kalman filtering (NRWKF) and trace random walk Kalman filtering (TRWKF) algorithms, respectively. The corresponding equivalent memory spans are

$$l_\infty^{\text{NRWKF}} = \frac{2n\sqrt{\text{tr}\{\Phi_o\}}}{\bar\kappa\,\text{tr}\{\Phi_o^{1/2}\}}\,,$$

$$l_\infty^{\text{TRWKF}} = \frac{2n\sqrt{\text{tr}\{\Phi_o\}}}{\tilde\kappa\sqrt{1-\lambda}\,\text{tr}\{\Phi_o^{1/2}\}}\,.$$

Even though they are invariant to scale modifications, both expressions depend on the covariance structure of regressors; the modified RWKF algorithms differ in this respect from the modified LMS algorithms described in Section 5.3.2.

Using the following inequality, valid for any positive sequence $\{x_i\}$:

$$\sum_{i=1}^n x_i^2 < \left(\sum_{i=1}^n x_i\right)^2 \le n\sum_{i=1}^n x_i^2,$$

one obtains

$$\frac{1}{\sqrt{n}} \leq \frac{\sqrt{\mathrm{tr}\{\Phi_o\}}}{\mathrm{tr}\{\Phi_o^{1/2}\}} < 1 \tag{7.41}$$

where the lower bound is achieved if and only if $\lambda_1(\Phi_o) = \ldots = \lambda_n(\Phi_o)$, i.e. for uncorrelated regressors.

Remark 2

Note that when the 'slow adaptation' condition (7.26) holds that

$$\mathrm{E}[\varphi^T(t)P(t-1)\varphi(t)] \cong \kappa \mathrm{E}[\varphi^T(t)\Phi_o^{-1/2}\varphi(t)] = \kappa \, \mathrm{tr}\left\{\Phi_o^{-1/2}\mathrm{E}[\varphi(t)\varphi^T(t)]\right\}$$

$$= \kappa \, \mathrm{tr}\{\Phi_o^{1/2}\} \ll 1,$$

hence $l_\infty \gg 2n$ from (7.38) and this brings us to the following conclusion, similar to Section 5.2. Slow adaptation takes place when the memory span of the estimation algorithm is much larger than the number of estimated coefficients – exactly as recommended by the principle of parsimony.

Tracking-oriented measure

To obtain the tracking-oriented measure of the estimation memory for an RWKF filter, one should compare expressions for the corresponding mean square parameter tracking errors. Since

$$\mathcal{T}_\infty^{\mathrm{RWKF}} = \lim_{t \mapsto \infty} \mathcal{T}[\hat{\theta}_{\mathrm{RWKF}}(t)] = \mathrm{tr}\left\{\mathrm{cov}[\hat{\theta}_{\mathrm{RWKF}}(t)]\right\} \cong \frac{\kappa \sigma_v^2 \, \mathrm{tr}\{\Phi_o^{-1/2}\}}{2}, \tag{7.42}$$

and

$$\mathcal{T}_\infty^{\mathrm{WLS}} = \lim_{t \mapsto \infty} \mathcal{T}[\hat{\theta}_{\mathrm{WLS}}(t)] \cong \frac{\mathrm{tr}\{\Phi_o^{-1}\}}{l_\infty}\sigma_v^2,$$

after setting

$$\mathcal{T}_\infty^{\mathrm{RWKF}} = \mathcal{T}_\infty^{\mathrm{WLS}}$$

one obtains

$$l_\infty^{\star \mathrm{RWKF}} = \frac{2 \, \mathrm{tr}\{\Phi_o^{-1}\}}{\kappa \, \mathrm{tr}\{\Phi_o^{-1/2}\}}. \tag{7.43}$$

Using the inequality

$$\left(\sum_{i=1}^n \frac{1}{x_i^2}\right)\left(\sum_{i=1}^n x_i\right) \geq n\left(\sum_{i=1}^n \frac{1}{x_i}\right),$$

valid for any sequence of positive numbers $\{x_i\}$ and following in a straightforward way from the generalized Chebyshev inequality, one obtains

$$\frac{l_\infty^\star}{l_\infty} = \frac{\mathrm{tr}\{\Phi_o^{-1}\}\mathrm{tr}\{\Phi_o^{1/2}\}}{n \, \mathrm{tr}\{\Phi_o^{-1/2}\}} \geq 1, \tag{7.44}$$

where equality holds if and only if $\lambda_1(\Phi_o) = \ldots = \lambda_n(\Phi_o)$. Since

$$\mathrm{tr}\{\Phi_o^{-1}\} \leq \frac{n}{\lambda_{\min}(\Phi_o)},$$

$$\mathrm{tr}\{\Phi_o^{1/2}\} \leq n\lambda_{\max}^{1/2}(\Phi_o),$$

$$\mathrm{tr}\{\Phi_o^{-1/2}\} \geq \frac{n}{\lambda_{\max}^{1/2}(\Phi_o)},$$

it is possible to obtain the following upper bound on the discrepancy between $l_\infty^\star$ and l_∞:

$$\frac{l_\infty^\star}{l_\infty} \leq \delta(\Phi_o)$$

where $\delta(\Phi_o)$ denotes the eigenvalue disparity index of Φ_o.

7.6 Dynamic characteristics of RWKF estimators

7.6.1 Impulse response associated with RWKF estimators

Consider a time-varying system obeying the conditions (B1.1) on stationarity and independence of regressors and (B2.2) on mutual independence of $\{\varphi(t)\}$, $\{v(t)\}$ and $\{\theta(t)\}$.

As argued in the previous section, if the adaptation gain κ is sufficiently small, the RWKF filter can be approximated by the following algorithm:

$$\widehat{\theta}(t) = \widehat{\theta}(t) + \kappa\Phi_o^{-1/2}\varphi(t)\epsilon(t). \tag{7.45}$$

Combining the system equation with (7.45) one arrives at

$$\widehat{\theta}(t) = \left(I_n - \kappa\Phi_o^{-1/2}\varphi(t)\varphi^T(t)\right)\widehat{\theta}(t-1) + \kappa\Phi_o^{-1/2}\varphi(t)\varphi^T(t)\theta(t) + \kappa\Phi_o^{-1/2}\varphi(t)v(t),$$

which leads, under (B1.1) and (B2.2), to

$$\bar{\theta}(t) = \left(I_n - \kappa\Phi_o^{1/2}\right)\bar{\theta}(t-1) + \kappa\Phi_o^{1/2}\theta(t). \tag{7.46}$$

The steady-state relationship between $\bar{\theta}(t)$ and $\theta(t)$ can be written down explicitly as

$$\bar{\theta}_{\mathrm{RWKF}}(t) \cong \sum_{i=0}^{\infty} H_{\mathrm{RWKF}}(i)\theta(t-i), \tag{7.47}$$

where

$$H_{\mathrm{RWKF}}(i) = \kappa\left(I_n - \kappa\Phi_o^{1/2}\right)^i \Phi_o^{1/2} \tag{7.48}$$

denotes the impulse response associated with the RWKF estimator.

Using the orthogonal transformation Q (cf. (5.17)) one gets

$$H_{\mathrm{RWKF}}(i) = \kappa Q(I_n - \kappa\Lambda_o^{1/2})^i\Lambda_o^{1/2}Q^T, \tag{7.49}$$

$$= \kappa Q\, \mathrm{diag}\left\{\lambda_1^{1/2}(\Phi_o)\left(1 - \kappa\lambda_1^{1/2}(\Phi_o)\right)^i, \ldots, \lambda_n^{1/2}(\Phi_o)\left(1 - \kappa\lambda_n^{1/2}(\Phi_o)\right)^i\right\}Q^T,$$

As in the case of LMS estimators (cf. (5.70) and (5.71)) it holds that

$$\sum_{i=0}^{\infty} H_{\text{RWKF}}(i) = \kappa Q \left[\sum_{i=0}^{\infty} \left(I_n - \kappa \Lambda_o^{1/2} \right)^i \right] \Lambda_o^{1/2} Q^T \tag{7.50}$$

$$= \kappa Q \left(\kappa \Lambda_o^{1/2} \right)^{-1} \Lambda_o^{1/2} Q^T = I_n,$$

and

$$\sum_{i=0}^{\infty} H_{\text{RWKF}}^2(i) = \kappa^2 Q \left[\sum_{i=0}^{\infty} \left(I_n - \kappa \Lambda_o^{1/2} \right)^{2i} \right] \Lambda_o Q^T$$

$$= \kappa Q \left(2I_n - \kappa \Lambda_o^{1/2} \right)^{-1} \Lambda_o^{1/2} Q^T \cong \frac{\kappa}{2} Q \Lambda_o^{1/2} Q^T = \frac{\kappa}{2} \Phi_o^{1/2},$$

leading to

$$\frac{1}{n} \text{tr} \left\{ \sum_{i=0}^{\infty} H_{\text{RWKF}}^2(i) \right\} \cong \frac{1}{l_\infty^{\text{RWKF}}} . \tag{7.51}$$

7.6.2   Frequency response associated with RWKF estimators

The (matrix) frequency response function associated with the RWKF estimator can be defined as the Fourier transform of the impulse response $H_{\text{RWKF}}(i)$:

$$H_{\text{RWKF}}(\omega) = \sum_{i=0}^{\infty} H_{\text{RWKF}}(i) e^{-j\omega i}. \tag{7.52}$$

Combining (7.52) with (7.49) one obtains

$$H_{\text{RWKF}}(\omega) =$$

$$Q \, \text{diag} \left\{ \frac{\kappa \lambda_1^{1/2}(\Phi_o)}{1 - \left(1 - \kappa \lambda_1^{1/2}(\Phi_o) \right) e^{-j\omega}}, \dots, \frac{\kappa \lambda_n^{1/2}(\Phi_o)}{1 - \left(1 - \kappa \lambda_n^{1/2}(\Phi_o) \right) e^{-j\omega}} \right\} Q^T. \tag{7.53}$$

The transfer matrix (7.53) is symmetric, i.e.

$$H_{\text{RWKF}}^T(\omega) = H_{\text{RWKF}}(\omega).$$

The average bias component of the mean square tracking error can be expressed analogously as the average bias component for the LMS estimator (5.74):

$$\overline{T}_b = \mathop{\text{E}}_{\Theta(t)} \left[\| \, \overline{\theta}(t) - \theta(t) \, \|^2 \right] \cong \frac{1}{\pi} \int_0^\pi \text{tr}\{E_{\text{RWKF}}(\omega) S_\theta(\omega)\} \, d\omega, \tag{7.54}$$

where

$$E_{\text{RWKF}}(\omega) = [I_n - H_{\text{RWKF}}(\omega)]^\dagger [I_n - H_{\text{RWKF}}(\omega)]. \tag{7.55}$$

7.7 Convergence and tracking performance of RWKF estimators

7.7.1 Initial convergence

Since no rigorous results quantifying the speed of initial convergence of RWKF algorithms seems to be available (in the noise-free case the exponential convergence rate was proved by Guo [66]) we will limit our analysis to heuristic arguments only. The WLS-like form of RWKF algorithms (7.18) allows one to conjecture that the RWKF filters should exhibit fast initial convergence typical of WLS filters. As in the case of the EWLS algorithm, the 'matrix stepsize' $\widetilde{\Sigma}(t|t)$ in the RWKF parameter update

$$\widehat{\theta}(t) = \widehat{\theta}(t-1) + K(t)\epsilon(t) = \widehat{\theta}(t-1) + \widetilde{\Sigma}(t|t)\varphi(t)\epsilon(t)$$

changes with time; it is larger (on the average) in the initial phase of convergence before it settles down around the steady-state value of $\kappa\Phi_o^{-1/2}$. This variable-stepsize strategy guarantees rapid initial convergence.

Practical evidence fully confirms the above statements; according to Goddard [60], under typical channel equalization operating conditions the RWKF algorithm needs approximately $2n$ time steps to reach its steady-state performance.

7.7.2 Tracking performance

Suppose that the RWKF algorithm is used to track system parameters varying according to the following random walk model:

$$\theta(t) = \theta(t-1) + w(t),$$

$$\text{cov}[w(t)] = W.$$

Let us stress the fact that W – the covariance matrix characterizing the true parameter variation – is not assumed to be identical with $\sigma_w^2 I_n$ – the hypothetical covariance matrix which led to a particular, parsimonious form of the RWKF algorithm. In other words, we do not assume that the Kalman tracker is based on a true model of parameter changes.

Once more we will refer to the results derived by Guo and Ljung [68]. If conditions which guarantee exponential stability of the RWKF filter are met, the evolution of the covariance matrix $\Pi(t)$ of the estimation error

$$\widetilde{\theta}(t) = \theta(t+1) - \widehat{\theta}(t)$$

can be approximated using the following difference equation:

$$\Pi(t) = (I_n - \kappa\Phi_o^{1/2})\Pi(t-1)(I_n - \kappa\Phi_o^{1/2})^T + \kappa^2\sigma_v^2 I_n + W, \qquad (7.56)$$

which is a special case of (5.93) obtained for $G(t) = \Phi_o^{1/2}$, $H(t) = I_n$ and $W(t) = W$ (cf. (5.94)). The steady-state solution of (7.56) can be obtained from

$$\Pi_\infty = (I_n - \kappa\Phi_o^{1/2})\Pi_\infty(I_n - \kappa\Phi_o^{1/2})^T + \kappa^2\sigma_v^2 I_n + W, \qquad (7.57)$$

or after neglecting the term proportional to $\kappa^2\Pi_\infty$ on the right-hand side of (7.57), from

$$\Phi_o^{1/2}\Pi_\infty + \Pi_\infty\Phi_o^{1/2} \cong \kappa\sigma_v^2 I_n + \frac{W}{\kappa}. \qquad (7.58)$$

Predictive ability

In order to evaluate the steady-state excess mean square prediction error

$$\mathcal{P}_\infty^{\mathrm{RWKF}} \cong \mathrm{tr}\{\Pi_\infty \Phi_o\},$$

multiply both sides of (7.58) by $\Phi_o^{1/2}$:

$$\Phi_o \Pi_\infty + \Phi_o^{1/2} \Pi_\infty \Phi_o^{1/2} \cong \kappa \sigma_v^2 \Phi_o^{1/2} + \frac{\Phi_o^{1/2} W}{\kappa}.$$

Since $\mathrm{tr}\{\Phi_o^{1/2} \Pi_\infty \Phi_o^{1/2}\} = \mathrm{tr}\{\Pi_\infty \Phi_o\} = \mathrm{tr}\{\Phi_o \Pi_\infty\}$ one obtains

$$\mathcal{P}_\infty^{\mathrm{RWKF}} \cong \frac{\kappa \sigma_v^2 \, \mathrm{tr}\{\Phi_o^{1/2}\}}{2} + \frac{\mathrm{tr}\{\Phi_o^{1/2} W\}}{2\kappa}, \tag{7.59}$$

or equivalently (cf. (7.38))

$$\mathcal{P}_\infty^{\mathrm{RWKF}} \cong \frac{n\sigma_v^2}{l_\infty} + \frac{\mathrm{tr}\{\Phi_o^{1/2} W\}\mathrm{tr}\{\Phi_o^{1/2}\}}{4n} l_\infty. \tag{7.60}$$

As in the case of EWLS/LMS estimators, (5.112) and (5.113), the first term on the right-hand side of (7.60) can be recognized as a variance component of the excess prediction error and the second term as its bias component. Since the first term is inversely proportional to l_∞ and the second term is proportional to l_∞, the estimation memory of the RWKF filter should be chosen so as to trade off the low variance requirement against the low bias requirement.

It is interesting to compare (7.60) with the analogous expressions derived for the EWLS and LMS filters. First of all, note that when $W = \sigma_w^2$ the bias term in (7.60) is equal to

$$\frac{\sigma_w^2}{4n} \left[\mathrm{tr}\{\Phi_o^{1/2}\}\right]^2 l_\infty.$$

The analogous expression for the EWLS and LMS filters is

$$\frac{\sigma_w^2}{4} \mathrm{tr}\{\Phi_o\} l_\infty.$$

According to (7.41)

$$\frac{1}{n} \left[\mathrm{tr}\{\Phi_o^{1/2}\}\right]^2 \le \mathrm{tr}\{\Phi_o\},$$

which means that in the case considered the RWKF filter will always (i.e. for *any* choice of l_∞) perform better than the 'equivalent' EWLS and LMS filters – a hardly surprising conclusion if one realizes that the adopted model of parameter changes almost perfectly matches the true model (the only unknown factor is the true value of κ). In particular, note that when κ is set to σ_w/σ_v the RWKF filter becomes the 'true' Kalman filter, i.e. under Gaussian assumptions the optimal parameter tracking algorithm.

For nondiagonal covariance matrices W the comparison is less conclusive. Since

$$\lambda_{\min}(\Phi_o^{1/2})\lambda_{\text{ave}}(\Phi_o^{1/2})\,\text{tr}\{W\} \leq \frac{\text{tr}\{\Phi_o^{1/2}W\}\text{tr}\{\Phi_o^{1/2}\}}{n}$$

$$\leq \lambda_{\max}(\Phi_o^{1/2})\lambda_{\text{ave}}(\Phi_o^{1/2})\,\text{tr}\{W\}$$

and

$$\lambda_{\min}(\Phi_o) \leq \lambda_{\min}(\Phi_o^{1/2})\lambda_{\text{ave}}(\Phi_o^{1/2}) \leq \lambda_{\text{ave}}(\Phi_o)$$

$$\leq \lambda_{\max}(\Phi_o^{1/2})\lambda_{\text{ave}}(\Phi_o^{1/2}) \leq \lambda_{\max}(\Phi_o), \tag{7.61}$$

the sensitivity of $\mathcal{P}_\infty^{\text{RWKF}}$ to the covariance structure of regressors places the RWKF algorithm somewhere in between the LMS algorithm, characterized by the middle term in (7.61), and the EWLS algorithm, characterized by the outer terms in (7.61); see expressions (5.114) and (5.115).

Minimization of (7.60) with respect to l_∞ results in

$$\left(l_\infty^{\text{RWKF}}\right)_{\text{opt}} = \frac{2n\sigma_v}{\sqrt{\text{tr}\{\Phi_o^{1/2}W\}\text{tr}\{\Phi_o^{1/2}\}}}, \tag{7.62}$$

$$\left(\mathcal{P}_\infty^{\text{RWKF}}\right)_{\text{min}} = \sigma_v\sqrt{\text{tr}\{\Phi_o^{1/2}W\}\text{tr}\{\Phi_o^{1/2}\}}. \tag{7.63}$$

Comparing (7.62) and (7.63) with expressions (5.116) to (5.119) describing the best achievable performance of the EWLS and LMS filters, one arrives at the same conclusions as those reached above.

Tracking ability

To evaluate the steady-state mean square parameter tracking error

$$\mathcal{T}_\infty^{\text{RWKF}} \cong \text{tr}\{\Pi_\infty\},$$

multiply both sides of (7.58) by $\Phi_o^{-1/2}$:

$$\Pi_\infty + \Phi_o^{-1/2}\Pi_\infty\Phi_o^{1/2} \cong \kappa\sigma_v^2\Phi_o^{-1/2} + \frac{\Phi_o^{-1/2}W}{\kappa}.$$

Since $\text{tr}\{\Phi_o^{-1/2}\Pi_\infty\Phi_o^{1/2}\} = \text{tr}\{\Phi_o\}$ one obtains

$$\mathcal{T}_\infty^{\text{RWKF}} \cong \frac{\kappa\sigma_v^2\,\text{tr}\{\Phi_o^{-1/2}\}}{2} + \frac{\text{tr}\{\Phi_o^{-1/2}W\}}{2\kappa}, \tag{7.64}$$

or equivalently (cf. (7.43))

$$\mathcal{T}_\infty^{\text{RWKF}} \cong \frac{\text{tr}\{\Phi_o^{-1}\}\sigma_v^2}{l_\infty^\star} + \frac{\text{tr}\{\Phi_o^{-1/2}W\}\text{tr}\{\Phi_o^{-1/2}\}}{4\,\text{tr}\{\Phi_o^{-1}\}}l_\infty^\star. \tag{7.65}$$

Comparing the bias term in (7.65) with the corresponding expressions derived earlier for the LMS algorithm (5.120) and the EWLS algorithm (5.121), one arrives at conclusions identical with those reached in the preceding subsection – the sensitivity of $\mathcal{T}_\infty^{\text{RWKF}}$ to the covariance structure of regressors places the RWKF filter 'between' the LMS filter and the EWLS filter.

Minimization of $\mathcal{T}_\infty^{\text{RWKF}}$ with respect to $l_\infty^\star$ yields

$$\left(l_\infty^{\star\text{RWKF}}\right)_{\text{opt}} = \frac{2\sigma_v \operatorname{tr}\{\Phi_o^{-1}\}}{\sqrt{\operatorname{tr}\{\Phi_o^{-1/2}W\}\operatorname{tr}\{\Phi_o^{-1/2}\}}}, \tag{7.66}$$

$$\left(\mathcal{T}_\infty^{\text{RWKF}}\right)_{\text{min}} = \sigma_v \sqrt{\operatorname{tr}\{\Phi_o^{-1/2}W\}\operatorname{tr}\{\Phi_o^{-1/2}\}}. \tag{7.67}$$

Analysis of (7.67) produces conclusions identical to those reached above.

Remark

The optimal values of l_∞ and $l_\infty^\star$ can be 'translated back' to the optimal values of κ and $\kappa^\star$:

$$\kappa_{\text{opt}} = \frac{\sqrt{\operatorname{tr}\{\Phi_o^{1/2}W\}}}{\sigma_v\sqrt{\operatorname{tr}\{\Phi_o^{1/2}\}}}, \tag{7.68}$$

$$\kappa_{\text{opt}}^\star = \frac{\sqrt{\operatorname{tr}\{\Phi_o^{-1/2}W\}}}{\sigma_v\sqrt{\operatorname{tr}\{\Phi_o^{-1/2}\}}}. \tag{7.69}$$

Note that when $W = \sigma_w^2 I_n$ it holds that

$$\kappa_{\text{opt}} = \kappa_{\text{opt}}^\star = \frac{\sigma_w}{\sigma_v},$$

i.e. the 'true' Kalman settings minimize both the prediction-oriented and tracking-oriented quality measures. The same statement is true when Φ_o is similar to an identity matrix, which takes place for FIR systems with uncorrelated inputs. Note that for $\Phi_o = \sigma_u^2 I_n$ one obtains

$$\kappa_{\text{opt}} = \kappa_{\text{opt}}^\star = \frac{\sqrt{\operatorname{tr}\{W\}/n}}{\sigma_v}.$$

This is not surprising as for $\Phi_o = \sigma_u^2 I_n$ the prediction-oriented and tracking-oriented quality measures coincide.

7.8 Parameter matching using the Kalman smoothing approach

Whenever feasible, the Kalman filtering framework allows one to obtain improved identification results by taking the process for estimating $\theta(t)$ and incorporating not only the 'past' measurements $\xi(t), i \le t$ but also a certain number of 'future' measurements: $\xi(t), i > t$. Recursive algorithms which provide such estimates are known as Kalman smoothers.

7.8.1   Fixed interval smoothing

The fixed interval smoothing approach can be used in all applications based on process segmentation. Consider an analysis frame $T = [1, \ldots, N]$ of length N and suppose that the evolution of process coefficients is governed by the state space model (7.1). Under Gaussian assumptions the minimum mean square estimate of $\theta(t)$ based on all available data $\Xi(N)$ is given by

$$\widehat{\theta}(t|N) = \mathrm{E}[\theta(t)|\Xi(N)],$$

and can be evaluated by postprocessing the estimates from the Kalman filtering algorithms (7.3) or (7.4):

$$\widehat{\theta}(t|N) = \widehat{\theta}(t|t) + A(t) \left[\widehat{\theta}(t+1|N) - \widehat{\theta}(t+1|t) \right], \tag{7.70}$$

$$t = N - 1, \ldots, 1$$

where

$$A(t) = \Sigma(t|t) F^T \Sigma^{-1}(t+1|t) = \widetilde{\Sigma}(t|t) F^T \widetilde{\Sigma}^{-1}(t+1|t). \tag{7.71}$$

Note that (7.70), known as the Rauch–Tung–Striebel smoother [125], is a recursive *backward* processing algorithm since the time index t must be decreased from $N-1$ down to 1.

The covariance matrix $\Sigma(t|N)$ for the smoothed estimate $\widehat{\theta}(t|N)$ can be obtained from

$$\Sigma(t|N) = \Sigma(t|t) + A(t) \left[\Sigma(t+1|N) - \Sigma(t+1|t) \right] A^T(t). \tag{7.72}$$

However, unlike the filtering run, $\Sigma(t|N)$ does *not* need to be calculated to generate the smoothed parameter estimates.

Another useful form of a Kalman smoother, proposed by Bryson and Frazier [125], allows one to avoid inversion of the covariance matrix $\Sigma(t+1|t)$ (or $\widetilde{\Sigma}(t+1|t)$) and has good numerical properties [201]:

$$\widehat{\theta}(t|N) = \widehat{\theta}(t|t) - \widetilde{\Sigma}(t|t) F^T L(t), \tag{7.73}$$

where

$$L(t-1) = \left[I_n - K(t)\varphi^T(t) \right] \left\{ F^T L(t) - \varphi(t) \left[y(t) - \varphi^T(t)\widehat{\theta}(t|t-1) \right] \right\} \tag{7.74}$$

and

$$K(t) = \widetilde{\Sigma}(t|t)\varphi(t)$$

is the gain vector of the (forward time) Kalman filter. The initial condition for calculation of the backward gain vector should be set to $L(N) = 0$.

Application of (7.70), (7.71) or (7.73), (7.74) to smoothing of RWKF estimates is straightforward ($F = I_n$). In order to smooth the IRWKF estimates, one should replace the quantities $\widehat{\theta}(t|t)$, $\widehat{\theta}(t|t-1)$, $\widetilde{\Sigma}(t|t)$, $\widetilde{\Sigma}(t|t-1)$, F and $\varphi(t)$, which appear in the smoothing formulas presented above, with their augmented counterparts $\widehat{\theta}_a(t|t)$, $\widehat{\theta}_a(t|t-1)$, $\widetilde{\Sigma}_a(t|t)$, $\widetilde{\Sigma}_a(t|t-1)$, F_a and $\varphi_a(t)$.

7.8.2   Fixed lag smoothing

Whenever decisions based on the results of parameter estimation can be postponed by a certain number of sampling intervals, say τ, the filtered estimates $\widehat{\theta}(t - \tau | t - \tau)$ can be replaced with more accurate smoothed estimates

$$\widehat{\theta}(t - \tau | t) = \mathrm{E}[\theta(t - \tau) | \Xi(t)].$$

Fixed lag smoothing can be realized using a Kalman filtering algorithm designed for an augmented state space model of parameter variation.

Let

$$\theta_a(t) = [\theta^T(t), \ldots, \theta^T(t - \tau)]^T.$$

Note that the estimate $\theta_a(t|t)$, yielded by the Kalman filter, can be written down as

$$\widehat{\theta}_a(t|t) = \mathrm{E}[\theta_a(t)|\Xi(t)] = \begin{bmatrix} \mathrm{E}[\theta(t)|\Xi(t)] \\ \vdots \\ \mathrm{E}[\theta(t - \tau)|\Xi(t)] \end{bmatrix} = \begin{bmatrix} \widehat{\theta}(t|t) \\ \vdots \\ \widehat{\theta}(t - \tau|t) \end{bmatrix}, \qquad (7.75)$$

i.e. the last n coordinates of $\widehat{\theta}_a(t|t)$ form a vector of the smoothed parameter estimates we are looking for:

$$\widehat{\theta}(t - \tau|t) = C_a \widehat{\theta}_a(t|t)$$

where

$$C_a = [O \ \ldots \ O \ I_n]_{n \times (\tau+1)n}.$$

For the system governed by (7.1) the augmented state equation is

$$\theta_a(t) = F_a \theta_a(t - 1) + G_a w(t),$$

where

$$F_a = \begin{bmatrix} F & O & \ldots & O & O \\ I_n & O & \ldots & O & O \\ & & \ddots & & \\ O & O & \ldots & I_n & O \end{bmatrix}_{(\tau+1)n \times (\tau+1)n}, \qquad G_a = \begin{bmatrix} G \\ O \\ \vdots \\ O \end{bmatrix}_{(\tau+1)n \times n}.$$

The corresponding output (process) equation is

$$y(t) = \varphi_a^T(t)\theta_a(t) + v(t),$$

where

$$\varphi_a(t) = [\varphi^T(t), 0^T, \ldots, 0^T]_{(\tau+1)n \times 1}^T.$$

For the integrated random walk model of order $k \leq \tau$ one gets

$$F_a = \begin{bmatrix} -f_1 I_n & \ldots & -f_{k-1}I_n & f_k I_n & O & \ldots & O & O \\ I_n & & O & O & O & \ldots & O & O \\ \vdots & & & & & & & \vdots \\ O & \ldots & O & I_n & O & \ldots & O & O \\ \vdots & & & & & & & \vdots \\ O & \ldots & O & O & O & \ldots & I_n & O \end{bmatrix}_{(\tau+1)n \times (\tau+1)n},$$

$$G_a = \begin{bmatrix} I_n \\ O \\ \vdots \\ O \\ \vdots \\ O \end{bmatrix}_{(\tau+1)n \times n}.$$

The estimate $\widehat{\theta}_a(t|t)$ can be obtained using the KF algorithm (7.4) provided that all system matrices and vectors are replaced with their augmented counterparts.

7.9 Computer simulations

Figures 7.1 to 7.10 show identification results obtained using the RWKF algorithm for the time-varying finite impulse response systems (FIR1 and FIR2) and autoregressive processes (AR1 and AR2) described in Section 2.9. Since for the RWKF filter the tracking-oriented and prediction-oriented memory spans are different, the first sequence of plots (Figures 7.1 to 7.5) correspond to the case where $l_\infty^\star = 50$ (the parameter tracking oriented analysis) and the second sequence of plots (Figures 7.6 to 7.10) correspond to the case where $l_\infty = 50$ (the prediction-oriented analysis).

Figures 7.1 and 7.6 show the output of the linear filter associated with the LMS estimator. For both FIR systems the 'theoretical' plots stay in very good agreement with the results of averaging (over 100 simulation runs) the estimated parameter trajectories (Figures 7.3 and 7.8). Estimation results obtained for a single realization of the identified process are shown in Figures 7.2 and 7.7. Note that the cross-coupling effect (which can be seen for a system with abrupt parameter changes) is much less emphasized than the analogous effect observed for LMS filters. The same applies to the difference in the parameter tracking behavior between RWKF filters with $l_\infty = 50$ and $l_\infty^\star = 50$.

Figures 7.4, 7.5 and 7.9, 7.10 show identification results obtained for the time-varying autoregressive processes AR1 and AR2. Since the tracking-oriented and prediction-oriented estimation memory spans of the LMS filter depend on the covariance structure of regressors, for a nonstationary AR process they are not constant; the stepsize μ was chosen so as to enforce $l_\infty = 50$ (or $l_\infty^\star = 50$) at the beginning of the analysis interval, i.e. for the starting values of the process coefficients. The 'variable memory' feature of the KF algorithm explains its faster response to parameter changes which can be observed for AR processes compared to their FIR counterparts.

Remark 1

All plots show the steady-state tracking behavior of the RWKF algorithm. To reach the steady state, parameter estimation was initialized at instant $t = -500$, i.e. before the analysis interval $T = [0, 500]$ started. In the initial convergence period the system parameters were constant $(a_1(t) = a_1(0), a_2(t) = a_2(0)$ for $t < 0)$.

Remark 2

The greater variance of the estimates in the AR case, compared to the FIR case, is mainly the consequence of a smaller signal-to-noise ratio.

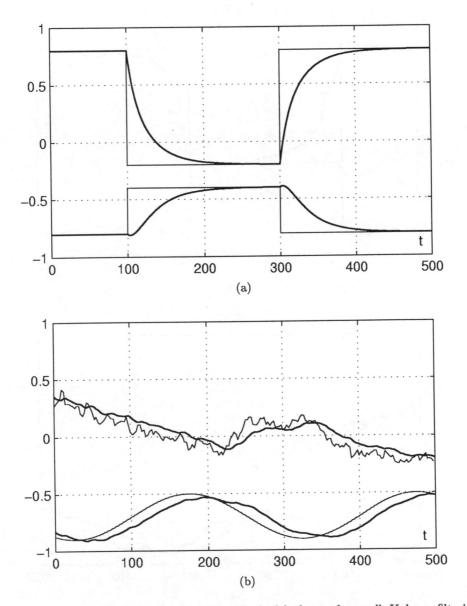

Figure 7.1 Output of a linear filter associated with the random walk Kalman filtering (RWKF) estimator for a system with jump parameter changes (a) and continuous parameter changes (b); the equivalent memory of the estimation algorithm was set to $l_\infty^\star = 50$ ($\kappa = 0.0447$).

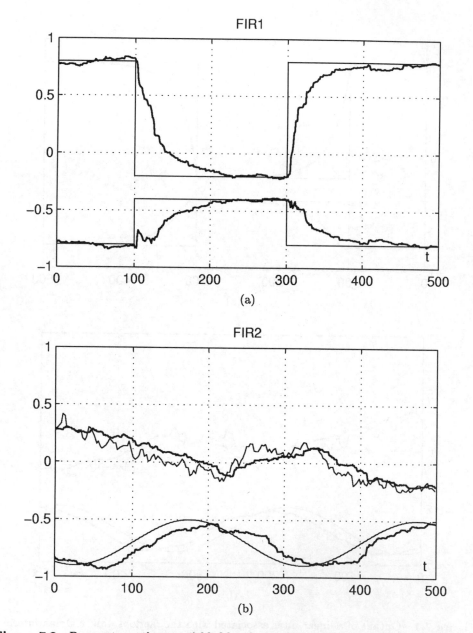

Figure 7.2 Parameter estimates yielded by the random walk Kalman filtering (RWKF) algorithm for a single realization of an FIR system with jump parameter changes (a) and continuous parameter changes (b); the equivalent memory of the estimation algorithm was set to $l_\infty^\star = 50$ ($\kappa = 0.0447$).

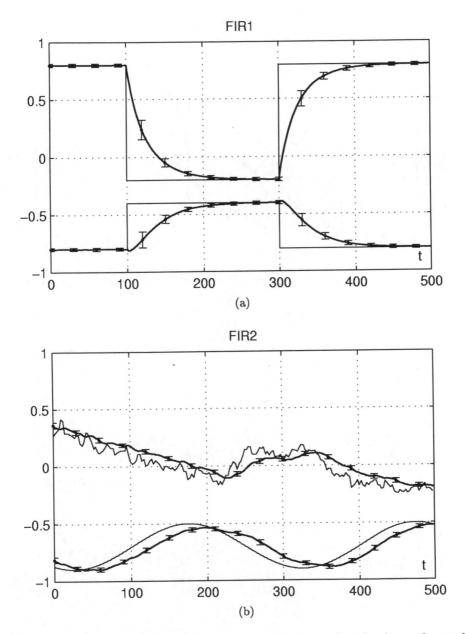

Figure 7.3   Average trajectories of parameter estimates yielded by the random walk Kalman filtering (RWKF) algorithm for an FIR system with jump parameter changes (a) and continuous parameter changes (b); vertical bars show standard deviation of the estimates.

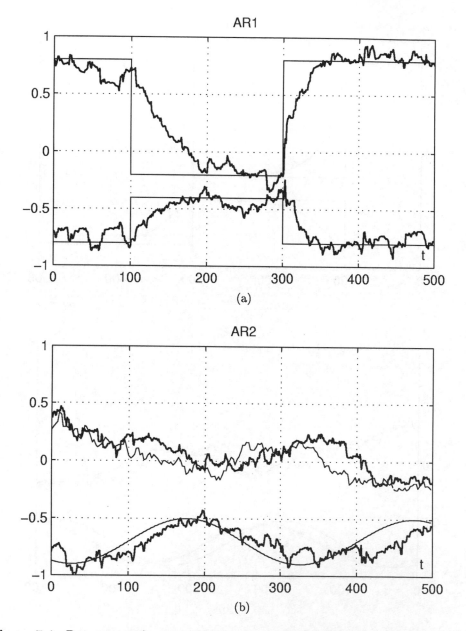

Figure 7.4 Parameter estimates yielded by the random walk Kalman filtering (RWKF) algorithm for a single realization of an AR system with jump parameter changes (a) and continuous parameter changes (b); the starting value of the equivalent memory of the estimation algorithm was set to $l_\infty^* = 50$.

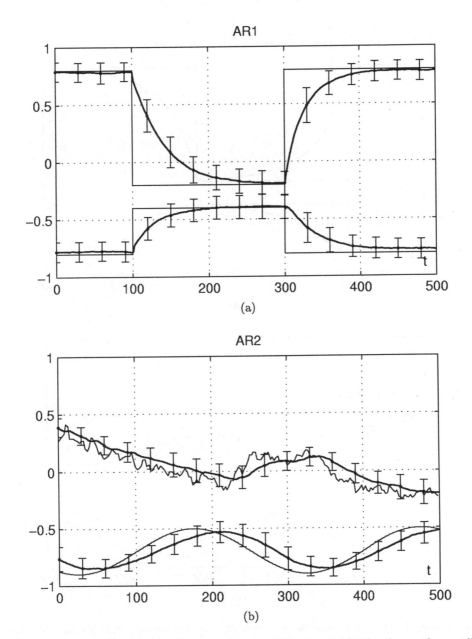

Figure 7.5   Average trajectories of parameter estimates yielded by the random walk
Kalman filtering (RWKF) algorithm for an AR system with jump parameter changes (a)
and continuous parameter changes (b); vertical bars show standard deviation of the
estimates.

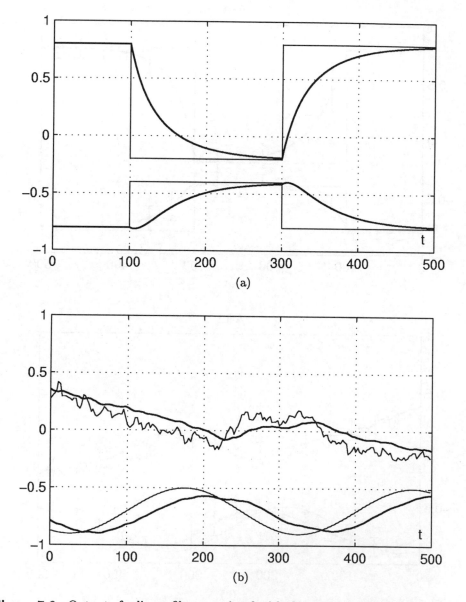

(a)

(b)

Figure 7.6 Output of a linear filter associated with the random walk Kalman filtering (RWKF) estimator for a system with jump parameter changes (a) and continuous parameter changes (b); the equivalent memory of the estimation algorithm was set to $l_\infty = 50$ ($\kappa = 0.0268$).

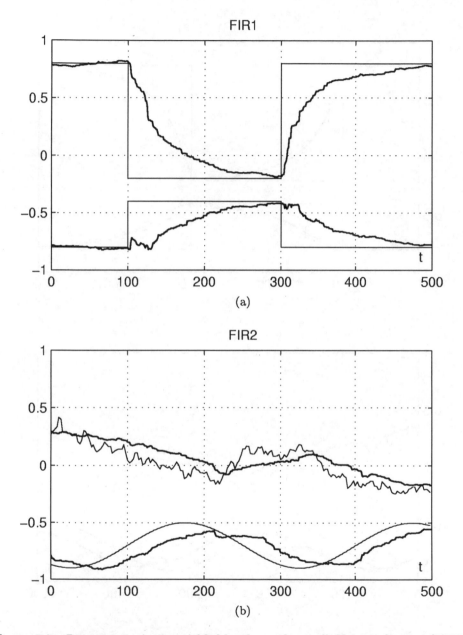

Figure 7.7 Parameter estimates yielded by the random walk Kalman filtering (RWKF) algorithm for a single realization of an FIR system with jump parameter changes (a) and continuous parameter changes (b); the equivalent memory of the estimation algorithm was set to $l_\infty = 50$ ($\kappa = 0.0268$).

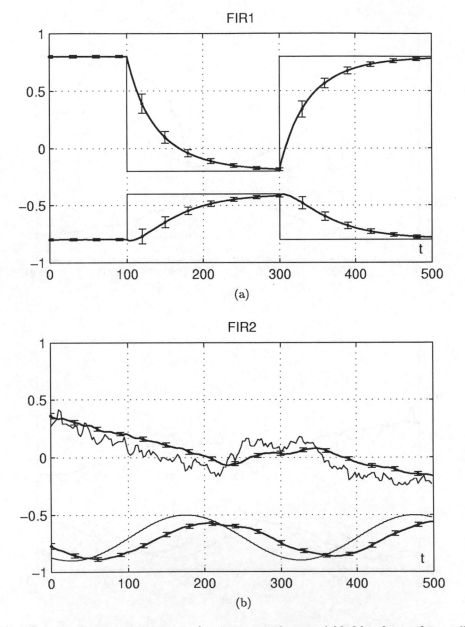

Figure 7.8 Average trajectories of parameter estimates yielded by the random walk
Kalman filtering (RWKF) algorithm for an FIR system with jump parameter changes (a)
and continuous parameter changes (b); vertical bars show standard deviation of the
estimates.

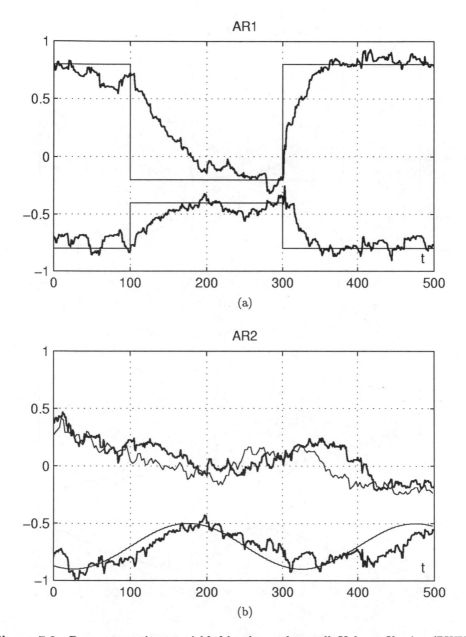

Figure 7.9 Parameter estimates yielded by the random walk Kalman filtering (RWKF) algorithm for a single realization of an AR system with jump parameter changes (a) and continuous parameter changes (b); the starting value of the equivalent memory of the estimation algorithm was set to $l_\infty = 50$.

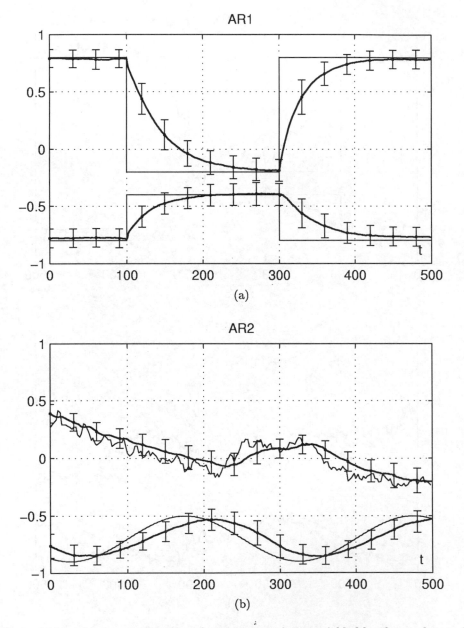

Figure 7.10 Average trajectories of parameter estimates yielded by the random walk Kalman filtering (RWKF) algorithm for an AR system with jump parameter changes (a) and continuous parameter changes (b); vertical bars show standard deviation of the estimates.

7.10 Extension to ARMAX processes

Consider a time-varying ARMA process governed by

$$y(t+1) = \sum_{i=1}^{n_A} a_i(t)y(t-i+1) + v(t+1) + \sum_{i=1}^{n_C} c_i(t)v(t-i+1)$$

$$= \varphi_y^T(t)\theta_y(t) + \varphi_v^T(t)\theta_v(t) + v(t+1), \tag{7.76}$$

where

$$\varphi_y(t) = [y(t), \ldots, y(t-n_A+1)]^T, \quad \varphi_v(t) = [v(t), \ldots, v(t-n_C+1)]^T,$$

$$\theta_y(t) = [a_1(t), \ldots, a_{n_A}(t)]^T, \quad \theta_v(t) = [c_1(t), \ldots, c_{n_C}(t)]^T.$$

Observe that there is a slight difference between the model (7.76) which, after changing the time coordinate from $t+1$ to t, can be rewritten in the form

$$y(t) = \varphi^T(t)\theta(t-1) + v(t),$$

$$\varphi(t) \quad = \quad [y(t-1), \ldots, y(t-n_A), v(t-1), \ldots, v(t-n_C)]^T,$$

$$\theta(t) \quad = \quad [a_1(t), \ldots, a_{n_A}(t), c_1(t), \ldots, c_{n_C}(t)]^T,$$

and the model

$$y(t) = \varphi^T(t)\theta(t) + v(t),$$

which has been used so far.

Suppose that variation of the parameter vector $\theta(t) = [\theta_y^T(t), \theta_v^T(t)]^T$ can be described by the random walk model

$$\theta(t+1) = \theta(t) + w(t+1). \tag{7.77}$$

Note that equations (7.76) and (7.77) can be written in the following (nonminimal) state space form:

$$x(t+1) \quad = \quad f[x(t)] + \nu(t+1),$$

$$y(t) \quad = \quad c_o^T x(t), \tag{7.78}$$

where

$$x(t) = [\varphi_y^T(t), \varphi_v^T(t), \theta_y^T(t), \theta_v^T(t)]^T$$

denotes the $2n$-dimensional state vector, $\nu(t)$ is the input noise

$$\nu(t) = \begin{bmatrix} hv(t) \\ w(t) \end{bmatrix}, \quad h = [1, 0, \ldots, 1, 0, \ldots, 0]^T,$$

$c_o = [1, 0, \ldots, 0]^T$ is the output vector and $f[x(t)]$ denotes a nonlinear state transition function

$$f[x(t)] = \begin{bmatrix} A_y & O & B_y[\varphi_y(t)] & B_v[\varphi_v(t)] \\ O & A_v & O & O \\ O & O & I_{n_A} & O \\ O & O & O & I_{n_C} \end{bmatrix} \begin{bmatrix} \varphi_y(t) \\ \varphi_v(t) \\ \theta_y(t) \\ \theta_v(t) \end{bmatrix}$$

where

$$A_y = \begin{bmatrix} 0 & \cdots & 0 & 0 \\ 1 & \cdots & 0 & 0 \\ & \ddots & 0 & 0 \\ 0 & \cdots & 1 & 0 \end{bmatrix}_{n_A \times n_A}, \quad A_v = \begin{bmatrix} 0 & \cdots & 0 & 0 \\ 1 & \cdots & 0 & 0 \\ & \ddots & 0 & 0 \\ 0 & \cdots & 1 & 0 \end{bmatrix}_{n_C \times n_C},$$

$$B_y[\varphi_y(t)] = \begin{bmatrix} 0 & \varphi_y^T(t) & & \\ & \cdots & 0 & 0 \\ \vdots & & \vdots & \\ 0 & \cdots & 0 & 0 \end{bmatrix}_{n_A \times n_A}, \quad B_v[\varphi_v(t)] = \begin{bmatrix} 0 & \varphi_v^T(t) & & \\ & \cdots & 0 & 0 \\ \vdots & & \vdots & \\ 0 & \cdots & 0 & 0 \end{bmatrix}_{n_A \times n_C},$$

According to (7.78) the problem of joint estimation of $\theta(t)$ and $\varphi_v(t)$ can be regarded as a nonlinear filtering problem in the state space. It can be solved using the extended Kalman filtering (EKF) approach.

Denote by $F[x(t)]$ the state transition matrix of a linearized system:

$$F[x(t)] = \nabla_x f[x(t)] = \begin{bmatrix} D_y[\theta_y(t)] & D_v[\theta_v(t)] & B_y[\varphi_y(t)] & B_v[\varphi_v(t)] \\ O & A_v & O & O \\ O & O & I_{n_A} & O \\ O & O & O & I_{n_C} \end{bmatrix}$$

where

$$D_y[\theta_y(t)] = \begin{bmatrix} 1 & \theta_y^T(t) & & \\ & \cdots & 0 & 0 \\ & \ddots & & \vdots \\ 0 & \cdots & 1 & 0 \end{bmatrix}_{n_A \times n_A}, \quad D_v[\theta_v(t)] = \begin{bmatrix} 0 & \theta_v^T(t) & & \\ & \cdots & 0 & 0 \\ \vdots & & \vdots & \\ 0 & \cdots & 0 & 0 \end{bmatrix}_{n_A \times n_C}$$

For the problem at hand, the extended Kalman filter (i.e. the ordinary Kalman filter designed for a linearized system) can be summarized as follows:

$$
\begin{aligned}
\hat{x}(t|t) &= \hat{x}(t|t-1) + K(t)\epsilon(t), \\
\hat{x}(t|t-1) &= f[\hat{x}(t-1|t-1)], \\
\epsilon(t) &= y(t) - c_o^T(t)\hat{x}(t|t-1), \\
K(t) &= \frac{\tilde{\Sigma}(t|t-1)c_o}{c_o^T\tilde{\Sigma}(t|t-1)c_o}, \\
\tilde{\Sigma}(t|t-1) &= F_o(t-1)\tilde{\Sigma}(t-1|t-1)F_o^T(t-1) + H_o, \\
\tilde{\Sigma}(t|t) &= \tilde{\Sigma}(t|t-1) - \frac{\tilde{\Sigma}(t|t-1)c_o c_o^T\tilde{\Sigma}(t|t-1)}{c_o^T\tilde{\Sigma}(t|t-1)c_o},
\end{aligned}
\qquad (7.79)
$$

where $F_o(t-1) = F[\hat{x}(t-1|t-1)]$ and

$$H_o = \begin{bmatrix} hh^T & O \\ O & \kappa^2 I_n \end{bmatrix}.$$

It is straightforward to show that

$$\widehat{\varphi}_y(t|t) = \varphi_y(t),$$

which is intuitively obvious since the vector of output measurements $\varphi_y(t)$ is *known* at instant t.

The EKF algorithm can be extended in a pretty straightforward manner to ARMAX systems; see Ljung [112]. The same reference provides convergence analysis of the EKF algorithm (for time-invariant linear systems subject to the output measurement noise).

The EKF algorithm (7.79) can be viewed as a combination of two Kalman filters coupled in a nonlinear fashion. The first filter recovers $\varphi_v(t)$, i.e. the missing moving average part of the regression vector associated with (7.76) and the second filter tracks variations of process parameters. A simplified, hence computationally more attractive, version of the EKF algorithm can be obtained by separating its inverse filtering part from the parameter tracking part. This can be achieved, for example, by using the 'certainty equivalence' projection technique proposed in [149]. Another possibility is to use the following ELS-like recursive scheme:

$$
\begin{aligned}
\widehat{\theta}(t) &= \widehat{\theta}(t-1) + K(t)\epsilon(t), \\
\epsilon(t) &= y(t) - \widetilde{\varphi}^T(t)\widehat{\theta}(t-1), \\
K(t) &= \frac{P(t-1)\widetilde{\varphi}(t)}{1 + \widetilde{\varphi}^T(t)P(t-1)\widetilde{\varphi}(t)}, \\
P(t) &= P(t-1) - \frac{P(t-1)\widetilde{\varphi}(t)\widetilde{\varphi}^T(t)P(t-1)}{1 + \widetilde{\varphi}^T(t)P(t-1)\widetilde{\varphi}(t)} + \kappa^2 I_n,
\end{aligned}
\tag{7.80}
$$

where, similar to the the ELS scheme,

$$\widetilde{\varphi}(t) = [y(t-1),\ldots,y(t-n_A), u(t-1),\ldots,u(t-n_B), \epsilon(t-1),\ldots,\epsilon(t-n_C)]^T.$$

Comments and extensions

Section 7.1

- The idea of using Kalman filters to identify a time-varying system goes back to Mayne [123] amd Lee [102]. The paper of Bohlin [29] is an excellent survey of the practical applications. Another good reference is the book of Kitagawa and Gersch [95].

Section 7.3

- The use of integrated random walk models was popularized by Norton [152] and Young [201].

Section 7.8

- Application of Kalman smoothing to the identification of time-varying systems goes back to Norton [152]. Interesting extensions of this approach can be found in the papers of Kitagawa [92] and Kitagawa and Gersh [93], [94].

8

Practical Issues

8.1 Numerical safeguards

So far our analysis of the properties of different adaptive filters has been constrained to the case where the sequence $\{\varphi(t)\}$ is persistently exciting, i.e. it spans the entire regression space. As explained in Chapter 3, persistence of excitation (which guarantees system identifiability) may be lost if the components of the regression vector become linearly dependent. In simple terms, nonidentifiability means that the incoming input/output data does not contain information about the *entire* parameter vector θ, i.e. it does not span the entire parameter space. Such 'information blackout', when combined with forgetting mechanisms typical of all finite-memory adaptive filters, leads unavoidably to estimation problems; since the 'old' data items are systematically replaced with the 'new' but noninformative items, all estimation routines discussed in Chapters 3 to 7 become numerically ill-conditioned and yield unreliable (and often unreasonable) results.

Basically, there are two ways of coping with identifiability problems. The first (active) approach focuses on changing defective experimental conditions, i.e. on removing the source of system nonidentifiability. As shown in Section 3.1.1, nonidentifiability of ARX systems is caused by an inadequate excitation: the input signal is either not rich enough to excite all system modes (open-loop identification) or it is linearly dependent on the remaining regression variables (closed-loop identification under a linear feedback). A simple way of restoring identifiability in both these cases is to add some *dither* to the input signal; dither is a low-variance random perturbation, usually a white noise sequence [11]. Dithering prevents estimation algorithms from numerical ill-conditioning but it also disturbs operation of the adaptive system, causing deterioration of its performance. For this reason it is seldom used in practice.

The second (passive) approach to resolving nonidentifiability problems is based on a different strategy. It accepts that adaptive systems may be forced, from time to time, to operate under poor excitation conditions and it attempts to modify the parameter estimation algorithms so as to make them insensitive to a temporary lack of excitation. The rest of this section is an overview of these passive techniques.

8.1.1 Least squares algorithms

When process identification is carried out under poor excitation conditions, the regression matrix inverted by all algorithms belonging to the least squares family

(WLS, BF, WBF) may become too close to a singular matrix, with obvious numerical consequences.

To better understand what happens when persistence of excitation is temporarily lost, we will study the behavior of the EWLS algorithm (4.8). First of all, consider the case where

$$\varphi(t) = 0, \quad \forall t > t_o, \tag{8.1}$$

i.e. the identified system is *not excited* after the time instant t_o. A trivial example which illustrates this case is an FIR system driven by an input signal which dies out to zero at instant $t_o - n$:

$$u(t) = 0, \quad \forall t > t_o - n.$$

A much more interesting example, where condition (8.1) holds asymptotically, comes from the area of adaptive control. Suppose that a self-tuning minimum variance regulator (1.67) is used to control a plant which under certain operating conditions may become time invariant ($\theta(t) = \theta_o, \forall t > t_o$) and noise-free ($E[v^2(t)] = 0, \forall t > t_o$); nearly deterministic nonlinear plants operated under infrequent set point changes usually fall into this category [11].

In the absence of measurement errors, parameter estimates rapidly converge to their true values,

$$\widehat{\theta}(t) \xrightarrow[t \to \infty]{} \theta_o,$$

which entails

$$y(t) \xrightarrow[t \to \infty]{} v(t) = 0$$

and consequently

$$u(t) \xrightarrow[t \to \infty]{} 0.$$

Combining the last two relationships, one obtains

$$\varphi(t) \xrightarrow[t \to \infty]{} 0. \tag{8.2}$$

Substituting (8.1) into (4.8) one arrives at

$$\begin{aligned} \widehat{\theta}(t) &= \widehat{\theta}(t-1), \\ P(t) &= \frac{1}{\lambda} P(t-1), \quad \forall t \geq t_o, \end{aligned}$$

or equivalently,

$$\begin{aligned} \widehat{\theta}(t) &= \widehat{\theta}(t_o), \\ P(t) &= \frac{1}{\lambda^{t-t_o}} P(t_o), \quad \forall t \geq t_o. \end{aligned} \tag{8.3}$$

The first relationship in (8.3) is pretty obvious; since no new information about the identified system arrives after time t_o the estimates remain unchanged. The second relationship allows one to identify the source of numerical problems. Divided at each time step by a quantity less than one, the matrix $P(t)$ grows exponentially fast with time. Since $P(t)$ plays the role of the matrix stepsize in the EWLS parameter update, for large values of $P(t)$ the estimation algorithm becomes extremely sensitive to noise and/or numerical errors. This means, for example, that even a relatively small disturbance or a set point change occurring after a long period with no excitation may cause instability of an adaptive control system. The effect is often known as a covariance blowup or an estimator windup, due to its similarity with windup in classical integral control.

The estimator windup takes place in all cases where the regression variables are linearly dependent. To justify this statement, consider a situation where

$$x^T \varphi(t) = 0, \quad \forall t > t_o, \tag{8.4}$$

for some nonzero vector x. Note that (8.1) is just a special case of (8.4). Since

$$P^{-1}(t) = \lambda P^{-1}(t-1) + \varphi(t)\varphi^T(t)$$

one easily arrives at

$$x^T P^{-1}(t)x = \lambda x^T P^{-1}(t-1)x, \quad \forall t > t_o,$$

which leads to

$$x^T P^{-1}(t)x = \lambda^{t-t_o} x^T P^{-1}(t_o)x, \quad \forall t > t_o. \tag{8.5}$$

Observe that

$$x^T P^{-1}(t)x \geq \|x\|^2 \, \lambda_{\min}(P^{-1}(t)) = \frac{\|x\|^2}{\lambda_{\max}(P(t))}, \tag{8.6}$$

and

$$x^T P^{-1}(t_o)x \leq \|x\|^2 \, \lambda_{\max}(P^{-1}(t_o)) = \frac{\|x\|^2}{\lambda_{\min}(P(t_o))}. \tag{8.7}$$

After combining (8.5), (8.6) and (8.7) one arrives at

$$\lambda_{\max}(P(t)) \geq \frac{1}{\lambda^{t-t_o}}\lambda_{\min}(P(t_o)), \tag{8.8}$$

which means that under (8.4) at least one eigenvalue of $P(t)$ will grow exponentially with time, leading to similar sensitivity problems as those described above. It is straightforward to show that the same happens when

$$x^T \varphi(t) \xrightarrow[t \mapsto \infty]{} 0,$$

provided that convergence takes place sufficiently fast, namely

$$\sum_{i=t_o+1}^{t} \lambda^{i-t_o} (x^T \varphi(i))^2 \xrightarrow[t \mapsto \infty]{} 0.$$

Square root filtering

Consider the EWLS filter (4.8). When the regression matrix, the inverse of which is propagated by the EWLS algorithm, becomes poorly conditioned, numerical errors (caused by finite precision of computation) may destroy nonnegative definiteness of $P(t)$ and yield 'unreasonable' parameter estimates. The main source of numerical problems is the covariance update equation in (4.8); since $P(t)$ is evaluated as a *difference* of two matrices, when roundoff errors become pronounced the result of subtraction may become indefinite or negative definite.

A pretty straightforward way of getting around the problem is to use the modified covariance update equation

$$P(t) = \frac{1}{\lambda} \left(I_n - K(t)\varphi^T(t)\right) P(t-1) \left(I_n - K(t)\varphi^T(t)\right)^T + K(t)K^T(t), \qquad (8.9)$$

known in the theory of Kalman filtering as Joseph's stabilized recursion [5]. Note that the modified recursion (mathematically identical with the original one) is subtraction-free, hence it guarantees positive definiteness of $P(t)$. There are, however, better ways of achieving the same goal.

Numerical robustness of the EWLS filter can be significantly improved if, instead of the matrix $P(t)$, its square root $S(t)$ is propagated:

$$P(t) = S(t)S^T(t). \qquad (8.10)$$

First, the nonnegative definite character of $P(t)$ is preserved by virtue of the fact that the product of any square matrix and its transpose is always a nonnegative definite matrix. Second, the numerical conditioning of $S(t)$ is generally much better than that of $P(t)$. Denote by

$$c(A) = \delta(AA^T) = \frac{\lambda_{\max}(AA^T)}{\lambda_{\min}(AA^T)}$$

the condition number of a square matrix A [76].

Observe that

$$c(P(t)) = c(S(t)S^T(t)) = c^2(S(t)),$$

i.e. the condition number of $S(t)$ is the square root of $c(P(t))$. Propagation of $S(t)$ does *not* prevent the corresponding EWLS algorithm from exhibiting the covariance blowup. However, while numerical operations with $P(t)$ may encounter difficulties when $c(P(t)) = 10^k$, the square root filter should function until $c(P(t)) = 10^{2k}$. This effective double precision of the square root filter allows it to cope better (and longer) with numerical problems caused by a temporary lack of persistence of excitation.

The matrix square roots are not uniquely determined – infinitely many matrices $S(t)$ exist which obey (8.10). Uniqueness can be enforced by insisting that the square root factors be symmetric or triangular (with positive diagonal elements). However, even in this case, square roots can be computed in many different ways. This explains the existence of a large number of different square root algorithms described in the literature [75].

One of the first square root algorithms, known to have good numerical properties, was proposed by Potter [17]. The derivation of Potter's algorithm is very simple. Using (8.10) the covariance update recursion of (4.8) can be rewritten in the form

$$S(t)S^T(t) = \frac{S(t-1)}{\sqrt{\lambda}} \left[I_n - \frac{r(t)r^T(t)}{\beta(t)}\right] \frac{S^T(t-1)}{\sqrt{\lambda}},$$

where

$$
\begin{aligned}
r(t) &= S^T(t-1)\varphi(t), \\
\beta(t) &= \lambda + r^T(t)r(t).
\end{aligned}
$$

It is straightforward to check that

$$\left[I_n - \frac{r(t)r^T(t)}{\beta(t)}\right] = \left[I_n - \frac{r(t)r^T(t)}{\alpha(t)}\right]\left[I_n - \frac{r(t)r^T(t)}{\alpha(t)}\right],$$

where

$$\alpha(t) = \beta(t) + \sqrt{\lambda\beta(t)}.$$

Hence the square root $S(t)$ can be updated using the following recursion:

$$S(t) = \frac{S(t-1)}{\sqrt{\lambda}}\left[I_n - \frac{r(t)r^T(t)}{\alpha(t)}\right].$$

Finally, note that the gain vector in (4.8) can be rewritten in the form

$$K(t) = \frac{P(t-1)\varphi(t)}{\lambda + \varphi^T(t)P(t-1)\varphi(t)} = \frac{S(t-1)r(t)}{\beta(t)}.$$

The Potter's square root version of the EWLS algorithm can be summarized as follows:

$$\begin{aligned}
\widehat{\theta}(t) &= \widehat{\theta}(t-1) + K(t)\epsilon(t), \\
\epsilon(t) &= y(t) - \varphi^T(t)\widehat{\theta}(t-1), \\
r(t) &= S^T(t-1)\varphi(t), \\
\beta(t) &= \lambda + r^T(t)r(t), \\
\alpha(t) &= \beta(t) + \sqrt{\lambda\beta(t)}, \\
K(t) &= \frac{S(t-1)r(t)}{\beta(t)}, \\
S(t) &= \frac{1}{\sqrt{\lambda}}\left[S(t-1) - \frac{\beta(t)}{\alpha(t)}K(t)r^T(t)\right].
\end{aligned} \tag{8.11}$$

To put the problem of square root filtering in a more general perspective, denote by T any $k \times k$ unitary matrix which transforms the $k \times l$ ($l \geq k$) matrix A, consisting of four blocks $A_{11}, A_{12}, A_{21}, A_{22}$, into a lower triangular (in a block sense) matrix B:

$$\left[\begin{array}{cc} A_{11} & A_{12} \\ A_{21} & A_{22} \end{array}\right] T = \left[\begin{array}{cc} B_{11} & 0 \\ B_{21} & B_{22} \end{array}\right].$$

The triangularization task can be achieved in many different ways, e.g. by using a sequence of elementary transformations such as rotations (circular or hyperbolic) [169], Householder transformations [86] or modified Gram–Schmidt operations [86].

Sayed and Kailath [168] observed that the square root of $P(t)$ (denoted by $S(t)$), the scaled gain vector ($\sqrt{\beta(t)/\lambda}K(t)$) and the scaling coefficient ($\sqrt{\beta(t)/\lambda}$) can be easily obtained by triangularization of the following $(n+1) \times (n+1)$ matrix:

$$\left[\begin{array}{cc} 1 & \frac{1}{\sqrt{\lambda}}\varphi^T(t)S(t-1) \\ 0 & \frac{1}{\sqrt{\lambda}}S(t-1) \end{array}\right], \tag{8.12}$$

namely

$$\left[\begin{array}{cc} 1 & \frac{1}{\sqrt{\lambda}}\varphi^T(t)S(t-1) \\ 0 & \frac{1}{\sqrt{\lambda}}S(t-1) \end{array} \right] T = \left[\begin{array}{cc} \sqrt{\frac{\beta(t)}{\lambda}} & 0^T \\ \sqrt{\frac{\beta(t)}{\lambda}}K(t) & S(t) \end{array} \right], \tag{8.13}$$

resulting in the following parameter update:

$$\widehat{\theta}(t) = \widehat{\theta}(t-1) + \left(\sqrt{\frac{\beta(t)}{\lambda}}K(t) \right) \left(\frac{\epsilon(t)}{\sqrt{\beta(t)/\lambda}} \right). \tag{8.14}$$

The verification of (8.13) is straightforward: 'square' both sides of (8.13) by multiplying the corresponding matrices with their transposes and note that

$$TT^T = I_{n+1};$$

the resulting set of equations is identical with the original EWLS recursions (4.8).

A very interesting class of square root algorithms can be obtained by applying circular plane (Givens) rotations. Denote by $T_{ij}(A) = [t_{ij}], j > i$, the $k \times k$ circular plane rotation matrix which can be used to *annihilate* (reduce to zero) the (i,j) element of an arbitrary $k \times l$, $l \geq k$, matrix $A = [a_{ij}]$:

$$T_{ij}(A) = \left[\begin{array}{ccccccc} 1 & & & & & & \\ & \ddots & & & & & \\ & & c & & -s & & \\ & & & \ddots & & & \\ & & & & 1 & & \\ & & s & & c & & \\ & & & & & \ddots & \\ & & & & & & 1 \end{array} \right]$$

where $t_{ii} = t_{jj} = c$, $t_{ji} = -t_{ij} = s$ and

$$c = \cos\alpha = \frac{1}{\sqrt{1+\rho^2}},$$

$$s = \sin\alpha = \frac{\rho}{\sqrt{1+\rho^2}},$$

$$\alpha = \text{arctg}\rho, \quad \rho = \frac{a_{ii}}{a_{jj}}.$$

It is straightforward to check that

$$[AT_{ij}]_{ij} = 0$$

and that

$$T_{ij}T_{ij}^T = I_k,$$

i.e. the rotation matrix is orthogonal.

Note that

1. Since T_{ij} differs from the identity matrix only in its $(i, i), (i, j), (j, i)$ and (j, j) components, only two columns of A (i and j) are modified when rotation is executed.
2. If the superdiagonal elements of the ith column of A are zero ($a_{1i} = \ldots = a_{i-1,i} = 0$), rotation preserves all zero elements of the jth column located above a_{ij} ($a_{1j}, \ldots, a_{i-1,j}$).

One can use both properties for sequential triangularization purposes, selecting the rotation matrix in the form (row sweep)

$$T = \prod_{i=1}^{k} \left(\prod_{j=i+1}^{l} T_{ij} \right) = T_r, \qquad (8.15)$$

or in the form (column sweep)

$$T = \prod_{j=2}^{l} \left(\prod_{i=1}^{\min(j-1,k)} T_{ij} \right) = T_c. \qquad (8.16)$$

Such transforms are equivalent to performing a sequence of Givens rotations, each of which annihilates a particular element of the upper triangle of the transformed matrix. Using this approach one gets (schematically)

$$\begin{bmatrix} x & x & x & x \\ x & x & x & x \\ x & x & x & x \end{bmatrix} T = \begin{bmatrix} x & 0 & 0 & 0 \\ x & x & 0 & 0 \\ x & x & x & 0 \end{bmatrix}.$$

In the case of 3×4 matrices

$$T_r = T_{12}T_{13}T_{14}T_{23}T_{24}T_{34}, \quad T_c = T_{12}T_{13}T_{23}T_{14}T_{24}T_{34}.$$

Remark

Caution is needed when interpreting (8.15) and (8.16). Strictly speaking, the operation

$$AT_{i_1 j_1} T_{i_2 j_2} \ldots T_{i_m j_m} = B$$

should be written in the form

$$\begin{aligned} B_1 &= AT_{i_1 j_1}(A), \\ B_2 &= B_1 T_{i_2 j_2}(B_1), \\ &\vdots \\ B_m &= B_{m-1} T_{i_m j_m}(B_{m-1}), \\ B &= B_m, \end{aligned}$$

i.e. the sine and cosine parameters of the orthogonal matrix $T_{i_l j_l}$, $1 \le l \le m$, cannot be evaluated before the first $l - 1$ transformations are actually performed. ∎

The block triangularization required by (8.13) can be obtained by applying a sequence of n Givens rotations (in any desired order), each of which annihilates a particular element of the first row of (8.12):

$$T = \prod_{1 < i \leq n+1} T_{1i}.$$

If the full triangularization of (8.12) (instead of the block triangularization) is performed, the square root $S(t)$ is a lower triangular matrix, i.e. it is a Cholesky factor of $P(t)$. The corresponding procedure is known as the inverse QR decomposition based RLS algorithm, or simply an inverse QR-RLS algorithm (the term 'QR decomposition' was coined in matrix algebra for a process of orthogonal triangularization and the word 'inverse' indicates the fact that factorization is applied to $P(t)$, i.e. to the inverse of the regression matrix $R(t)$).

The inverse QR-RLS algorithm has very interesting properties. Due to its highly modular and highly parallel structure, it lends itself to implementation in the form of a systolic array, i.e. an array of individual (but repeatable) processing cells arranged in a regular pattern [4], namely

1. A triangular array which performs (8.13)
2. A linear array which performs (8.14)

Systolic arrays can be used for fast hardware implementations of square root algorithms.

Dead zone or stopping

Paradoxically, identifiability problems occur in many adaptive systems when the system performance becomes too good, i.e. when further identification of the process dynamics is (at least temporarily) not needed. As already argued, in adaptive control systems this may happen when the measurement noise vanishes. It is interesting to note that the same problem may arise in open-loop systems. Consider the behavior of an adaptive predictor

$$\hat{y}(t+1|t) = \sum_{i=1}^{n_A} \hat{a}_i(t) y(t+1-i) \qquad (8.17)$$

in the case where the analyzed process is a mixture of a sinusoidal signal and white measurement noise $v(t)$:

$$y(t) = \sin \omega t + v(t).$$

As long as the variance of $v(t)$ is nonzero, all elements of the regression vector associated with (8.17)

$$\varphi(t) = [y(t-1), \ldots, y(t-n_A)]^T$$

are linearly independent (for arbitrarily large n_A) which guarantees system identifiability. If, however, the measurement noise is temporarily absent ($v(t) = 0$, $\forall t \geq t_o$) and the number of estimated coefficients is larger than two ($n_A > 2$), the components of $\varphi(t)$ become linearly dependent (see Example 3.2) and this inevitably leads to numerical problems.

In both cases discussed above, identifiability problems arise when

$$\epsilon(t) \xrightarrow[t \to \infty]{} 0,$$

so the magnitude of the prediction errors can serve as a simple measure for the richness of system excitation. When prediction errors fall beyond a certain threshold δ, one might simply stop the process of parameter estimation, preventing the covariance matrix from blowing up. The corresponding estimation algorithm can be written as follows:

$$\epsilon(t) = y(t) - \varphi^T(t)\widehat{\theta}(t-1),$$

$$\text{if } |\epsilon(t)| > \delta > 0:$$

$$
\begin{aligned}
\widehat{\theta}(t) &= \widehat{\theta}(t-1) + K(t)\epsilon(t), \\
K(t) &= \frac{P(t-1)\varphi(t)}{\lambda + \varphi^T(t)P(t-1)\varphi(t)}, \\
P(t) &= \frac{1}{\lambda}\left[P(t-1) - \frac{P(t-1)\varphi(t)\varphi^T(t)P(t-1)}{\lambda + \varphi^T(t)P(t-1)\varphi(t)} \right],
\end{aligned}
\qquad (8.18)
$$

$$\text{if } |\epsilon(t)| \leq \delta > 0:$$

$$
\begin{aligned}
\widehat{\theta}(t) &= \widehat{\theta}(t-1), \\
P(t) &= P(t-1).
\end{aligned}
$$

It can be readily extended to all kinds of LS estimators.

From a practical viewpoint it is very important to choose an appropriate value of δ, otherwise the covariance blowup may take place before the prediction error has permanently contracted to within the dead zone set by δ. If the variance σ_v^2 of the 'nominal' measurement noise is known a priori, one can set $\delta = \eta\sigma_v$ where $\eta \in [1,2]$.

To increase numerical safety, the prediction error test is often augmented with some additional identifiability checks. For example, in the commercially available adaptive regulators (Novatune from Asea Brown Boveri and Firstloop from First Control Systems) parameter estimation is suspended when the changes in the control signal and the process output are less than their prescribed values – see Åström and Wittenmark [11].

Variable forgetting factors

So far we have considered constant-memory WLS filters only. An obvious way of preserving identifiability conditions is via appropriate memory scheduling. When the amount of incoming information drops, the process of discarding 'old' measurements should be limited and eventually stopped; this will prevent the regression matrix from getting singular. In other words, the memory of an adaptive filter should be inversely proportional to the average information content of the incoming data.

In the exponentially weighted least squares framework, the variable-memory scheme can easily be realized by replacing the time-invariant forgetting constant λ with a time-varying forgetting factor $\lambda(t)$. This results in

$$
\begin{aligned}
\widehat{\theta}(t) &= R^{-1}(t)s(t), \\
R(t) &= \lambda(t)R(t-1) + \varphi(t)\varphi^T(t), \\
s(t) &= \lambda(t)s(t-1) + y(t)\varphi(t),
\end{aligned}
\tag{8.19}
$$

or equivalently $(P(t) = R^{-1}(t))$

$$
\begin{aligned}
\widehat{\theta}(t) &= \widehat{\theta}(t-1) + K(t)\epsilon(t), \\
\epsilon(t) &= y(t) - \varphi^T(t)\widehat{\theta}(t-1), \\
K(t) &= \frac{P(t-1)\varphi(t)}{\lambda(t) + \varphi^T(t)P(t-1)\varphi(t)}, \\
P(t) &= \frac{1}{\lambda(t)}\left[P(t-1) - \frac{P(t-1)\varphi(t)\varphi^T(t)P(t-1)}{\lambda(t) + \varphi^T(t)P(t-1)\varphi(t)}\right].
\end{aligned}
\tag{8.20}
$$

It is easy to show that the recursive estimation formulas given above can be written in the following explicit form:

$$
\widehat{\theta}(t) = \arg\min_{\theta} \sum_{i=0}^{t-1} w_t(i)\left(y(t-i) - \varphi^T(t-i)\theta\right)^2
$$

$$
= \left(\sum_{i=0}^{t-1} w_t(i)\varphi(t-i)\varphi^T(t-i)\right)^{-1} \left(\sum_{i=0}^{t-1} w_t(i)y(t-i)\varphi(t-i)\right),
\tag{8.21}
$$

where

$$
w_t(0) = 1,
$$

$$
w_t(i) = \prod_{j=0}^{i-1} \lambda(t-j), \quad 0 < i \le t-1.
$$

Since

$$
w_t(i) = \lambda(t)w_{t-1}(i-1), \quad i = 1.\dots,t,
$$

the effective memory of the EWLS algorithm with variable forgetting factor

$$
k_t = \sum_{i=0}^{t-1} w_t(i)
$$

can be evaluated recursively:

$$
k_t = \lambda(t)k_{t-1} + 1.
\tag{8.22}
$$

Unlike the case of a constant forgetting rate, note that algorithm (8.20), which is a generalized version of (4.8), is *not identical* with the following algorithm:

$$
\begin{aligned}
\widehat{\theta}(t) &= \widehat{\theta}(t-1) + K(t)\epsilon(t), \\
\epsilon(t) &= y(t) - \varphi^T(t)\widehat{\theta}(t-1), \\
K(t) &= \frac{\widetilde{P}(t-1)\varphi(t)}{1 + \varphi^T(t)\widetilde{P}(t-1)\varphi(t)}, \\
\widetilde{P}(t) &= \frac{1}{\lambda(t)}\left[\widetilde{P}(t-1) - \frac{\widetilde{P}(t-1)\varphi(t)\varphi^T(t)\widetilde{P}(t-1)}{1 + \varphi^T(t)\widetilde{P}(t-1)\varphi(t)}\right],
\end{aligned}
\tag{8.23}
$$

which is a generalized version of (4.9). This means that the parameter estimate obtained from (8.23) is not an exact solution to the least squares problem (8.21).

Fortescue et al. [50] suggested that $\lambda(t)$ can be chosen so as to stabilize the weighted sum of residual errors

$$
q(t) = \sum_{i=0}^{t-1} w_t(i)\left(y(t-i) - \varphi^T(t-i)\widehat{\theta}(t)\right)^2,
\tag{8.24}
$$

where $\widehat{\theta}(t)$ is given by (8.21).

Note that for a time-invariant system $(\theta(t) = \theta)$ and fixed-rate forgetting $(\lambda(t) = \lambda_o, \ 1 - \lambda_o \ll 1)$ it holds that $\widehat{\theta}(t) \cong \theta$, hence

$$
y(t-i) - \varphi^T(t-i)\widehat{\theta}(t) \cong v(t-i),
$$

leading to

$$
\lim_{t\to\infty} \mathrm{E}[q(t)] \cong k_o\sigma_v^2 = q_o,
$$

where

$$
k_o = \sum_{i=0}^{\infty} \lambda_o^i = \frac{1}{1-\lambda_o}
$$

denotes the effective memory of the corresponding filter. This means that if $q(t)$ is stabilized at q_o and the intensity of the measurement noise is constant, the corresponding values of $\lambda(t)$ will fluctuate around

$$
\lambda_o = 1 - \frac{1}{k_o}.
$$

If, however, the noise activity drops below its nominal level, the effective memory of the filter will be increased above the nominal value of k_o. In particular, in the noise-free case $(\mathrm{E}[v^2(t)] = 0)$ discussed before, the forgetting constant $\lambda(t)$ will tend to 1, i.e. the process of forgetting old data will eventually be stopped.

To solve the stabilization problem, we will refer to the recursive formula describing the evolution of $q(t)$:

$$
\begin{aligned}
q(t) &= \lambda(t)q(t-1) + \left[1 - \varphi^T(t)P(t)\varphi(t)\right]\epsilon^2(t) \\
&= \lambda(t)q(t-1) + \left[1 - K^T(t)\varphi(t)\right]\epsilon^2(t) \\
&= \lambda(t)q(t-1) + \frac{\lambda(t)\epsilon^2(t)}{\lambda(t) + \varphi^T(t)P(t-1)\varphi(t)},
\end{aligned}
\tag{8.25}
$$

which can be derived in an analogous way to (3.20). Setting $q(t) = q(t-1) = q_o$ and

$$\gamma(t) = \varphi^T(t)P(t-1)\varphi(t),$$

one obtains from (8.25) the following equation:

$$q_o = \lambda(t)q_o + \frac{\lambda(t)\epsilon^2(t)}{\lambda(t) + \gamma(t)}, \qquad (8.26)$$

which can easily be solved for $\lambda(t)$:

$$\lambda(t) = \frac{\delta(t) + \sqrt{\delta^2(t) + 4\gamma(t)}}{2} \qquad (8.27)$$

where

$$\delta(t) = 1 - \gamma(t) - \frac{\epsilon^2(t)}{q_o}.$$

As suggested by Ydstie [199], [200] the exact solution (8.27) can be replaced with the following approximation:

$$\lambda(t) = \frac{1}{1 + \epsilon^2(t)/[(1+\gamma(t))q_o]}, \qquad (8.28)$$

which can be obtained from (8.26) after setting $\lambda(t) + \gamma(t) \cong 1 + \gamma(t)$. Approximation (8.28) yields slightly larger values of $\lambda(t)$ than (8.27).

The forgetting factor $\lambda(t)$ obtained from (8.27) or (8.28) can take any value between 0 and 1. In practice it may be dangerous to adopt values too far from one. Suppose that $k_{t-1} = 100$. According to (8.22), if λ is set to 0.5 at the instant t, the memory of the EWLS algorithm will be halved ($k_t = 51$). However, even if forgetting is switched off ($\lambda = 1$) at instants $t+1$, $t+2,\ldots$ it will take at least 49 time steps to restore the initial value of the effective memory. This clearly demonstrates that if no lower threshold λ_{min} for $\lambda(t)$ were set, the variable forgetting factor EWLS algorithm would be extremely sensitive to impulsive disturbances.

Here is a summary of the variable forgetting factor (VFF) algorithm with the safety jacket mentioned above:

$$\begin{aligned}
\widehat{\theta}(t) &= \widehat{\theta}(t-1) + K(t)\epsilon(t), \\
\epsilon(t) &= y(t) - \varphi^T(t)\widehat{\theta}(t-1), \\
\gamma(t) &= \varphi^T(t)P(t-1)\varphi(t), \\
\lambda(t) &= \max\left\{\lambda_{min}, \frac{1}{1 + \epsilon^2(t)/[(1+\gamma(t))q_o]}\right\}, \\
K(t) &= \frac{P(t-1)\varphi(t)}{\lambda(t) + \gamma(t)}, \\
P(t) &= \frac{1}{\lambda(t)}\left[I_n - K(t)\varphi^T(t)\right]P(t-1).
\end{aligned} \qquad (8.29)$$

The value of λ_{min} is usually set to 0.8.

The original VFF algorithm, presented in the frequently cited paper of Fortescue et al. [50] has a slightly different structure than (8.29). The stabilized quantity is computed recursively from

$$
\begin{aligned}
\tilde{q}(t) &= \lambda(t)\tilde{q}(t-1) + \left[1 - \tilde{K}^T(t)\varphi(t)\right]\epsilon^2(t) \\
&= \lambda(t)\tilde{q}(t-1) + \frac{\lambda(t)\epsilon^2(t)}{\lambda(t) + \varphi^T(t)\tilde{P}(t-1)\varphi(t)},
\end{aligned} \tag{8.30}
$$

where $\tilde{K}(t)$ and $\tilde{P}(t-1)$ are generated by the modified EWLS algorithm (8.23). Setting $\tilde{q}(t) = \tilde{q}(t-1) = q_o$ and $\tilde{\gamma}(t) = \varphi^T(t)\tilde{P}(t-1)\varphi(t)$ in (8.30) one obtains the formula

$$
\lambda(t) = 1 - \frac{\epsilon^2(t)}{(1+\tilde{\gamma}(t))q_o}, \tag{8.31}
$$

which for small values of $\epsilon(t)$ resembles (8.28). Note that for small values of ϵ it is possible to use the following approximation $1/(1+\epsilon) \cong 1 - \epsilon$.

The corresponding VFF algorithm is as follows

$$
\begin{aligned}
\hat{\theta}(t) &= \hat{\theta}(t-1) + \tilde{K}(t)\epsilon(t), \\
\epsilon(t) &= y(t) - \varphi^T(t)\hat{\theta}(t-1), \\
\tilde{\gamma}(t) &= \varphi^T(t)\tilde{P}(t-1)\varphi(t), \\
\tilde{K}(t) &= \frac{\tilde{P}(t-1)\varphi(t)}{\lambda(t) + \tilde{\gamma}(t)}, \\
\lambda(t) &= \max\left\{\lambda_{min}, 1 - \frac{\epsilon^2(t)}{(1+\tilde{\gamma}(t))q_o}\right\}, \\
\tilde{P}(t) &= \frac{1}{\lambda(t)}\left[I_n - \tilde{K}(t)\varphi^T(t)\right]\tilde{P}(t-1).
\end{aligned} \tag{8.32}
$$

The theoretical justification of (8.31) is less convincing than the theoretical justification of (8.27) or (8.28). When λ varies with time, $\tilde{q}(t)$ is not identical with the exponentially weighted residual sum of squares $q(t)$, although it behaves in a similar way. Furthermore, since the forgetting constant $\lambda(t)$ is evaluated when the gain vector $\tilde{K}(t)$ is already set, for large values of $\epsilon(t)$ the formula (8.31) may yield negative values of $\lambda(t)$. Despite the differences mentioned above, algorithm (8.32) behaves in a very similar way to algorithm (8.29), i.e. it increases the memory length of the EWLS filter if prediction errors are consistently smaller than expected.

Both these VFF algorithms are heuristic and neither guarantees complete anti-windup protection; note that if the components of $\varphi(t)$ are nonzero but linearly dependent then neither (8.28) nor (8.31) will force $\lambda(t)$ to converge to 1.

Remark 1

To obtain smooth variations of $\lambda(t)$ one can replace $\epsilon^2(t)$ in (8.29) or (8.32) with the average value of the recently observed squared prediction errors [23]:

$$
\frac{1}{M}\sum_{i=0}^{M-1}\epsilon^2(t-i).
$$

Remark 2

Since the maximum eigenvalue of $\widetilde{P}(t)$ obeys the inequality

$$\lambda_{\max}(\widetilde{P}) \leq \sum_{i=1}^{n} \lambda_i(\widetilde{P}) = \text{tr}\{\widetilde{P}(t)\},$$

an easy way of keeping control over the size of the matrix $\widetilde{P}(t)$ is by stabilizing its trace. Note that, according to (8.23)

$$\text{tr}\{\widetilde{P}(t)\} = \frac{1}{\lambda(t)} \left[\text{tr}\{\widetilde{P}(t-1)\} - \frac{\|\widetilde{\eta}(t)\|^2}{1+\widetilde{\gamma}(t)} \right], \tag{8.33}$$

where

$$\widetilde{\eta}(t) = \widetilde{P}(t-1)\varphi(t).$$

Hence, after setting $\text{tr}\{\widetilde{P}(t)\} = \text{tr}\{\widetilde{P}(t-1)\} = c_o$ in (8.33), one obtains

$$\lambda(t) = 1 - \frac{\|\widetilde{\eta}(t)\|^2}{(1+\widetilde{\gamma}(t))c_o} . \tag{8.34}$$

The EWLS algorithm equipped with (8.34) is often known as a constant trace EWLS filter. In principle the same technique can be used for stabilization of $P(t)$ in (8.20) but the corresponding expressions for $\lambda(t)$ are more complicated than (8.34), so they are seldom used in practice.

Regularization

A number of ad hoc modifications of the RLS covariance update were proposed to keep control over the size of the matrix $P(t)$. This technique is usually called covariance regularization. One of the simplest regularization algorithms is based on a trace normalization [62]:

$$
\begin{aligned}
\widehat{\theta}(t) &= \widehat{\theta}(t-1) + K(t)\varphi(t)\epsilon(t), \\
\epsilon(t) &= y(t) - \varphi^T(t)\widehat{\theta}(t-1), \\
K(t) &= \alpha\frac{\overline{P}(t-1)\varphi(t)}{1+\varphi^T(t)\overline{P}(t-1)\varphi(t)}, \\
P(t) &= \overline{P}(t-1) - \alpha\frac{\overline{P}(t-1)\varphi(t)\varphi^T(t)\overline{P}(t-1)}{1+\varphi^T(t)\overline{P}(t-1)\varphi(t)}, \\
\overline{P}(t) &= c_o\frac{P(t)}{\text{tr}\{P(t)\}},
\end{aligned}
\tag{8.35}
$$

where α is a positive gain coefficient, typically $\alpha \in [0.1, 0.5]$ and $c_o = \text{tr}\{\overline{P}(t))\}$ is the desired value of the trace.

A more sophisticated regularization, combining the elements of exponential data forgetting and exponential covariance resetting, was proposed by Salgado et al. [167]:

$$\widehat{\theta}(t) = \widehat{\theta}(t-1) + K(t)\varphi(t)\epsilon(t),$$

$$\epsilon(t) = y(t) - \varphi^T(t)\widehat{\theta}(t-1),$$

$$K(t) = \alpha\frac{P(t-1)\varphi(t)}{1 + \varphi^T(t)P(t-1)\varphi(t)},$$

$$P(t) = \frac{1}{\lambda}P(t-1) - \alpha\frac{P(t-1)\varphi(t)\varphi^T(t)P(t-1)}{1 + \varphi^T(t)P(t-1)\varphi(t)} + \beta I_n - \delta P^2(t-1). \quad (8.36)$$

The general guidelines for choosing the positive constants α, β, λ and δ are as follows:

1. α adjusts the gain of the least squares algorithm; typically $\alpha \in [0.1, 0.5]$.
2. β is a small constant related to the minimum eigenvalue of $P(t)$; typically $\beta \in [0, 0.01]$.
3. λ is the exponential forgetting factor; typically $\lambda \in [0.9, 0.99]$.
4. δ is a small constant related to the maximum eigenvalue of $P(t)$; typically $\delta \in [0, 0.01]$.

It can be shown [167] that if the above algorithm is used and if the constants α, β, λ and δ obey some extra (easy to fulfill) technical conditions, both the covariance matrix $P(t)$ and its inverse $P^{-1}(t)$ are bounded from above.

Directional forgetting

Directional forgetting is another interesting concept; it allows one to substantially reduce the risk of encountering the estimation windup problems.

With conventional exponential forgetting

$$P^{-1}(t) = \lambda P^{-1}(t-1) + \varphi(t)\varphi^T(t),$$

which means in geometrical terms that the matrix $P^{-1}(t-1)$ is deflated in all directions of the regression space and inflated in a direction indicated by $\varphi(t)$. Hence, if $\{\varphi(t)\}$ is not persistently exciting, the size of $P^{-1}(t)$ (measured in terms of the diameters of the ellipsoid of concentration $x^T P^{-1}(t)x = $ const) systematically diminishes in the directions that are not excited, resulting in a blowup of $P(t)$.

The idea behind directional forgetting is to deflate $P^{-1}(t-1)$ in the direction of the incoming information only. If no forgetting is applied, the regression matrix can be written in the form

$$P^{-1}(t) = P^{-1}(t-1) + \varphi(t)\varphi^T(t)$$

$$= \left[P^{-1}(t-1) - \frac{1}{\gamma(t)}\varphi(t)\varphi^T(t)\right] + \left[\left(1 + \frac{1}{\gamma(t)}\right)\varphi(t)\varphi^T(t)\right], \quad (8.37)$$

where

$$\gamma(t) = \varphi^T(t)P(t-1)\varphi(t).$$

Using the identity

$$\det(A + \mu xx^T) = (1 + \mu x^T A^{-1}x)\det(A),$$

valid for any nonsingular matrix A, any $n \times 1$ vector x and any scalar μ, one arrives at

$$\det \left[P^{-1}(t-1) - \frac{1}{\gamma(t)} \varphi(t) \varphi^T(t) \right] = 0,$$

which means that if the matrix $P(t-1)$ is positive definite, the first matrix on the right-hand side of (8.37) has rank $(n-1)$. If $\| \varphi(t) \| \neq 0$ then the second matrix on the right-hand side of (8.37) is a dyad of rank 1.

The dyadic decomposition (8.37) identifies the part of $P^{-1}(t)$ which can be solely attributed to excitation acting in the direction of $\varphi(t)$. To forget information only in this particular direction, one can replace (8.37) with

$$P^{-1}(t) = \left[P^{-1}(t-1) - \frac{1}{\gamma(t)} \varphi(t) \varphi^T(t) \right] + \lambda \left[\left(1 + \frac{1}{\gamma(t)} \right) \varphi(t) \varphi^T(t) \right]$$

$$= P^{-1}(t-1) + \frac{\varphi(t) \varphi^T(t)}{\delta(t)} , \tag{8.38}$$

where

$$\delta(t) = \frac{\gamma(t)}{\lambda \gamma(t) + \lambda - 1} .$$

According to (3.10)

$$P(t) = P(t-1) - \frac{P(t-1) \varphi(t) \varphi^T(t) P(t-1)}{\delta(t) + \varphi^T(t) P(t-1) \varphi(t)} ,$$

provided that $\gamma(t) \neq 0$, i.e. $\| \varphi(t) \| \neq 0$. The constant λ in (8.38) can be interpreted as the exponential forgetting rate along the direction of the incoming information. Note that the weighting coefficient $\delta(t)$ is data dependent and that its range is not restricted to positive numbers.

A complete directional forgetting (DF) algorithm, proposed by Kulhavy and Karny [97], can be summarized as follows:

$$\text{if } \| \varphi(t) \| \neq 0:$$

$$\begin{aligned}
\widehat{\theta}(t) &= \widehat{\theta}(t-1) + K(t) \epsilon(t), \\
\epsilon(t) &= y(t) - \varphi^T(t) \widehat{\theta}(t-1), \\
\gamma(t) &= \varphi^T(t) P(t-1) \varphi(t), \\
\delta(t) &= \frac{\gamma(t)}{\lambda \gamma(t) + \lambda - 1} , \\
K(t) &= \frac{P(t-1) \varphi(t)}{1 + \gamma(t)} , \\
P(t) &= P(t-1) - \frac{P(t-1) \varphi(t) \varphi^T(t) P(t-1)}{\delta(t) + \gamma(t)} ,
\end{aligned} \tag{8.39}$$

$$\text{if } \| \varphi(t) \| = 0:$$

$$\begin{aligned}
\widehat{\theta}(t) &= \widehat{\theta}(t-1), \\
P(t) &= P(t-1).
\end{aligned}$$

8.1.2   Gradient algorithms

In LS adaptive filters, numerical problems are caused by singularity or near-singularity of the inverted regression matrix. Since the classic LMS filter is division-free it is certainly not endangered by the blowup phenomenon. It does not mean, however, that the lack of an appropriate excitation does not negatively affect the gradient-type estimation algorithms. If the sequence $\{\varphi(t)\}$ is not persistently exciting, all the gradient filters in Chapter 5 suffer from a negative phenomenon known as parameter drift. After a long operation under poor excitation conditions, parameter estimates can attain arbitrarily large values despite the boundedness of the process (input/output) and noise variables. The origin of an unbounded drift of parameter estimates is less explicit than for the LS filters. Consider the error equation for a time-invariant system:

$$\tilde{\theta}(t) = \left(I_n - \mu\varphi(t)\varphi^T(t)\right)\tilde{\theta}(t-1) - \mu\varphi(t)v(t). \tag{8.40}$$

Note that in the case of no excitation (8.1) one obtains from (8.40)

$$\tilde{\theta}(t) = \tilde{\theta}(t_o), \quad \forall t > t_o,$$

or equivalently

$$\hat{\theta}(t) = \hat{\theta}(t_o), \quad \forall t > t_o,$$

which means that the LMS filter should operate safely under such trivially defective experimental conditions. The same conclusion can be reached if a more general nonidentifiability constraint (8.4) is imposed instead of (8.1). Can anything go wrong under these circumstances?

The key observation is that when $\{\varphi(t)\}$ is not persistently exciting, the stationary point $\tilde{\theta} = 0$ of the error equation (8.40) is not exponentially stable but only marginally stable. It is well known that marginally stable systems are not robust to perturbations; they tend to integrate errors caused by various departures from the ideal scheme, e.g. departures caused by numerical errors. Following this line, Weiss and Mitra [191] have shown that if the ideal LMS computational scheme

$$\begin{aligned}
\hat{\theta}(t) &= \hat{\theta}(t-1) + \mu\varphi(t)\epsilon(t), \\
\epsilon(t) &= y(t) - \varphi^T(t)\hat{\theta}(t-1)
\end{aligned}$$

is replaced with the perturbed scheme

$$\begin{aligned}
\hat{\theta}(t) &= \hat{\theta}(t-1) + \mu[\varphi(t) + \eta_1(t)]\epsilon(t), \\
\epsilon(t) &= y(t) - \varphi^T(t)[\hat{\theta}(t-1) + \eta_2(t)]
\end{aligned}$$

where $\{\eta_1(t)\}$ and $\{\eta_2(t)\}$ denote bounded roundoff and/or quantization errors, the lack of persistence of excitation results in unbounded drift of parameter estimates. The work of Weiss and Mitra is an elegant theoretical explanation of instability observed earlier in some hardware implementations of LMS filters [43].

Summing up, the instability mechanism observed for gradient filters in the case of insufficient excitation is more tricky than the mechanism for LS filters. With LS filters, accumulation of noninformative data leads directly to numerical problems, whether the computations are performed with finite or infinite accuracy. With gradient filters it manifests itself in the lack of numerical robustness of the recursive estimation algorithm. It should be stressed that even though less spectacular than the covariance blowup, the parameter drift can be equally harmful to an adaptive system. In particular, if the estimation algorithm is part of a closed-loop adaptive system, drift in the parameter estimates due to slight nonidealities (e.g. roundoff noise or unmodeled dynamics) in the absence of persistence of excitation can ultimately destabilize the system and cause its output to *burst*. Thorough discussions of this bursting phenomenon have been given by Anderson [6] for adaptive control and by Macchi [117] for adaptive equalization.

Dead zone or stopping

A simple way of eliminating, or at least reducing, the drift phenomenon is by stopping the estimation when the prediction error is adequately small:

$$\epsilon(t) = y(t) - \varphi^T(t)\widehat{\theta}(t-1),$$

$$\text{if } |\epsilon(t)| > \delta > 0:$$

$$\widehat{\theta}(t) = \widehat{\theta}(t-1) + \mu\varphi(t)\epsilon(t), \tag{8.41}$$

$$\text{if } |\epsilon(t)| \leq \delta:$$

$$\widehat{\theta}(t) = \widehat{\theta}(t-1).$$

Note that (8.41) can be rewritten in a more compact form as

$$\widehat{\theta}(t) = \widehat{\theta}(t-1) + \mu\varphi(t)D[\epsilon(t)],$$

where $D[\cdot]$ is the dead-zone transformation

$$D[\epsilon(t)] = \begin{cases} 0 & \text{if } |\epsilon(t)| \leq \delta \\ \epsilon(t) & \text{if } |\epsilon(t)| > \delta \end{cases}$$

For less abrupt transitions when crossing the dead-zone boundary [157] one can use

$$D[\epsilon(t)] = \begin{cases} \epsilon(t) - \delta & \text{if } \epsilon(t) > \delta \\ 0 & \text{if } |\epsilon(t)| \leq \delta \\ \epsilon(t) + \delta & \text{if } \epsilon(t) < -\delta \end{cases}$$

As in the case of the LS estimators, the dead-zone threshold δ can be expressed in terms of the expected measurement noise variance.

Leakage

A simple but conceptually appealing technique of drift elimination is known as *leakage* [37]. Suppose that the instantaneous quality measure (5.3), which was a basis for deriving the LMS algorithm, is replaced with the following modified measure:

$$J_t(\theta) = v^2(t,\theta) + \alpha \, \|\theta\|^2, \tag{8.42}$$

where α is a small positive constant. Since the modified measure explicitly penalizes large excursions of $\|\theta\|$ from zero, its minimization should yield a drift-free gradient algorithm.

Substituting (8.42) into (5.2) one obtains the so-called leaky LMS algorithm

$$\widehat{\theta}(t) = (1 - \beta)\widehat{\theta}(t - 1) + \mu\varphi(t)\epsilon(t), \tag{8.43}$$

where $\beta = \alpha\mu$. The value of β should be chosen so as to fulfill

$$0 < \beta < 1. \tag{8.44}$$

In order to better understand the behavior of leaky LMS, consider the error equation associated with (8.43). For a time-invariant system $(\theta(t) = \theta)$ one obtains

$$\widetilde{\theta}(t) = \left[(1 - \beta)I_n - \mu\varphi(t)\varphi^T(t)\right]\widetilde{\theta}(t - 1) - \mu\varphi(t)v(t) + \beta\theta. \tag{8.45}$$

Note that even if $\varphi(t) = 0$, $\forall t$, i.e. the sequence $\{\varphi(t)\}$ is trivially not persistently exciting, the error system is exponentially stable.

Due to the presence of the constant forcing term $\beta\theta$ on the right-hand side of (8.45), the leaky LMS error system cannot achieve the asymptotic zero-state stability even if $\{\varphi(t)\}$ is persistently exciting. This means that the estimates generated by (8.43) are biased. The amount of offset introduced by leakage can be easily evaluated under the independence assumption (B1.1). Observe that for a time-invariant system, algorithm (8.43) can be rewritten as

$$\widehat{\theta}(t) = \left[(1 - \beta)I_n - \mu\varphi(t)\varphi^T(t)\right]\widehat{\theta}(t - 1) + \mu\varphi(t)\varphi^T(t)\theta + \mu\varphi(t)v(t). \tag{8.46}$$

Hence, after taking expectations of both sides of (8.46), one obtains

$$E[\widehat{\theta}(t)] = \left[(1 - \beta)I_n - \mu\Phi_o\right]E[\widehat{\theta}(t - 1)] + \mu\Phi_o\theta,$$

which, for sufficiently small μ, has the following steady-state solution:

$$\lim_{t \to \infty} E[\widehat{\theta}(t)] = \left[I_n + \frac{\beta}{\mu}\Phi_o^{-1}\right]\theta. \tag{8.47}$$

Therefore, in order to keep the bias small, the leakage coefficient β should be chosen so as to guarantee that, in addition to (8.44), the following inequality holds:

$$\frac{\beta}{\mu}\Phi_o^{-1} \ll I_n,$$

or equivalently

$$\beta I_n \ll \mu\Phi_o. \tag{8.48}$$

Note that for FIR systems (8.48) is equivalent to $\beta \ll \mu\sigma_u^2$.

Remark

An interesting alternative to (8.43), the error-weighted leakage, was proposed by Narendra and Annaswamy [130]. The proposed estimation algorithm is

$$\widehat{\theta}(t) = \left(1 - \frac{\beta|\epsilon(t)|}{\delta + \beta|\epsilon(t)|}\right)\widehat{\theta}(t - 1) + \mu\varphi(t)\epsilon(t), \tag{8.49}$$

where β and δ are the appropriately chosen positive constants. Note that the leakage in (8.49) is controlled by the magnitude of the prediction errors; if $\epsilon(t)$ is reduced to zero then so is the amount of leakage.

8.1.3 Kalman filter algorithms

Suppose that the KF algorithm (7.4) is operated under the no-excitation condition (8.1). Substituting (8.1) into (7.4) one arrives at

$$
\begin{aligned}
\widehat{\theta}(t|t) &= \widehat{\theta}(t|t-1), \\
\widehat{\theta}(t|t-1) &= F\widehat{\theta}(t-1|t-1), \\
\widetilde{\Sigma}(t|t-1) &= F\widetilde{\Sigma}(t-1|t-1)F^T + G\widetilde{W}G^T, \\
\widetilde{\Sigma}(t|t) &= \widetilde{\Sigma}(t|t-1).
\end{aligned}
\tag{8.50}
$$

Combining the last two updates gives

$$
\widetilde{\Sigma}(t+1|t) = F\widetilde{\Sigma}(t|t-1)F^T + G\widetilde{W}G^T,
$$

which is exponentially stable if and only if all eigenvalues of F have magnitudes less than one.

The stability constraint formulated above is not met for the random walk model and its generalizations. For the RWKF algorithm (7.9) one obtains

$$
\begin{aligned}
\widehat{\theta}(t) &= \widehat{\theta}(t-1), \\
P(t) &= P(t-1) + \kappa^2 I_n, \quad \forall t > t_o,
\end{aligned}
$$

or equivalently

$$
\begin{aligned}
\widehat{\theta}(t) &= \widehat{\theta}(t_o), \\
P(t) &= P(t_o) + (t - t_o)\kappa^2 I_n, \quad \forall t > t_o.
\end{aligned}
\tag{8.51}
$$

Comparing (8.51) with (8.3) reveals that the RWKF algorithm, although vulnerable to estimator windup, is considerably more robust to excitation failures than the EWLS algorithm. According to (8.51) the covariance blowup in an RWKF algorithm has a *linear* growth rate, whereas the analogous rate in the EWLS algorithm is *exponential*. For this reason the RWKF algorithm is sometimes known as an adaptive filter with linear forgetting.

Remark

It is relatively easy to derive the deterministic upper bound on the rate of growth of $P(t)$, valid for *any* sequence of regressors. First, observe that under the random walk hypothesis

$$
\widetilde{\Sigma}^{-1}(t|t) = \widetilde{\Sigma}^{-1}(t|t-1) + \varphi(t)\varphi^T(t) \geq \widetilde{\Sigma}^{-1}(t|t-1),
$$

hence

$$
\lambda_{\max}(\widetilde{\Sigma}(t|t)) \leq \lambda_{\max}(\widetilde{\Sigma}(t|t-1)).
\tag{8.52}
$$

Second, it holds that

$$\lambda_{max}(\tilde{\Sigma}(t+1|t)) = \lambda_{max}\left[\tilde{\Sigma}(t|t) + \kappa^2 I_n\right] = \lambda_{max}(\tilde{\Sigma}(t|t)) + \kappa^2. \qquad (8.53)$$

Combining (8.52), (8.53) and $\tilde{\Sigma}(t+1|t) = P(t)$ gives

$$\lambda_{max}(P(t)) \le \lambda_{max}(P(t-1)) + \kappa^2,$$

which leads to

$$\lambda_{max}(P(t)) \le \lambda_{max}(P(t_o)) + (t - t_o)\kappa^2. \qquad (8.54)$$

Since (8.51) entails

$$\lambda_{max}(P(t)) = \lambda_{max}(P(t_o)) + (t - t_o)\kappa^2,$$

the least favorable excitation under which the upper bound in (8.54) is achieved corresponds to $\varphi(t) = 0$, i.e. to the no-excitation case considered earlier.

■

Square root filtering

To avoid ill-conditioning of the a posteriori matrix $\tilde{\Sigma}(t|t)$, Joseph's stabilized version of the covariance update can be used:

$$\tilde{\Sigma}(t|t) = \left(I_n - K(t)\varphi^T(t)\right)\tilde{\Sigma}(t|t-1)\left(I_n - K(t)\varphi^T(t)\right)^T + K(t)K^T(t), \qquad (8.55)$$

instead of the recursion found in (7.4).

To improve numerical robustness of the RWKF algorithm one can propagate the square roots of the covariance matrices $\tilde{\Sigma}(t|t-1)$ and $\tilde{\Sigma}(t|t)$, denoted here by $S(t|t-1)$ and $S(t|t)$, respectively:

$$\begin{aligned}
\tilde{\Sigma}(t|t-1) &= S(t|t-1)S^T(t|t-1), \\
\tilde{\Sigma}(t|t) &= S(t|t)S^T(t|t).
\end{aligned}$$

To obtain the square root version of the time update recursions $\tilde{\Sigma}(t-1|t-1) \mapsto \tilde{\Sigma}(t|t-1)$ of the general KF algorithm (7.4), one should solve the following triangularization problem:

$$\left[\ FS(t-1|t-1)\quad G\widetilde{W}^{1/2}\ \right] T = \left[\ S(t|t-1)\quad 0\ \right], \qquad (8.56)$$

where T denotes a $2n \times 2n$ unitary matrix.

The measurement update recursions $\tilde{\Sigma}(t|t-1) \mapsto \tilde{\Sigma}(t|t)$ can be replaced either by Potter's recursions (which are, in fact, a specialized version of (8.11) obtained for $\lambda = 1$):

$$\begin{aligned}
r(t) &= S^T(t|t-1)\varphi(t), \\
\beta(t) &= 1 + r^T(t)r(t), \\
\alpha(t) &= \beta(t) + \sqrt{\beta(t)}, \\
K(t) &= \frac{S(t|t-1)r(t)}{\beta(t)}, \\
S(t|t) &= S(t|t-1) - \frac{\beta(t)}{\alpha(t)}K(t)r^T(t), \qquad (8.57)
\end{aligned}$$

or by the following triangularization update (which is also a special case of the EWLS update (8.13)):

$$\begin{bmatrix} 1 & \varphi^T(t)S(t|t-1) \\ 0 & S(t|t-1) \end{bmatrix} T = \begin{bmatrix} \sqrt{\beta(t)} & 0^T \\ \sqrt{\beta(t)}K(t) & S(t|t) \end{bmatrix}. \tag{8.58}$$

The square root counterpart of the covariance update in the RWKF algorithm $P(t-1) \mapsto P(t)$ can be obtained by solving

$$\begin{bmatrix} 1 & \varphi^T(t)S(t-1) & 0^T \\ 0 & S(t-1) & \kappa^2 I_n \end{bmatrix} T = \begin{bmatrix} \sqrt{\beta(t)} & 0^T & 0^T \\ \sqrt{\beta(t)}K(t) & S(t) & O \end{bmatrix}, \tag{8.59}$$

where $S(t)$ is a square root of $P(t)$: $P(t) = S(t)S^T(t)$.

Fading memory

Numerical problems which arise when KF parameter trackers operate under insufficiently rich excitation can be considered as a natural consequence of the adopted model of parameter variation. Since neither the random walk model nor its extensions are stable, from the viewpoint of information theory, the unlimited growth of the filter covariance matrices is perfectly understandable. If the process parameters are expected to drift away to infinity and the collected data does not contain information allowing for their unambiguous estimation, the covariance matrices $\widetilde{\Sigma}(t|t-1)$ and $\widetilde{\Sigma}(t|t)$, which quantify the accuracy of the KF estimates, must blow up. Divergence of Kalman filtering algorithms can be easily stopped if the original state space model (7.1) is replaced by the following modified model:

$$\theta(t) = \lambda F \theta(t-1) + G w(t) \tag{8.60}$$

where λ, $0 < \lambda < 1$, denotes the forgetting constant.

The corresponding KF algorithm can easily be obtained after replacing the matrix F in (7.4) with λF. For the modified random walk model

$$\theta(t) = \lambda \theta(t-1) + w(t), \tag{8.61}$$

one obtains these recursions:

$$\begin{aligned}
\widehat{\theta}(t) &= \lambda \widehat{\theta}(t-1) + K(t)\epsilon(t), \\
\epsilon(t) &= y(t) - \varphi^T(t)\widehat{\theta}(t-1), \\
K(t) &= \frac{P(t-1)\varphi(t)}{1 + \varphi^T(t)P(t-1)\varphi(t)}, \\
P(t) &= \lambda^2 \left[P(t-1) - \frac{P(t-1)\varphi(t)\varphi^T(t)P(t-1)}{1 + \varphi^T(t)P(t-1)\varphi(t)} \right] + \kappa^2 I_n, \tag{8.62}
\end{aligned}$$

which constitute the leaky or fading RWKF algorithm. Note that when $\lambda = 1$, algorithm (8.62) becomes identical with (7.9).

One can easily show that if $P(t)$ evolves according to (8.62) it is upper bounded, namely

$$P(t) \leq \frac{\kappa^2}{1 - \lambda^2} I_n, \tag{8.63}$$

for any value of t and any sequence of regression vectors. Hence the covariance blowup is prevented. Note, however, that for small values of κ and $1 - \lambda$ the right-hand side of (8.63) may take very large values.

Another form of fading filter may be produced by modifying only the time update recursion of the KF algorithm (7.4),

$$\widetilde{\Sigma}(t|t - 1) = \lambda^2 F \widetilde{\Sigma}(t - 1|t - 1) F^T + G \widetilde{W} G^T, \tag{8.64}$$

while the remaining Kalman recursions remain unchanged, including the parameter update equations

$$\begin{aligned} \widehat{\theta}(t|t) &= \widehat{\theta}(t|t - 1) + K(t)\epsilon(t), \\ \widehat{\theta}(t|t - 1) &= F\widehat{\theta}(t - 1|t - 1). \end{aligned} \tag{8.65}$$

It can be shown that the resulting algorithm minimizes the exponentially weighted version of the error criterion which underlies the ordinary Kalman filter [5].

8.2 Optimization

8.2.1 Memory optimization

To improve the performance of adaptive filters, their design parameters should be chosen so they take into account the way the process parameters change as well as how quickly they change. The most important parameter of an adaptive filter is its estimation memory. The general guidelines for the filter memory optimization are pretty straightforward; the memory span of an adaptive filter should be inversely proportional to the speed of parameter variation. When the identified system undergoes rapid changes, the memory of the filter should be shortened so as to allow for fast parameter resetting; when the system changes are slow, the memory should be increased to make the parameter estimates more accurate. We have already shown that the memory of an adaptive filter should trade off between the variance and bias components of the performance measure. In a special case where process parameters drift according to the random walk model, the optimal values of l_∞ can be obtained analytically: EWLS case (5.117), (5.123); LMS case (5.116), (5.122); and RWKF case (7.62), (7.66).

Even though interesting from the theoretical viewpoint, all the expressions mentioned above have very limited practical importance; not only were they derived for a very specific model of system time variation (the random walk hypothesis), they also require a detailed knowledge of the covariance structure of regressors Φ_o, the measurement noise variance σ_v^2 and the statistics of parameter changes W. In practice at least some of these quantities are either not known a priori or they are time dependent.

If the prior knowledge about the identified time-varying system is not available then the filter memory, or the memory-related parameters, can be adjusted adaptively. A number of such memory optimization procedures will be revised below.

The memory tuning procedures can be divided into sequential and parallel. The first case uses a single tracking algorithm, equipped with an adjustable memory-controlling parameter. In the second case several algorithms, with different memory settings, are run in parallel and compared according to their predictive abilities.

Sequential estimation

Denote by $\Delta(t)$ the local measure of the rate of system parameter variation and denote by γ the gain of an adaptive filter. A simple gain scheduling technique is based on the relationship

$$\gamma(t) = f(\Delta(t)), \tag{8.66}$$

where $f(\cdot)$ is a nondecreasing positive function.

Quite obviously, the speed of parameter variation cannot be measured directly; some indirect nonstationarity measures have to be used instead. The local prediction error statistic

$$\Delta(t) = \sum_{i=0}^{M-1} \epsilon^2(t-i), \tag{8.67}$$

where M is the length of the local analysis window, is the most frequently used indirect nonstationarity index [104], [50]; in particular note that the VFF rules (8.28) and (8.31), developed as the anti-windup devices, can alternatively be thought of as the memory optimization procedures.

Gain scheduling has some obvious limitations. First, if memory adjustments are based on monitoring prediction errors, any increase in the prediction error variance is regarded as a sign of faster parameter variation and therefore shortens the estimation memory. Note however, that the increase in the prediction error variance may be caused either by the parameter changes or by the increase in the noise variance σ_v^2, due to the set point change or a spurious disturbance. Even though decreasing the filter memory in the second case does not make any sense at all, it is not possible to differentiate between the two situations when using the simple strategy described above.

Second, even if the direction of necessary memory adjustments is determined correctly, there is no obvious way of relating the changes in the local nonstationarity index to the appropriate changes of the filter gain (memory). This is because gain scheduling is an *open-loop* technique.

Some limitations of memory scheduling can be overcome if the optimal gain is looked for using a gradient search technique. Consider, for example, the problem of optimizing a stepsize μ in an LMS algorithm

$$\begin{aligned}
\widehat{\theta}(t) &= \widehat{\theta}(t-1) + \mu\varphi(t)\epsilon(t), \\
\epsilon(t) &= y(t) - \varphi^T(t)\widehat{\theta}(t-1).
\end{aligned}$$

Since we are interested in choosing μ so as to minimize prediction errors, it is natural to adopt

$$J(\mu) = \mathrm{E}\left[\epsilon^2(t)\right] \tag{8.68}$$

as our local performance measure (note that the prediction error $\epsilon(t)$ is an *implicit* function of μ). Furthermore, since in practical situations the ensemble average cannot be evaluated (only one process realization is available) one can replace (8.68) with the following instantaneous quality measure:

$$J_t(\mu) = \epsilon^2(t). \tag{8.69}$$

The recursive gradient search of the optimal value of μ can be realized using this algorithm (cf. (5.4)):

$$\mu(t) = \mu(t-1) - \frac{1}{2}\delta\frac{\partial J_t(\mu)}{\partial \mu} \tag{8.70}$$

where δ denotes a small positive constant.
Note that

$$\frac{\partial J_t(\mu)}{\partial \mu} = -2\varphi^T(t)\frac{\partial \widehat{\theta}(t-1)}{\partial \mu}\epsilon(t)$$

and

$$\frac{\partial \widehat{\theta}(t)}{\partial \mu} = \frac{\partial \widehat{\theta}(t-1)}{\partial \mu} - \mu\varphi(t)\varphi^T(t)\frac{\partial \widehat{\theta}(t-1)}{\partial \mu} + \varphi(t)\epsilon(t). \tag{8.71}$$

Hence the LMS algorithm with adaptive stepsize adjustment can be summarized as follows [21]:

$$
\begin{aligned}
\widehat{\theta}(t) &= \widehat{\theta}(t-1) + \mu(t-1)\varphi(t)\epsilon(t), \\
\epsilon(t) &= y(t) - \varphi^T(t)\widehat{\theta}(t-1), \\
\mu_o(t) &= \mu(t-1) + \delta\varphi^T(t)\psi(t-1)\epsilon(t), \\
\mu(t) &= \begin{cases} 0 & \text{if } \mu_o(t) < 0 \\ \mu_o(t) & \text{if } 0 < \mu_o(t) \le \mu_{\max} \\ \mu_{\max} & \text{if } \mu_o(t) > \mu_{\max} \end{cases} \\
\psi(t) &= \left[I_n - \mu(t-1)\varphi(t)\varphi^T(t)\right]\psi(t-1) + \varphi(t)\epsilon(t), \tag{8.72}
\end{aligned}
$$

where $\psi(t)$ denotes an approximation of $\partial\widehat{\theta}(t)/\partial\mu$ obtained when the recursion (8.71) is used with arbitrary (e.g. zero) initial conditions. Note that in order to guarantee stability of the LMS filter, the stepsize update $\mu_o(t)$ is projected back into the admissible region $[0, \mu_{\max}]$ each time its value becomes negative or too large. The theoretical analysis of the convergence and tracking behavior of (8.72) is provided in the paper by Kushner and Yang [100].

The gradient search technique described above can be easily extended to other adaptive filters. For example, the following self-optimizing version of the EWLS algorithm, proposed by Haykin [75], was derived in an analogous way to (8.72):

$$
\begin{aligned}
\widehat{\theta}(t) &= \widehat{\theta}(t-1) + K(t)\epsilon(t), \\
\epsilon(t) &= y(t) - \varphi^T(t)\widehat{\theta}(t-1), \\
K(t) &= \frac{P(t-1)\varphi(t)}{\lambda(t-1) + \varphi^T(t)P(t-1)\varphi(t)}, \\
P(t) &= \frac{1}{\lambda(t-1)}\left[P(t-1) - \frac{P(t-1)\varphi(t)\varphi^T(t)P(t-1)}{\lambda(t-1) + \varphi^T(t)P(t-1)\varphi(t)}\right], \\
\lambda_o(t) &= \lambda(t-1) + \delta\varphi^T(t)\psi(t-1)\epsilon(t),
\end{aligned}
$$

$$\lambda(t) \;=\; \begin{cases} 1 & \text{if } \lambda_o(t) > 1 \\ \lambda_o(t) & \text{if } \lambda_{\min} < \lambda_o(t) \le 1 \\ \lambda_{\min} & \text{if } \lambda_o(t) < \lambda_{\min} \end{cases}$$

$$\psi(t) \;=\; \left[I_n - K(t)\varphi^T(t) \right] \psi(t-1) + Z(t)\varphi(t)\epsilon(t),$$

$$Z(t) \;=\; \frac{1}{\lambda(t)} \left[I_n - K(t)\varphi^T(t) \right] Z(t-1) \left[I_n - K(t)\varphi^T(t) \right]$$

$$+ \frac{1}{\lambda(t)} \left[K(t)K^T(t) - P(t) \right]. \tag{8.73}$$

The $n \times n$ matrix $Z(t)$ in (8.73) is an approximation of the sensitivity matrix $\partial P(t)/\partial \lambda$.

Both self-optimizing algorithms presented above are subject to limitations typical of all sequential gradient-based search procedures; the process of adaptation is either fast but erratic (for 'large' values of δ) or precise but slow (for 'small' values of δ).

When system parameters change abruptly, the memory tuning algorithm should quickly forget past data; this can be achieved by assigning relatively large values to δ in (8.72) or (8.73). But when δ is large, the tuning procedure becomes overly sensitive to noise; a spurious disturbance causing a short-term increase of prediction errors, but not affecting system parameters, may significantly but unnecessarily reduce the filter memory, i.e. essentially, restart its operation.

Parallel estimation

Most of the limitations of the sequential memory tuning procedures can be overcome if the parallel estimation schemes are applied. The derivation of the corresponding formulas is based on Bayesian reasoning.

Consider a bank of RWKF filters corresponding to I different values of κ, $\{\kappa_i, i = 1, \dots, I\}$. Then

$$\widehat{\theta}_i(t) \;=\; \widehat{\theta}_i(t-1) + K_i(t)\epsilon_i(t),$$

$$\epsilon_i(t) \;=\; y(t) - \varphi^T(t)\widehat{\theta}_i(t-1),$$

$$K_i(t) \;=\; \frac{\widetilde{\Sigma}_i(t|t-1)\varphi(t)}{1 + \varphi^T(t)\widetilde{\Sigma}_i(t|t-1)\varphi(t)},$$

$$\widetilde{\Sigma}_i(t|t-1) \;=\; \widetilde{\Sigma}_i(t-1|t-1) + \kappa_i^2 I_n,$$

$$\widetilde{\Sigma}_i(t|t) \;=\; \widetilde{\Sigma}_i(t|t-1) - \frac{\widetilde{\Sigma}_i(t|t-1)\varphi(t)\varphi^T(t)\widetilde{\Sigma}_i(t|t-1)}{1 + \varphi^T(t)\widetilde{\Sigma}_i(t|t-1)\varphi(t)}, \tag{8.74}$$

$$i = 1, \dots, I.$$

Denote by $T(t) = [t, t-1, \dots, t-M+1]$ the local analysis interval covering the M most recent time instants. We would like to combine, in a rational way, the results given by such a bank of tracking algorithms. Since different values of κ correspond to different hypothetical rates of variation of the process parameters, the general problem we are trying to address is how to make the identification results less sensitive to the unknown and/or time-varying 'rate of nonstationarity' for the analyzed system. Local analysis is performed to account for possible changes in the modes of process variation.

Suppose one would like to find the optimal-local one-step-ahead predictor; that is, the predictor which minimizes the following local measure of performance:

$$E\left[y(t+1) - \widehat{y}(t+1|t)\right]^2,\tag{8.75}$$

where the expectation is taken over $y(t+1)$, i (the number of a hypothetical 'true' model of parameter variation) and all possible data sets $\Xi_T(t) = \{\xi(j), j \in T(t)\}$. The quantities i and σ_v^2 will be regarded as unknown but constant within the analysis interval $T(t)$; they will be modeled as random variables with assigned prior distributions.

The predictor which minimizes (8.75) is given by

$$\widehat{y}(t+1|t) = \sum_{i=1}^{I} \mu_i(t)\widehat{y}_i(t+1|t)\tag{8.76}$$

where

$$\widehat{y}_i(t+1|t) = \varphi^T(t+1)\widehat{\theta}_i(t),$$

and

$$\mu_i(t) = p(i|\Xi_T(t)).$$

The solution obtained has some interesting properties. The optimal predictor is given in the form of a linear combination of competitive predictors $\widehat{y}_i(t+1|t)$ obtained for different hypothetical rates of parameter variation. The weights $\mu_i(t)$, which in the Bayesian framework have the meaning of the posterior probabilities given the data set $\Xi_T(t)$, will be called the *model credibility coefficients*.

Note that (8.76) can be rewritten in the form

$$\widehat{y}(t+1|t) = \varphi^T(t+1)\widehat{\theta}(t),\tag{8.77}$$

where

$$\widehat{\theta}(t) = \sum_{i=1}^{I} \mu_i(t)\widehat{\theta}_i(t).\tag{8.78}$$

According to (8.77) the optimal predictor can be alternatively regarded as the one corresponding to the *averaged system model*, characterized by an averaged vector of parameter estimates $\widehat{\theta}(t)$. Assigning noninformative prior distributions to i and σ_v^2 [156],

$$\pi(i) = \frac{1}{I}, \quad \pi(\sigma_v^2|i) = \pi(\sigma_v^2) \propto \frac{1}{\sigma_v^2},\tag{8.79}$$

which is a way of saying that nothing is known a priori about the true values of i and σ_v^2, one can derive the following expression for the model credibility coefficients [142]:

$$\mu_i(t) = \frac{\eta_i(t)}{\sum_{i=1}^{I} \eta_i(t)}\tag{8.80}$$

where

$$\eta_i(t) = \left(\prod_{j=0}^{M-1} \beta_i(t-j)\right)^{-1/2} \left(\sum_{i=1}^{M-1} \frac{\epsilon_i^2(t-j)}{\beta_i(t-j)}\right)^{-M/2}\tag{8.81}$$

and

$$\beta_i(t) = 1 + \varphi^T(t)\widetilde{\Sigma}_i(t|t-1)\varphi(t). \tag{8.82}$$

Remark 1

The noninformative prior distributions are designed so as to reflect our complete ignorance of unknown variables and hence to make the Bayesian analysis less subjective. The main advantage of this approach is that one arrives at the results which depend entirely upon the experimental data (i.e. they do not contain any subjectively determined quantities) but which still preserve the genuine structure of the Bayesian solution [156]. If some prior knowledge about the system changes is available, one can easily take it into account. For example, if different modes of parameter variation are not equiprobable, one can set $\pi(i)$ to the observed or anticipated relative frequency of occurrence for each mode. If the prior probabilities of different modes are not identical, the right-hand side of (8.81) should be multiplied by $\pi(i)$.

Remark 2

Note that the prior distribution of σ_v^2 adopted in (8.79) is *improper* (the density does not integrate to 1). This nonstandard choice follows from Jeffrey's maximum entropy rule for selecting noninformative priors [83].

∎

An interesting approach to statistical reasoning was suggested by Dawid [40]. According to his prequential (predictive + sequential) principle, analogous to Fisher's likelihood principle, different models should be assessed on their empirically confirmed predictive capabilities. The prequential framework allows one to extend the above results to banks of arbitrary adaptive filters, not necessarily RWKF filters.

Following Dawid, under Gaussian assumptions the prequential likelihood associated with the ith algorithm can be defined as

$$p(\Xi_T(t)|\sigma_\epsilon^2, i) = (2\pi\sigma_\epsilon^2)^{-M/2} \exp\left\{-\frac{1}{\sigma_\epsilon^2}\sum_{j=0}^{M-1}\epsilon_i^2(t-j)\right\}, \tag{8.83}$$

where $\{\epsilon_i(t)\}$ denotes the observed sequence of prediction errors of unknown but constant (and data-independent) variance σ_ϵ^2. According to (8.83) prediction errors yielded by different algorithms are treated as if they resulted from an optimal predictor; on this basis we make judgements on the (relative) credibility of the corresponding models.

Observe that

$$\mu_i(t) = p(i|\Xi_T(t)) = \frac{p(i, \Xi_T(t))}{\sum_{i=1}^{I} p(i, \Xi_T(t))} = c\int_0^\infty p(\Xi_T(t), \sigma_\epsilon^2|i)\pi(i)\, d\sigma_\epsilon^2, \tag{8.84}$$

where c does not depend on i. Furthermore

$$p(\Xi_T(t), \sigma_\epsilon^2|i) = p(\Xi_T(t)|\sigma_\epsilon^2, i)\pi(\sigma_\epsilon^2|i).$$

Carrying out the integration in (8.84) for noninformative priors,

$$\pi(i) = \frac{1}{I}, \quad \pi(\sigma_\epsilon^2|i) = \pi(\sigma_\epsilon^2) \propto \frac{1}{\sigma_\epsilon^2},$$

one obtains $\mu_i(t)$ in the form (8.80) with

$$\eta(i) = \left(\sum_{j=0}^{M-1} \epsilon_i^2(t-j) \right)^{-M/2} ; \qquad (8.85)$$

see [142] for more details.

Note that under the slow adaptation condition (7.26)

$$\beta_i(t) \cong 1, \quad \forall i, t,$$

hence the formula (8.81), derived for RWKF filters, will yield almost identical results as (8.85). The advantage of the prequential approach is that it has a much wider range of applicability; the weights evaluated according to (8.85) can be used to combine parameter estimates yielded by *any* set of adaptive filters differing in type (WLS, LMS, KF, WBF), order and/or estimation memory.

For large values of M, which corresponds to long analysis intervals, the weighted estimation formula (8.78) will de facto reduce itself to

$$\widehat{\theta}(t) = \widehat{\theta}_{i^*(t)}(t), \qquad (8.86)$$

where

$$i^*(t) = \arg \min_{1 \le i \le I} \sum_{j=0}^{M-1} \epsilon_i^2(t-j), \qquad (8.87)$$

which is the local (sliding window) version of Rissanen's predictive least squares (PLS) principle [163], [164]. The criterion proposed by Rissanen (originally for the purpose of model order selection only) makes its choice on the basis of comparing predictive abilities of different models or algorithms but it does not take into account the distribution of the accumulated prediction error over the set of competitive models. The prequential approach offers a natural way of bringing the notion of significance into comparison of predictive abilities of various models. For long analysis intervals, even small differences in the accumulated prediction error statistics will be significant, i.e. they will produce large differences in the values of the corresponding credibility coefficients. Consequently, the major contribution to $\widehat{\theta}(t)$ in (8.78) is likely to be due to the 'recently the best' estimator $\widehat{\theta}_{i^*(t)}(t)$. For small values of M, however, the difference between (8.78) and (8.86) will be more emphasized.

When the parallel estimation schemes are employed, the limitations of the sequential approach, mentioned previously, are circumvented in a pretty natural way.

Since decisions are based on comparing prediction error statistics obtained for different algorithms, i.e. on relative prediction errors rather than their absolute values, the results are not sensitive to variations in σ_v^2. For example, if σ_v^2 undergoes a sudden change, the prediction errors yielded by all algorithms are affected in the same way, that is, they all increase or all decrease at exactly the same rate. This allows one to differentiate between parameter changes and changes in the measurement noise statistics.

Similarly, since switching between different filters is based on their short-term performance, not on the instantaneous values of prediction errors, for reasonable values of M the memory adjustments are both fast and reliable.

Remark

The idea of using multiple filters for parameter tracking was pioneered by Andersson [8]. The parallel estimation scheme proposed by Andersson, called adaptive forgetting through multiple models (AFMM), is based on the concept of the finite-Gaussian sum approximation, a technique used to obtain approximate solutions to nonlinear filtering problems [5].

Assuming that the process parameters obey the random jump model

$$\theta(t) = \theta(t-1) + w(t),$$

where

$$\text{cov}[w(t)] = \begin{cases} W & \text{w.p.} \quad q \\ O & \text{w.p.} \quad 1-q \end{cases}$$

and q denotes the (small) jump probability, one can show that the optimal solution to the tracking problem is yielded by a bank of Kalman filters with an exponentially growing number of components. Since at each time step a parameter jump may or may not occur, the number of possible variants which should be considered at instant t is equal to 2^t. The idea behind the finite-Gaussian sum approximation is to prune, at each time instant t, the growing decision tree of possible variants by retaining only its I most likely components. The resulting suboptimal estimation scheme is based on a 'population' of Kalman tracking algorithms which are born, survive and die in an evolutionary process guided by Bayesian rules.

Choice of design parameters

When designing a multiple-model adaptive scheme, one is interested in selecting the memory spans of competing adaptive filters so as to increase the robustness of a predictor (8.76), namely to decrease its sensitivity to unknown and/or time-varying degrees of nonstationarity in the analyzed process.

Suppose the system parameters vary according to the random walk model and suppose a bank of I adaptive filters of the same type (EWLS, LMS, RWKF) but with different memory settings $l_\infty^i, i = 1, \ldots, I$ are used for parameter tracking. Recall that dependence of the excess mean square prediction error Π_∞ on the filter memory l_∞ can be written in the following form (cf. (5.116), (5.117) and (7.5)):

$$\Pi_\infty(l_\infty) = \frac{a}{l_\infty} + bl_\infty. \tag{8.88}$$

The value l_∞^{opt}, which minimizes (8.88), can be regarded as an optimal trade-off between the variance and bias components of Π_∞. Denote by $\delta = 1 + \epsilon$, where ϵ

is a small positive constant, the multiplier which allows one to specify what is meant by an 'insignificant increase' of the excess prediction error. For example, if insignificant changes are defined as those not exceeding 5% of the optimal value of $\Pi_\infty^{opt} = \Pi_\infty(l_\infty^{opt})$, one should set $\delta = 1.05$. Requiring that

$$\Pi_\infty(l_\infty) \leq \delta\Pi_\infty^{opt}, \tag{8.89}$$

one can find the minimum $\underline{l}_\infty$ and maximum $\overline{l}_\infty$ values of l_∞ which define the 'insensitivity' interval $\Delta_\delta = [\underline{l}_\infty, \overline{l}_\infty]$ (Figure 8.1). In practice the coefficients a and

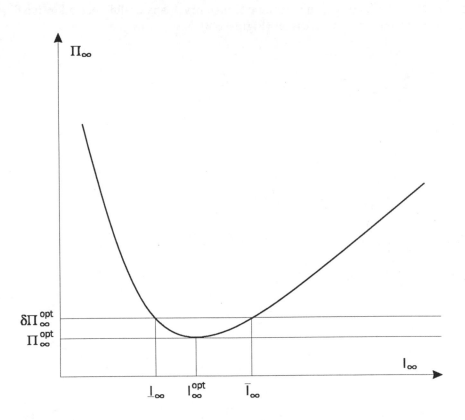

Figure 8.1   A typical dependence of the excess prediction error on the adaptation gain.

b in (8.88) are not known or they vary with time. Hence it may be very difficult, or even impossible, to choose l_∞ so that the condition $l_\infty \in \Delta_\delta$ holds at all time instants. We will show that the width of the insensitivity interval can be significantly increased if several adaptive filters are used instead of one algorithm. To get insight into the memory scheduling problem, we will study properties of the error characteristics associated with the 'ideal' (error-free) switching rule

$$\overline{\Pi}_\infty(l_\infty) = \min_{1 \leq i \leq I} \{\Pi_\infty^i(l_\infty)\}, \tag{8.90}$$

where

$$\Pi_\infty^{i+1}(l_\infty) = \Pi_\infty^i(\gamma_i l_\infty), \quad \gamma_i > 1, \tag{8.91}$$

and Π_∞^i is the error characteristic of the ith algorithm. According to (8.91) the competitive filters are arranged in the increasing memory order, namely

$$l_\infty^{i+1} = \gamma_i l_\infty^i, \quad i = 1, \ldots, I - 1.$$

Using simple geometrical arguments one can show that in order to satisfy (8.89) it is sufficient to choose $\gamma_1 = \ldots = \gamma_{I-1} = \gamma$, provided that γ is adjusted so as to match terminal points

$$\bar{l}_\infty^i = \underline{l}_\infty^{i+1}, \quad i = 1, \ldots, I - 1.$$

This results in the following rule of thumb: memory spans of different adaptive filters should form a geometric progression (Figure 8.2).

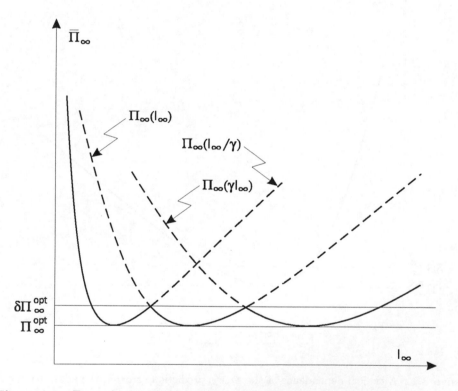

Figure 8.2 Error characteristic corresponding to the 'ideal' switching rule ($I = 3$)

It is not difficult to relate the multiplier γ to the accuracy level δ. It turns out that, *irrespective* of the values taken by a and b in (8.88), condition (8.89) is met if

$$\gamma = \frac{\delta + \sqrt{\delta^2 - 1}}{\delta - \sqrt{\delta^2 - 1}} = \left(\delta + \sqrt{\delta^2 - 1}\right)^2. \tag{8.92}$$

Practical experience shows that formula (8.92) is a good guideline for choosing γ, despite the fact it was derived for a specific model of parameter variation. In particular, the memory-doubling rule, which can be obtained from (8.92) for a 6% error bound, seems to work very well in practice.

The number of filters comprising the parallel structure depends on the required degree of robustness; in most applications two filters (short memory and long memory) or three filters (short memory, nominal memory and long memory) allow one to significantly reduce sensitivity of the estimation scheme to unknown or time-varying nonstationarity in the analyzed process.

Example 8.1

The second-order AR signal was generated according to

$$y(t) = a_1(t)y(t-1) + a_2(t)y(t-2) + v(t),$$

where $v(t) \sim \mathcal{N}(0,1)$ and time-dependent AR coefficients were varying in a stepwise manner according to

$$a_1(t) = 0.4\sin\alpha(t), \quad a_2(t) = 0.4\cos\alpha(t),$$

where

$$\alpha(t) = \begin{cases} \alpha(t-1) & \text{w.p.} \quad 0.99 \\ \alpha(t-1) + \Delta\alpha & \text{w.p.} \quad 0.01 \end{cases}$$

and $\alpha(0) = 0$, $\Delta\alpha = 2\pi/10$.

Identification of the nonstationary AR process described above was carried out using a bank of three EWLS filters working in parallel. The memory scaling coefficient γ was equal to 2 and the length of the analysis interval M was set to 20. The plots of the excess mean square prediction errors (averaged over 1000 time steps) for different values of the nominal filter memory (corresponding to the middle filter in the memory ordering) are shown in Figure 8.3 along with the plots obtained for all three component filters.

The results obtained clearly show the advantages of using the multiple filter structure. Note that in a certain range of values for the nominal memory, the multiple-model algorithm worked better than any of its component fixed-memory filters.

8.2.2 Other optimization issues

Some of the finite-memory algorithms described in Chapters 4 to 7 offer more degrees of freedom than memory tuning.

The WLS scheme allows one to influence the tracking characteristics of an adaptive filter by shaping the weighting sequence $\{w(i)\}$. All algorithms based on the method of basis functions (BF, WBF) require selection of the number and type of basis sequences. Finally, the algorithms based on Kalman filtering allow fine tuning of their tracking performance by selecting the state transition matrix F and the driving noise covariance matrix W in the state space model of parameter variation.

We will show that such 'secondary' parameters of adaptive filters can also be selected in a rational way.

Optimization of window shape in WLS estimation

If there is some prior knowledge on process parameter variation, one can attempt to optimize the *shape* of the window used in WLS estimation. An example of this analysis

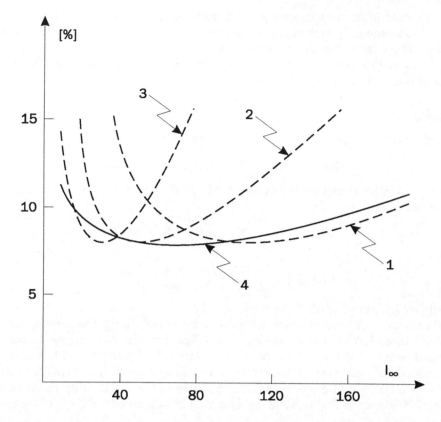

Figure 8.3 Relative excess prediction error as a function of the nominal memory for an adaptive filter bank: (1) short memory predictor, (2)nominal memory predictor, (3) long memory predictor and (4) weighted predictor.

is given in [134].

Assuming the parameters are subject to mixed-mode variations, slow drift with occasional jumps (2.35), and

$$\mathrm{cov}[\Delta\theta(t)] = W,$$

one can show that

$$\mathcal{P}_{\infty}^{\mathrm{WLS}} \cong \frac{n\sigma_v^2}{l_\infty} + \mathrm{tr}\{\Phi_o W\} \sum_{i=0}^{\infty}\sum_{j=0}^{\infty} h_{\mathrm{WLS}}(i) h_{\mathrm{WLS}}(j) \min(i,j), \qquad (8.93)$$

where

$$h_{\mathrm{WLS}}(i) = \frac{w(i)}{k_\infty} \qquad (8.94)$$

is the window-dependent impulse response associated with the WLS estimator. The optimization problem can be stated as follows: among all windows of the same equivalent width l_∞ find the one which minimizes the second (bias) term on the right-hand side of (8.93):

$$\sum_{i=0}^{\infty}\sum_{j=0}^{\infty} h_{\text{WLS}}(i)h_{\text{WLS}}(j)\min(i,j) \mapsto \min. \tag{8.95}$$

Since

$$\sum_{i=0}^{\infty} h_{\text{WLS}}(i) = 1, \quad \sum_{i=0}^{\infty} h_{\text{WLS}}^2(i) = \frac{1}{l_{\infty}}, \tag{8.96}$$

one arrives at a typical constrained optimization problem. Using the concept of integral approximations, introduced in Section 4.5.2, one can obtain the following continuous-time counterpart of (8.95): find $\tilde{h}_{\text{WLS}}(t)$ satisfying

$$\int_0^{\infty}\int_0^{\infty} \tilde{h}_{\text{WLS}}(s)\tilde{h}_{\text{WLS}}(t)\min(s,t)\,ds\,dt \mapsto \min, \tag{8.97}$$

subject to

$$\int_0^{\infty} \tilde{h}_{\text{WLS}}(s)\,ds = 1, \quad \int_0^{\infty} \tilde{h}_{\text{WLS}}^2(s)\,ds = \frac{1}{l_{\infty}}. \tag{8.98}$$

After a simple change of variables, both (8.97) and (8.98) can be transformed into a standard isoperimetric problem in the calculus of variations [134].
Solving (8.97) and (8.98) one obtains

$$\tilde{h}_{\text{WLS}}^{\text{opt}}(s) = \gamma e^{-\gamma s}, \quad s \geq 0,$$

which means that for the class of parameter variations considered, the widely used exponentially decaying weighting sequence is actually the optimal one.

Similar arguments lead to the conclusion that when process coefficients vary according to (2.35) the *two-sided* exponential window is the optimal one for parameter matching applications.

If the autocorrelation function of process parameter changes is known and if the analysis is restricted to finite-duration windows of length N, it is relatively easy to perform the joint window width/shape optimization. Lin et al. [107] performed such optimization for a fading telecommunication channel.

Denote by $S_\theta(\omega)$ the common Doppler power spectrum for every path of a fading channel. Likewise, let $r_\theta(i)$ be the corresponding autocorrelation function of the channel coefficient changes. If fading characteristics are determined by the Jakes model, $S_\theta(\omega)$ and $r_\theta(i)$ are given by (2.30) and (2.29), respectively.

The bias component of the steady-state excess mean square prediction error $\mathcal{P}_\infty$ can be expressed in the following form (cf. (4.27)):

$$E\left[\left(\varphi^T(t)(\theta(t) - \bar{\theta}(t))\right)^2\right] \cong \text{tr}\left\{\text{cov}\left[\theta(t) - \sum_{i=0}^{N-1} h_{\text{WLS}}(i)\theta(t-i)\right]\Phi_o\right\}. \tag{8.99}$$

Assuming that parameter changes in each path are uncorrelated ($\Phi_o = I_n\sigma_u^2$) one can express (8.99) in the form

$$n\sigma_u^2\left[r_\theta(0) - 2g_\theta^T h + h^T R_\theta h\right], \tag{8.100}$$

where
$$h = [h_{\mathrm{WLS}}(0), \ldots, h_{\mathrm{WLS}}(N-1)]^T,$$

is the vector of optimal window-dependent impulse response coefficients and

$$R_\theta = \begin{bmatrix} r_\theta(0) & \cdots & r_\theta(N-1) \\ \vdots & \ddots & \vdots \\ r_\theta(N-1) & \cdots & r_\theta(0) \end{bmatrix},$$

$$g_\theta = [r_\theta(1), \ldots, r_\theta(N)]^T.$$

Using (4.30) the variance term of $\mathcal{P}_\infty$ can be written in the form

$$n\sigma_v^2 h^T h. \tag{8.101}$$

Combining (8.100) and (8.101) one obtains the expression

$$\mathcal{P}_\infty \cong n\sigma_u^2 \left[r_\theta(0) - 2g_\theta^T h + h^T \left(R_\theta + \frac{\sigma_v^2}{\sigma_u^2} I_n \right) h \right], \tag{8.102}$$

which should be minimized with respect to h under the constraint

$$\sum_{i=0}^{N-1} h_{\mathrm{WLS}}(i) = h^T e = 1,$$

where $e = [1, \ldots, 1]$.

The constrained minimization problem can be easily solved using the technique of Lagrange multipliers:

$$h_{\mathrm{opt}} = D_\theta(g_\theta + \gamma e), \tag{8.103}$$

where

$$D_\theta = \left[R_\theta + \frac{\sigma_v^2}{\sigma_u^2} I_n \right]^{-1}, \qquad \gamma = \frac{1 - e^T D_\theta g_\theta}{e^T D_\theta e}.$$

As expected, the optimal window coefficients determined by (8.103) depend on the SNR of the received signal (σ_v^2/σ_u^2), statistical knowledge of the channel (R_θ, g_θ) and the number of estimated coefficients (n).

Remark 1

The optimal window derived above minimizes the mean square one-step-ahead prediction error, i.e. it optimizes the predictive and tracking performance of the corresponding adaptive filter. When a finite block of data is received, the interpolation/matching characteristics are more important. Note that the bias component of the mean square interpolation error, evaluated at position t of the received sequence of length N, can be written in the form

$$\mathrm{E}\left[(\varphi^T(t)(\theta(t) - \bar{\theta}(N)))^2 \right] \cong \mathrm{tr}\left\{ \mathrm{cov}\left[\theta(t) - \sum_{i=0}^{N-1} h_{\mathrm{WLS}}(i)\theta(N-i) \right] \Phi_\circ \right\}$$

and its variance component is given by (8.101), just as before. The minimization of mean square interpolation error can be carried out in an identical way to the minimization of mean square prediction error. Quite obviously, it yields different sets of optimal window coefficients for different positions within the block of received data [107].

Remark 2

In practice the spectral density $S_\theta(\omega)$ of the parameter variations is seldom known exactly; for example, the normalized Doppler frequency ω_n in (2.30) depends on the speed of the mobile unit and hence may change from frame to frame. For this reason in applications such as adaptive channel equalization one may attempt to design robust windows instead of optimal windows. Robust windows guarantee reduced sensitivity of the corresponding WLS estimators to departures from the assumed parameter and noise statistics, e.g. to variations in the Doppler spectrum. Lin et al. [107] present two procedures, called the average robust window design and the minimax robust window design. In both cases the shape of the window is determined in the frequency domain, based on the following expression for P_∞ (equivalent to (8.102); see Section 4.5):

$$P_\infty \cong \frac{n\sigma_v^2}{\pi} \int_0^\pi |H_{\mathrm{WLS}}(\omega)|^2\, d\omega + \frac{n\sigma_v^2}{\pi} \int_0^\pi |1 - H_{\mathrm{WLS}}(\omega)|^2 S_\theta(\omega)\, d\omega,$$

Optimization of the number and type of basis functions

When the number and type of basis functions used in the BF approach are not known a priori, some objective procedures for their selection are needed. The rational choice of the number of basis functions is particularly important since the BF approach is 'by nature' not parsimonious; each system parameter is represented by a set of k auxiliary coefficients, i.e. the overall number of degrees of freedom of the BF model may easily become too large.

Selection of the most appropriate basis can be performed using Akaike's information criterion (3.21). Consider I competitive BF estimators

$$\widehat{\theta}_i(t), \quad t \in T = [1, N],$$

$$i = 1, \ldots, I,$$

corresponding to different sets of basis functions. Let k_i denote the number of basis functions comprising each set and let $\widehat{\sigma}_i^2(T)$ be the corresponding estimate of the residual noise variance (cf. (4.3)):

$$\widehat{\sigma}_i^2(T) = \frac{1}{N}\sum_{t=1}^N [y(t) - \psi^T(t)\widehat{\gamma}_i(T)]^2 = \frac{1}{N}\sum_{t=1}^N [y(t) - \varphi(t)^T(t)\widehat{\theta}_i(t)]^2.$$

Selection can be based on minimizing

$$\mathrm{AIC}_i^\star = N \ln \widehat{\sigma}_i^2(T) + 2(nk_i + 1).$$

If selection is constrained to one basis set, i.e. only the number of basis functions has to be determined, the order-recursive estimation procedures can be applied and this substantially reduces the overall computational complexity [57].

When the weighted basis function approach is used for online tracking of parameter variations, the number and/or type of basis functions can be determined using the predictive least squares principle, i.e. by minimizing the accumulated prediction error statistics

$$\sum_{j=0}^{M-1} \epsilon_i^2(t-j),$$

where

$$\epsilon_i(t) = y(t) - \varphi^T(t)\widehat{\theta}_i(t).$$

Optimization of the Kalman filter estimation algorithms

The parameter tracking and matching properties of KF estimation algorithms depend on the choice of the system matrices F and G (appearing in the state space description of parameter changes) and the covariance matrix $\widehat{W} = W/\sigma_v^2$ (determining the relative speed of parameter variation).

The adopted value of $\widehat{W}$ is of primary importance as it governes the estimation memory of the KF filter. A number of procedures were proposed for adaptive (data-dependent) selection of $\widehat{W}$. In many practical applications one can safely assume that the rate of parameter change does not significantly change over time, which means that in long analysis intervals the 'true' value of $\widehat{W}$ can be regarded as unknown but constant. Suppose that the analysis is carried out in the interval $T = [1, N]$. Assuming the matrices F and G are known exactly and assuming the processes $\{w(t)\}$ and $\{v(t)\}$ are normally distributed, the likelihood function of σ_v^2 and W, based on the available set of measurements $\Xi(N)$ can be obtained [28] from

$$L(\sigma_v^2, W) = -\frac{1}{2}\sum_{t=1}^{N} \ln 2\pi\sigma_\epsilon^2(t) - \frac{1}{2}\sum_{t=1}^{N}\frac{\epsilon^2(t)}{\sigma_\epsilon^2(t)}, \qquad (8.104)$$

where $\sigma_\epsilon^2(t)$ denotes the variance of the prediction error $\epsilon(t)$:

$$\sigma_\epsilon^2(t) = \sigma_v^2 + \varphi^T(t)\Sigma(t|t-1)\varphi(t).$$

Note that (8.104) can be expressed alternatively as a function of σ_v^2 and the normalized covariance matrix $\widehat{W} = W/\sigma_v^2$ (which is more adequate when using the modified KF algorithm (7.4)):

$$L(\sigma_v^2, \widehat{W}) = -\frac{N}{2}\ln 2\pi\sigma_v^2 - \frac{1}{2}\sum_{t=1}^{N}\ln \beta(t) - \frac{1}{2\sigma_v^2}\sum_{t=1}^{N}\frac{\epsilon^2(t)}{\beta(t)}, \qquad (8.105)$$

where

$$\beta(t) = 1 + \varphi^T(t)\widetilde{\Sigma}(t|t-1)\varphi(t).$$

Maximization of (8.105) with respect to σ_v^2 is straightforward and results in the following ML estimate of the measurement noise variance:

$$\widehat{\sigma}^2(T) = \frac{1}{N}\sum_{t=1}^{N}\frac{\epsilon^2(t)}{\beta(t)}. \qquad (8.106)$$

In order to arrive at the ML estimate of $\widetilde{W}$, our primary interest, we should maximize the following concentrated log likelihood function:

$$L_c(\widetilde{W}) = \max_{\sigma_v^2} L(\sigma_v^2, \widetilde{W}) = L(\hat{\sigma}^2(T), \widetilde{W})$$

$$= -\frac{N}{2} \ln 2\pi \hat{\sigma}^2(T) - \frac{N}{2} - \frac{1}{2} \sum_{t=1}^{N} \ln \beta(t), \tag{8.107}$$

which is equivalent to *minimization* of the following statistics:

$$l_c(\widetilde{W}) = N \ln \hat{\sigma}^2(T) + \sum_{t=1}^{N} \ln \beta(t). \tag{8.108}$$

Since the quantities $\epsilon(t)$ and $\beta(t)$ are not explicit functions of $\widetilde{W}$, the minimization problem cannot be solved analytically; the ML estimate has to be searched iteratively by repetitive processing of the analyzed data segment. Alternatively, if several hypothetical values $\{\widetilde{W}_i, i = 1, \ldots, I\}$ are considered, one should pick the algorithm which minimizes (8.108).

Note that in the 'slow adaptation' case where $\beta(t) \cong 1$ then

$$l_c(\widetilde{W}) \cong N \ln \left(\sum_{t=1}^{N} \epsilon^2(t) \right) - N \ln N,$$

i.e. minimization of (8.108) is approximately equivalent to minimization of the accumulated error statistics (which is the cornerstone of Rissanen's predictive least squares principle [164]).

Remark

Even though estimation of the measurement noise variance σ_v^2 is not necessary – all that we need to run the KF algorithm (7.4) is the value of $\widetilde{W}$ – it may be beneficial in many prediction-oriented applications. Note that to obtain the measures of accuracy $\sigma_\epsilon^2(t)$, $\Sigma(t|t-1)$ and $\Sigma(t|t)$ for the KF estimates, one needs the true value of σ_v^2 or its estimate. Such measures are used, for example, in the detection of outliers (pops, clicks and scratches) in digital audio restoration systems [148]. If the measurement noise variance is known to vary slowly with time, one can use the following exponentially weighted estimate instead of (8.106):

$$\hat{\sigma}^2(t) = \frac{1 - \lambda}{1 - \lambda^t} \sum_{i=0}^{t-1} \lambda^i \frac{\epsilon^2(t-i)}{\beta(t-i)}. \tag{8.109}$$

Note that (8.109) is asymptotically equivalent to

$$\hat{\sigma}^2(t) = \lambda \hat{\sigma}^2(t-1) + (1 - \lambda) \left(\frac{\epsilon^2(t)}{\beta(t)} - \hat{\sigma}^2(t-1) \right).$$

■

If generalized random walk models of different orders $k_i, i = 1, \ldots, I$ are considered (Section 7.3), selection of the best algorithm can be based on minimization of

$$\mathrm{AIC}_i^\star = N \ln \widehat{\sigma}_i^2(T) + \sum_{t=1}^{N} \ln \beta_i(t) + 2(nk_i + 2);$$

see Kitagawa [92] for more details.

Iterative search is not convenient for tracking purposes. For this reason several recursive procedures for online estimation of W were proposed in the literature by Belanger [18], Brewer and Leondes [32] and Isaksson [80]. The corresponding algorithms are computationally very demanding since the number of estimated covariance parameters, equal to $n(n + 1)/2$, grows rapidly with the dimension of θ. Computational complexity is increased by the fact that at each step of the algorithm the update $\widehat{W}(t)$ must be guaranteed to be nonnegative definite. As the number n of the estimated system parameters becomes large, such adaptive procedures become increasingly inefficient unless the search is restricted to a few important covariance parameters, e.g. by imposing a diagonal structure on W.

Estimation of the state space transition matrix F in (7.1) can be based on a technique known as extended Kalman filtering [5]. An interesting alternative, estimation based on higher-order statistics of the input/output data, was proposed and verified in the adaptive equalization context by Tsatsanis et al. [187].

Comments and extensions

Section 8.1

- The problem of the memory length optimization was considered by many authors, e.g. Widrow and Walach [193], Eleftheriou and Falconer [44] and Macchi [118].

- For FIR systems and for purely AR processes the time shift invariance property of the input data can be taken advantage of when developing LS recursive estimation algorithms. The resulting fast transversal filter (FTF) or fast least squares (FLS) algorithms eliminate the covariance matrix recursion which is the most computationally involved step in (4.8); the gain $K(t)$ is updated using a smaller, compared to (4.8), set of auxiliary (state) variables [111], [36]. This allows one to reduce the computational complexity from $O(n^2)$ to $O(n)$ multiplication/addition operations per time update. If the number of estimated coefficients is large, the resulting computational savings may be essential (in some applications, such as echo canceling, the number of filter taps may reach several hundred [128]). Unfortunately, most of the FLS algorithms proposed in the literature are not numerically robust; they tend to accumulate errors due to finite-precision arithmetic, causing the parameter estimates to diverge. For this reason, in their original versions, the FTF/FLS filters are not suitable for tracking purposes. Using the concept of error feedback, Slock and Kailath [172] designed a stabilized FTF algorithm free from this drawback (though computationally more involved).

- In order to guarantee numerical robustness, recursive estimation algorithms should have the *exponential stability* property: under exact arithmetic the effects of an isolated perturbation, added to the algorithm's state vector, should decay to zero exponentially fast [114]. Both conventional and square root versions of the EWLS and KF parameter trackers are exponentially stable in the sense described above. The algorithms which are not exponentially stable, such as FTF/FLS, may suffer from numerical problems even if the identified process is persistently excited.

- The concept of *backward consistency* (closely related to exponential stability) is very useful for establishing error propagation properties of recursive estimation algorithms; it is based on backward error analysis - see Slock [173].

- A thorough analysis and comparison of numerical aspects of different EWLS and KF implementations can be found in Bierman [24], Verhaegen and Van Dooren [188], Verhaegen [189] and Slock [173], among many others.

- Stability analysis of adaptive fading memory Kalman filters was provided by Lee [103].

- Extensions of the VFF rule (8.30) and the constant trace rule (8.34) to the case of vector forgetting (4.10) can be found in the paper of Saelid and Foss [166].

- The original derivation of the DF algorithm (8.39) was based on Bayesian arguments. A similar, though conceptually different, approach was suggested by Hägglund [70]. Extensions of directional forgetting to the vector forgetting case can be found in [98] and its adaptive versions in [99] and [23]. A highly modular square root implementation of the DF algorithm was proposed by Kadlec [85].

- An interesting decomposition of parameter space into the persistently excited, decreasingly excited, otherwise excited and unexcited subspaces was proposed by Sethares et al. [170]. The same paper describes a very simple drift scenario: it can be shown that an unbounded drift of parameter estimates occurs if the LMS algorithm is used to identify an unknown coefficient of a first-order plant subject to a slower than exponentially decaying excitation and a constant output measurement noise.

Section 8.2

- For time-invariant systems the parallel estimation techniques were developed by Lainiotis [101].

- In the competing multiple model (CMM) scheme, described in Section 8.2.1, all Kalman filters comprising the adaptive filter bank work independently of each other. An alternative scheme, called interacting multiple models (IMM) was proposed as a solution to the problem of tracking maneuvering targets [27], [13]. The term 'interacting models' refers to the fact that at the beginning of each cycle of model-conditioned Kalman filtering, the state vectors and error covariance matrices of component filters are mixed appropriately. The mixing step is equivalent to 'hypothesis merging' and makes the filter bank behave more consistently compared to the no-mixing case. The parallel tracking scheme based on the concept of interacting multiple models was proposed in [146]. When process parameters are subject to jump changes the IMM scheme provides better tracking performance than the CMM scheme.

Epilogue

Practically oriented readers may be looking for advice on how to carry out identification of time-varying systems, or seeking answers to these simple questions:

- Is it possible to track successfully arbitrarily fast parameter variations?
- Which of the presented estimation methods is the best one?
- Which design parameters of tracking algorithms are more important and which are less important as far as the performance of an adaptive system is concerned?
- Is there a rational way to fit the estimation approach to the application at hand?

First of all, from the tracking viewpoint, nonstationary systems can be crudely divided into slowly varying and fast varying. The degree of nonstationarity, which determines this division, depends not only on the rate of parameter variation but also on the signal-to-noise ratio (i.e. on the relative size of the regression variables compared to the magnitude of measurement noise) and on the number of identified coefficients (the larger the number of unknown coefficients, the more difficult the corresponding tracking problem).

When the system is slowly varying, all tracking algorithms described in Chapters 3 to 7 yield good results provided that their estimation memory is chosen appropriately. In particular, all adaptive filters based on an implicit (weighted least squares approach, least mean squares approach) or explicit (random walk Kalman filtering approach) slow parameter drift hypothesis can be used and all give comparable results for identical memory settings.

Identification of fast-varying nonstationary processes can be handled successfully *only* in the case of structured nonstationarities, i.e. when the nonstationarities are at least partially known. The specific prior knowledge on the way system parameters vary with time can be incorporated in such model-based parameter tracking schemes as the Kalman filtering approach (stochastic model of parameter variation) and the basis function approach (deterministic model of parameter variation). Generally, the faster the system parameters change with time, the more detailed should be the prior knowledge on their variation to guarantee good tracking results. The corresponding adaptive filters then become more sensitive to modeling errors, i.e. to differences between the assumed parameter changes and the observed parameter changes.

None of the identification algorithms described in this book is universally the best. Approaches differ in terms of the amount of prior knowledge required, the speed of initial convergence, the sensitivity of tracking performance to the covariance structure of the regression variables, the sensitivity to variations in the design parameters, the robustness to a lack of proper excitation, and the computational complexity. The parsimony principle, formulated as a general guideline for selecting the complexity of instrumental models, remains good advice when selecting the right parameter tracking algorithm. When choosing among different adaptive filters, one should pick the algorithm which is as simple as possible, or more adequately, not more complex than necessary (complexity is understood here as the number of degrees of freedom of the parameter tracking algorithm). The price paid for superfluous complexity of an adaptive filter is in decreased robustness, i.e. increased sensitivity to various settings and/or assumptions.

The most important design parameter of every adaptive filter is its estimation memory. Short-memory algorithms are 'fast' (yield small tracking bias) but 'inaccurate' (yield large tracking variance) whereas the long-memory algorithms are 'slow' but 'accurate'. The best results are obtained if the estimation memory of a tracking algorithm is selected so as to match the degree of nonstationarity of the identified process, trading off the bias and variance error components. Optimization of the memory settings is possible using sequential or parallel estimation techniques. The first case uses a single tracking algorithm, equipped with an adjustable memory-controlling parameter. The second case takes several algorithms, with different memory settings, runs them in parallel and compares them according to their predictive abilities. Since the sequential procedures are slow and have constrained adaptation capabilities, the parallel estimation schemes should be preferred whenever possible; they work better and are more reliable than schemes using a single estimation algorithm.

References

[1] Akaike H. (1978). On newer statistical approaches to parameter estimation and structure determination. *Proc. 8th IFAC World Congress*, Helsinki, Finland.

[2] Akaike H. (1974). A new look at the statistical model identification. *IEEE Trans. Automat. Contr.*, vol. AC-19, pp. 716–723.

[3] Albert A.E. and Gardner L.S. Jr. (1967). *Stochastic approximation and Nonlinear Regression*. MIT Press, Cambridge MA.

[4] Alexander S.T. and Ghirnikar A.L. (1993). A method for recursive least-squares filtering based upon an inverse QR decomposition. *IEEE Trans. Signal Process.*, vol. 41, pp. 20–30.

[5] Anderson B.D.O. and Moore J.B. (1979). *Optimal Filtering*. Prentice Hall, Englewood Cliffs NJ.

[6] Anderson B.D.O. (1985). Adaptive systems, lack of persistency of excitation and bursting phenomena. *Automatica*, pp. 247–258.

[7] Anderson B.D.O. et al. (1986). *Stability of Adaptive Systems: Passivity and Averaging Analysis*. MIT Press, Cambridge MA.

[8] Andersson P. (1985). Adaptive forgetting in recursive identification through multiple models. *Int. J. Control*, vol. 42, pp. 1175–1193.

[9] Åström K.J., Borisson U., Ljung L. and Wittenmark B. (1977). Theory and application of self-tuning regulators. *Automatica*, vol. 13, pp. 457–476.

[10] Åström K.J., Kallström C., Thorell N., Eriksson, J. and Sten L. (1979). Adaptive autopilots for tankers. *Automatica*, vol. 15, pp. 241–254.

[11] Åström K.J. and Wittenmark B. (1989). *Adaptive Control*. Addison-Wesley, Reading MA.

[12] Bai E.W., Fu L.C. and Sastry S. (1988). Averaging analysis for discrete time and sampled data adaptive systems. *IEEE Trans. Circuits Syst.*, vol. CAS-35, pp. 137–148.

[13] Bar-Shalom Y. and Blom H.A.P. (1989). Tracking of a maneuvering target using input estimation versus the interacting multiple model algorithm. *IEEE Trans. Aerospace and Electron. Syst.*, vol. AES-25, pp. 296–300.

[14] Bartlett M.S. (1946). On theoretical specification of sampling properties of autocorrelated time series. *J. R. Statist. Soc. B*, vol. 8, pp. 27–41.

[15] Basseville M. (1988). Detecting changes in signals and systems - a survey. *Automatica*, vol. 24, pp. 309–326.

[16] Basseville M. and Nikiforov I. (1993). *Detecting of Abrupt Changes – Theory and Applications*. Prentice Hall, Englewood Cliffs NJ.

[17] Battin R.H. (1964). *Astronautical Guidance*. McGraw-Hill, New York.

[18] Bélanger P.R. (1974). Estimation of noise covariance matrices for a linear time-varying stochastic process. *Automatica*, vol. 10, pp. 267–275.

[19] Benveniste A. and Ruget G. (1982). A measure of the tracking capability of recursive stochastic algorithms with constant gains. *IEEE Trans. Automat. Contr.*, vol. AC-25, pp. 788–794.

[20] Benveniste A. (1987). Design of adaptive algorithms for the tracking of time-varying systems. *Int. J. Adaptive Contr. Signal Process.*, vol. 1, pp. 3–29.

[21] Benveniste A., Metivier M. and Priouret P. (1990). *Adaptive Algorithms for Stochastic Approximation*. Springer-Verlag, New York.

[22] Berger H. (1929). Über das Elektroenzephalogramm des Menschen. *Arch. Psychiatr. Nervenkrankheiten*, vol. 87, pp. 527–570.

[23] Bertin D., Bittanti S. and Bolzern P. (1986). Tracking of nonstationary systems by means of different prediction-error directional forgetting techniques. *Proc. 2nd IFAC Workshop on Adaptive Systems in Control and Signal Processing*, Lund, Sweden, pp. 91–96.

[24] Bierman G.J. (1977). *Factorization Methods for Discrete Sequential Estimation*. Academic Press, New York.

[25] Bitmead R.R. and Anderson B.D.O. (1980). Lyapunov techniques for the exponential stability of linear difference equations with random coefficients. *IEEE Trans. Automat. Contr.*, vol. AC-25, pp. 782–787.

[26] Bitmead R.R. and Anderson B.D.O. (1980). Performance of adaptive estimation algorithms in dependent random environments. *IEEE Trans. Automat. Contr.*, vol. AC-25, pp. 788–794.

[27] Blom H.A.P. and Bar-Shalom Y. (1988). The interacting multiple model algorithm for systems with Markovian switching coefficients. *IEEE Trans. Automat. Contr.*, vol. AC-33, pp. 780–783.

[28] Bohlin T. (1970). Information patterns for linear discrete-time models with stochastic coefficients. *IEEE Trans. Automat. Contr.*, vol. AC-15, pp. 104–106.

[29] Bohlin T. (1976). Four cases of identification of changing systems. In Mehra P.K. and Lainiotis D.G. (eds.), System Identification: *Advances and Case Studies*. Academic Press, New York, pp. 441–518.

[30] Borisson U. and Syding R. (1976). Self-tuning control of an ore crusher. *Automatica*, vol. 12, pp. 1–7.

[31] Box, G.E.P. and Jenkins G.M. (1970). *Time Series Analysis - Forecasting and Control*. Holden-Day, London.

[32] Brewer H.W. and Leondes C.T. (1977). Least squares estimation of nonstationary covariance parameters in linear systems. *Automatica*, vol. 13, pp. 265–277.

[33] Brewer J.W. (1978). Kronecker products and matrix calculus in system theory. *IEEE Trans. Circuits Syst.*, vol. CAS-25, pp. 772–781.

[34] Burg J.P. (1967). Maximum entropy spectral analysis. *Proc. 37th Annual Meeting of the Society of Exploration Geophysics*, Oklahoma City, OK.

[35] Castellini G., Conti F., Del Re E. and Pierucci L. (1997). A continuously adaptive MLSE receiver for mobile communications: algorithm and performance. *IEEE Trans. Commun.*, vol. 45, pp. 80–89.

[36] Cioffi J.M. and Kailath T. (1984).Fast recursive least squares transversal filters for adaptive filtering. *IEEE Trans. Acoust. Speech Signal Process.*, vol. ASSP-32, pp. 304–337.

[37] Cioffi J.M. (1987). Limited-precision effects in adaptive filtering. *IEEE Trans. Circuits Syst.*, vol. CAS-34, pp. 821–833.

[38] Cramer H. (1961). On some classes of non-stationary stochastic processes. Proc. 4th Berkeley Symp. on Math. Statist. and Probability, vol. 2, Univ. of California Press, Los Angeles CA.

[39] Czyżewski A., Kostek B. and Zieliński S.K. (1996). A new approach to synthesis of organ pipe sound based on simplified physical models. *Archives of Acoustics*, vol. 21, pp. 131–147.

[40] Dawid A.P. (1984). Present position and potential developments: some personal views, statistical theory, the prequential approach. *J. R. Statist. Soc. A*, vol. 147, pp. 278–292.

[41] De Bruin G. and Van Walstijn M. (1995). Physical models of wind instruments: a generalized excitation coupled with a modular tube simulation platform. *J. New Music Research*, vol. 24, pp. 148–163.

[42] Deller J.R. Jr, Proakis J.G. and Hansen J.H.L. (1993). *Discrete-Time Processing of Speech Signals*. Macmillan, New York.

[43] Duttweiler D.L. (1978). A twelve-channel digital echo canceller. *IEEE Trans. Commun.*, vol. COM-26, pp. 647–653.

[44] Eleftheriou E. and Falconer D. (1986). Tracking properties and steady-state performance of RLS adaptive filter algorithms. *IEEE Trans. Acoust. Speech Signal Process.*, vol. ASSP-34, pp. 1097–1109 (see also *Proc. Int. Conf. Acoust. Speech Signal Process.*, 1985).

[45] Falconer D.D. and Ljung L. (1978). Application of fast Kalman estimation to adaptive equalization. *IEEE Trans. Commun.*, vol. COM-26, pp. 1439–1446.

[46] Falconer D.D. and Gitlin R.D. (1978), Optimization reception of digital data signals in the presence of timing-phase hits. *Bell Syst. Tech. J*, vol. 57, pp. 3181–3208.

[47] Fasol K.H. and Jörgl H.P. (1980). Principles of model building and identification. *Automatica*, vol. 16, pp. 505–518.

[48] Feuer A. and Berman N. (1986). Performance analysis of the smoothed least mean square (SLMS) algorithm. *Signal Process.*, vol. 11, pp. 265–276.

[49] Forney G.D. Jr (1973). The Viterbi algorithm. *Proc. IEEE*, vol. 61, pp. 268–278.

[50] Fortescue T.R., Kershenbaum L.S. and Ydstie B.E. (1981). Implementation of self-tuning regulators with variable forgetting factors. *Automatica*, vol. 17, pp. 831–835.

[51] Gardner W.A. (1984). Learning characteristics of stochastic-gradient-descent algorithms: a general study, analysis and critique. *Signal Process.*, vol. 6, pp. 113–133.

[52] Gardner W.A. (1987). Nonstationary learning characteristics of the LMS algorithm. *IEEE Trans. Circuits Syst.*, vol. CAS-34, pp. 1199–1207.

[53] Gardner W.A. (1988). Correlation estimation and time-series modeling for nonstationary processes. *Signal Process.*, vol. 15, pp. 31–41.

[54] Gardner W.A. (1988). *Statistical Spectral Analysis: A Nonprobabilistic Theory*. Prentice Hall, Englewood Cliffs NJ.

[55] Gersch W. (1971). Spectral analysis of EEGs by autoregressive decomposition of time series. *Math. Biosci.*, vol. 7, pp. 205–222.

[56] Gersch W. and Yonemoto J. (1977). Parametric time series models for multivariate EEG analysis. *Comput. Biomed. Res.*, vol. 10, pp. 113–125.

[57] Gersch W. and Kitagawa G. (1983). A multivariate time-verying autoregressive modelling of nonstationary econometric time series. *Am. Statist. Assoc, Proc. Business Econ. Statist.*, pp. 399–404.

[58] Giannakis G.B. and Tepedelenlioğlu (1998). Basis expansion models and diversity techniques for blind identification and equalization of time-varying channels. *Proc. IEEE*, vol. 86, pp. 1969–1986.

[59] Giannakis G.B. (1998). Cyclostationary signal analysis. In Madisetti V.K. and Williams D.B. (eds), *The Digital Signal Processing Handbook*. CRC Press, Boca Raton FL, pp. 17/1–17/31.

[60] Goddard D. (1974). Channel equivalization using a Kalman filter for fast data transmission. *IBM J. Res. Develop.*, vol. 18, pp. 267–273.

[61] Goodwin G.C. and Sin K.S. (1984). *Adaptive Filtering, Prediction and Control*. Prentice Hall, Englewood Cliffs NJ.

[62] Goodwin G.C., Hill D.J. and Palaniswami M. (1985). Towards an adaptive robust controller. *Proc. 7th IFAC/IFORS Symp. on Identification and System Parameter Estimation*, York, UK.

[63] Grenier Y. (1981). Rational non-stationary spectra and their estimation. *Proc. 1st ASSP Workshop on Spectral Estimation*, Hamilton, Canada, pp. 6.8.1–6.8.8.

[64] Grenier Y. (1981). Time-dependent ARMA modeling of non-stationary signals. *IEEE Trans. Acoust. Speech Signal Process.*, vol. ASSP-31, pp. 899–911.

[65] Grillenzoni C. (1990). Modeling of time-varying systems. *J. Am. Statist. Association*, vol.85, pp. 499–507.

[66] Guo L. (1990). Estimating time-varying parameters by the Kalman filter based algorithm: stability and convergence. *IEEE Trans. Automat. Contr.*, vol. 35, pp. 141–147.

[67] Guo L. (1994). Stability of recursive stochastic tracking algorithms. *SIAM J. Contr. Optimiz.*, vol. 32, pp. 1195–1225.

[68] Guo L. and Ljung L. (1995). Performance analysis of general tracking algorithms. *IEEE Trans. Automat. Contr.*, vol. 40, pp. 1388–1402.

[69] Guo L., Ljung L. and Wang G.J. (1997). Necessary and sufficient conditions for stability of LMS. *IEEE Trans. Automat. Contr.*, vol. 42, pp. 761–770.

[70] Hägglund T. (1984). Adaptive control of systems subject to large parameter changes. *Proc. 9th IFAC World Congress, Budapest*, Hungary, pp. 993–998.

[71] Hall M., Oppenheim A.V. and Willsky A.S. (1983). Time-varying parametric modeling of speech. *Signal Process.*, vol. 5, pp. 267–285 (see also *Proc. 16th IEEE Conf. on Decision and Control*, New Orleans, pp. 1085–1091, 1977).

[72] Hannan E.J. (1970). *Multiple Time Series*. Wiley, New York.

[73] Harris F.J. (1978). On the use of windows for harmonic analysis with the discrete Fourier transform. *Proc. IEEE*, vol. 66, pp. 51–83.

[74] Hassibi B., Sayed A.H. and Kailath T. (1996). H^∞ optimality of the LMS algorithm. *IEEE Trans. Signal Process.*, vol. 44, pp. 267–280.

[75] Haykin S. (1996). *Adaptive Filter Theory*. Prentice Hall, Englewood Cliffs NJ.

[76] Householder A.S. (1964). *The Theory of Matrices in Numerical Analysis*. Blaisdell, London.

[77] Hsia T.C. (1983). Convergence analysis of LMS and NLMS adaptive algorithms. *Proc. ICASSP*, Boston, pp. 667–670.

[78] Isaksson L. and Wennberg A. (1975). An EEG simulator – a means of objective clinical interpretation of EEG. *Electroencephalogr. Clin. Neurophysiol.*, vol. 39, pp. 313–320.

[79] Isaksson L., Wennberg A. and Zetterberg L. (1981). Computer analysis of EEG signals with parametric models. *Proc. IEEE*, vol. 69, pp. 451–461.

[80] Isaksson A. (1988). Identification of time-varying systems through adaptive Kalman filtering. *Proc. 10th IFAC World Congress*, Munich, Germany, pp. 306–311.

[81] Jakes W.C. (1974). *Microwave Mobile Communications*. Wiley, New York.

[82] Jazwinski A.H. (1970). *Stochastic Processes and Filtering Theory*. Academic Press, New York.

[83] Jeffreys H. (1961). *Theory of Probability*. Clarendon Press, Oxford.

[84] Kaczmarz S. (1937). Angenäherte Auflosung von Systemen linearer gleichungen. *Bull. Int. Acad. Polon. Sci. Lett. Cl. Sci. Math. Nat. A*.

[85] Kadlec J. (1991). A recursive modified Gram-Schmidt identification with directional tracking of parameters. *Proc. 9th IFAC/IFORS Symp. on Identification and System Parameter Estimation*, Budapest, Hungary, pp. 1707–1712.

[86] Kaminski P.G., Bryson, A.E. Jr and Schmidt S.F. (1971). Discrete square root filtering: a survey of current techniques. *IEEE Trans. Automat. Contr.*, vol. AC-16, pp. 727–735.

[87] Kashyap R.L. and Ramachandra Rao A. (1976). *Dynamic Stochastic Models from Empirical Data*. Academic Press, New York.

[88] Katkovnik V.Y. (1994). Identification of physical rapidly time-varying parameters. *Proc. SYSID*, Copenhagen, Denmark, pp. 349–354.

[89] Kay S.M. (1987). *Modern Spectrum Analysis*. Prentice Hall, Englewood Cliffs NJ.

[90] Kay S.M. and Marple S.L. Jr (1981). Spectrum analysis – a modern perspective. *Proc. IEEE*, vol. 69, pp. 1380–1419.

[91] Kitagawa G. and Akaike H. (1978). A procedure for the modeling of non-stationary time series. *Ann. Inst. Statist. Math.*, vol. 30, part B, pp. 351–363.

[92] Kitagawa G. (1983). Changing spectrum estimation. *J. Sound Vibration*, vol. 89, pp. 433–445.

[93] Kitagawa G. and Gersch W. (1985). A smoothness priors time-varying AR coefficient modelling of nonstationary covariance time series. *IEEE Trans. Automat. Contr.*, vol. AC-30, pp. 48–56.

[94] Kitagawa G. and Gersch W. (1985). A smoothness priors long AR model method for spectral estimation. *IEEE Trans. Automat. Contr.*, vol. AC-30, pp. 57–65.

[95] Kitagawa G. and Gersch W. (1996). *Smoothness Priors Analysis of Time Series*. Springer-Verlag, New York.

[96] Kubo H., Murakami K. and Fujino T. (1994). An adaptive maximum likelihood sequence estimator for fast time-varying intersymbol interference channels. *IEEE Trans. Commun.*, vol. 42, pp. 1872–1880.

[97] Kulhavy R. and Karny M. (1984). Tracking of slowly varying parameters by directional forgetting. Proc. 9th IFAC World Congress, Budapest, Hungary, pp. 78–83.

[98] Kulhavy R. (1986). Directional tracking of regression-type model parameters. *Proc. 2nd IFAC Workshop on Adaptive Systems in Control and Signal Processing*, Lund, Sweden, pp. 97–101.

[99] Kulhavy R. (1987). Restricted exponential forgetting in real-time identification. *Automatica*, vol. 23, pp. 589–600.

[100] Kushner H.J. and Yang J. (1995). Analysis of adaptive step size SA algorithms for parameter tracking. *IEEE Trans. Automat. Contr.*, vol. AC-40, pp. 1403–1410.

[101] Lainiotis D.G. (1976). Partitioning: a unifying framework for adaptive systems - Part I: Estimation", *Proc. IEEE*, pp. 1126–1143.

[102] Lee R.C.K. (1964). *Optimal Identification, Estimation and Control*. MIT Press, Cambridge MA.

[103] Lee T.S. (1988). Theory and application of adaptive fading memory Kalman filters. *IEEE Trans. Circuits Syst.*, vol. CAS-35, pp. 474–477.

[104] Li S. and Dickinson B.W. (1987). Jump detection and fast parameter tracking for picewise AR processes using adaptive lattice filters. *Proc. Int. Conf. Acoust. Signal and Speech Processing*, Dallas, pp. 328–331.

[105] Li Z. (1986). On line identification for time-varying systems. *Proc. 25th IEEE Conf. on Decision and Control*, Athens, Greece, pp. 1648–1652.

[106] Li Z. (1987). Discrete-time adaptive control of deterministic fast time-varying systems. *IEEE Trans. Automat. Contr.* vol. AC-32, 1648–1652.

[107] Lin J., Proakis J.G., Ling F. and Lev-Ari H. (1995). Optimal tracking of time-varying channels: a frequency domain approach for known and new algorithms. *IEEE Trans. Selected Areas in Commun.*, vol. 13, pp. 141–154.

[108] Lindbom L., Sternad M. and Ahlén A. (1991). A Viterbi detector based on sinusoid modeling of fading mobile radio channels: an illustration of the utility of deterministic models of time-variation in adaptive systems. *Proc. STU Workshop on Digital Communication*, Gothenburg, Sweden.

[109] Ling F. and Proakis J.G. (1984). Nonstationary learning characteristics of least squares adaptive estimation algorithms. *Proc. Int. Conf. Acoust. Signal Speech Process.*, San Diego CA, pp. 3.7.1–3.7.4.

[110] Liporace J.M. (1975). Linear estimation of non-stationary signals. *J. Acoust. Soc. Am.*, vol. 58, pp. 1288–1295.

[111] Ljung L., Morf M. and Falconer D.D. (1978). Fast calculation of gain matrices for recursive estimation schemes. *Int. J. Control*, vol. 27, pp. 1–19.

[112] Ljung L. (1979). Asymptotic behavior of the extended Kalman filter as a parameter estimator for linear systems. *IEEE Trans. Automat. Contr.*, vol. AC-24, pp. 36–50.

[113] Ljung L. and Söderström T. (1983). *Theory and Practice of Recursive Identification.* MIT Press, Cambridge MA.

[114] Ljung S. and Ljung L. (1985). Error propagation properties of recursive least-squares adaptation algorithms. *Automatica*, vol. 21, pp. 157–167.

[115] Ljung L. and Gunnarsson S. (1990). Adaptative tracking in system identification - a survey. *Automatica*, vol. 26, pp. 7–22.

[116] Loynes R.M. (1968). On the concept of spectrum for nonstationary processes. *J. R. Statist. Soc. B*, vol. 30, pp. 1–30.

[117] Macchi O. (1995). *Adaptive processing: the least mean squares approach with applications in transmission.* Wiley, New York.

[118] Macchi O. (1996). Optimization of adaptive optimization for time-varying filters. *IEEE Trans. Automat. Contr.*, vol. AC-31, pp. 283–287.

[119] Majkowski J. (1991). *The Electroencephalographic Atlas.* PZWL, Warsaw (in Polish).

[120] Makhoul J.I. and Cosell L.K. (1981). Adaptive lattice analysis of speech. *IEEE Trans. Circuits Syst.*, vol. CAS-28, pp. 494–499.

[121] Marple S.L., Jr. (1987). *Digital Spectral Analysis with Applications.* Prentice Hall, Englewood Cliffs NJ.

[122] Martin-Sanchez J.M. and Shah S.L. (1984). Multivariable adaptive control of a binary distillation column. *Automatica*, vol. 20, pp. 607–620.

[123] Mayne D.Q. (1963). Optimal nonstationary estimation of the parameters of a linear system with Gaussian inputs. *J. Electron. Contr.*, vol. 14, pp. 101.

[124] Medaugh R.S. and Griffiths L.J. (1981). A comparison of two linear predictors. *Proc. Int. Conf. Acoust. Signal Speech Process.*, Atlanta GA, pp. 293–296.

[125] Meditch J.S. (1973). A survey of data smoothing for linear and nonlinear dynamic systems. *Automatica*, vol. 9, pp. 151–162.

[126] Mendel J.M. (1973). *Discrete techniques of parameter estimation: the equation error formulation*. Dekker, New York.

[127] Moustakides G.V. (1997). Study of the transient phase of the forgetting factor RLS. *IEEE Trans. Signal Process.*, vol. 45, pp. 2468–2476.

[128] Murano K. et al. (1990). Echo cancellation and applications. *IEEE Trans. Commun.*, vol. 28, pp. 49–55.

[129] Nagumo J.I. and Noda A. (1967). A learning method for system identification. *IEEE Trans. Automat. Contr.*, vol. AC-12, pp. 282–287.

[130] Narendra K.S. and Annaswamy M. (1987). A new adaptive law for robust adaptive control without persistent excitation. *IEEE Trans. Automat. Contr.*, vol. AC-32, pp. 134–145.

[131] Niedźwiecki M. (1980). On the extension of Akaike's final prediction error criterion to systems with slowly varying coefficients. *Proc. 1st European Signal Processing Conference EUSIPCO-80*, Lausanne, Switzerland.

[132] Niedźwiecki M. (1984). On the localized estimators and generalized Akaike's criteria. *IEEE Trans. Automat. Contr.*, vol. AC-29, pp. 970–983 (see also *Proc. 20th IEEE Conf. on Decision and Control*, San Diego CA, pp. 65–61, 1981).

[133] Niedźwiecki M. (1985). On time and frequency characteristics of weighted least squares estimators applied to nonstationary system identification. *Proc. 24th IEEE Conf. on Decision and Control*, Fort Lauderdale FL, pp. 225–230.

[134] Niedźwiecki M. (1986). Optimization of the window shape in weighted least squares identification of a class of nonstationary systems. *Proc. 7th Int. Conf. on Analysis and Optimization of Systems*, Antibes, France. Springer-Verlag, New York, pp. 889–901.

[135] Niedźwiecki M. (1987). Functional series modeling identification of nonstationary stochastic systems – the clipping technique. *Proc. 1st IAESTED Symp. on Signal Processing and its Applications*, Brisbane, Australia, pp. 321–326.

[136] Niedźwiecki M. (1987). Recursive functional series modeling estimators for adaptive control of time-varying plants. *Proc. 26th IEEE Conf. on Decision and Control*, Los Angeles CA, pp. 1239–1244.

[137] Niedźwiecki M. (1988). First-order tracking properties of weighted least squares estimators. *IEEE Trans. Automat. Contr.*, vol. AC-33, pp. 94–96.

[138] Niedźwiecki M. (1988). On tracking characteristics of weighted least squares estimators applied to nonstationary system identification. *IEEE Trans. Automat. Contr.*, vol. AC-33, pp. 96–98.

[139] Niedźwiecki M. (1988). Functional series modeling approach to identification of nonstationary stochastic systems. *IEEE Trans. Automat. Contr.*, vol. AC-33, pp. 955–961.

[140] Niedźwiecki M. (1989). Steady-state and parameter tracking properties of self-tuning minimum variance regulators. *Automatica*, vol. 25, pp. 597–602.

[141] Niedźwiecki M. (1990). Recursive functional series modeling approach to identification of time-varying plants – more bad news than good?. *IEEE Trans. Automat. Contr.*, vol. AC-35, pp. 610–616.

[142] Niedźwiecki M. (1990). Identification of nonstationary stochastic systems using parallel estimation schemes. *IEEE Trans. Automat. Contr.*, vol. AC-35, pp. 329–334.

[143] Niedźwiecki M. (1990). Identification of time-varying systems using combined parameter estimation and filtering. *IEEE Trans. Acoust. Speech Signal Process.*, vol. ASSP-38, pp. 679–686.

[144] Niedźwiecki M. and Guo L. (1991). Nonasymptotic results for finite-memory WLS filters. *IEEE Trans. Automat. Contr.*, vol. AC-36, pp. 515–522.

[145] Niedźwiecki M. (1993). Statistical reconstruction of multivariate time series. *IEEE Trans. Signal Process.*, vol. 41, pp. 451–457.

[146] Niedźwiecki M. and Suchomski P. (1996). On parallel estimation approach to adaptive filtering. *Bull. Polish Acad. Sci., Tech. Sci.*, vol. 44, pp. 385–397.

[147] Niedźwiecki M. and Wasilewski A. (1996). Application of adaptive filtering to dynamic weighing of vehicles. *Control Engng. Practice*, vol. 4, pp. 635–664.

[148] Niedźwiecki M. and Cisowski K. (1996). Adaptive scheme for elimination of broadband noise and impulsive disturbances from AR and ARMA signals. *IEEE Trans. Signal Process.*, vol. 44, pp. 528–537.

[149] Niedźwiecki M. (1997). Identification of time-varying processes in the presence of measurement noise and outliers. *Proc. 11th IFAC Symposium on System Identification*, Kitakyushu, Japan, pp. 1765–1770.

[150] Niedźwiecki M. and Kłaput T. (2000). Fast algorithms for identification of rapidly varying channels. *Proc. 10th European Signal Processing Conference EUSIPCO-2000*, Tampere, Finland.

[151] Nieto M.M. (1972). *The Titius–Bode Law of Planetary Distances: Its History and Theory*. Pergamon Press, Oxford.

[152] Norton J.P. (1975). Optimal smoothing in the identification of linear time-varying systems. *Proc. IEE*, vol. 122, pp. 663–668.

[153] Ohtsu K., Horigome H. and Kitagawa G. (1979). A new ship's autopilot design through a stochastic model. *Automatica*, vol. 15, pp. 255–268.

[154] Ozaki T. and Tong H. (1975). On the fitting of nonstationary autoregressive models in time series analysis. *Proc. 8th Hawaii International Conf. on System Sciences*, Honolulu HI, pp. 224–246.

[155] Parzen E. (1969). Multiple time series modeling. In *Multivariable Analysis II*. Academic Press, New York.

[156] Peterka V. (1981). Bayesian system identification. *Automatica*, vol. 17, pp. 41–53.

[157] Peterson B.B. and Narendra K.S. (1982). Bounded error adaptive control. *IEEE Trans. Automat. Contr.*, vol. AC-27, pp. 1161–1168.

[158] Porat B. (1985). Second-order equivalence of rectangular and exponential windows in least-squares estimation of Gaussian autoregressive processes. IEEE Trans. Acoust. Speech Signal Process., vol. ASSP-33, pp. 1209–1212.

[159] Priestley M.B. (1981). *Spectral Analysis and Time Series.* Academic Press, New York (2 vols).

[160] Proakis J.G. (1989). *Digital Communications.* McGraw-Hill, New York.

[161] Rabiner L.R. and Schaffer R.W. (1978). *Digital Processing of Speech Signals.* Prentice Hall, Englewood Cliffs NJ.

[162] Raheli R., Polydoros A. and Tzou C.-K. (1991). The principle of per-survivor processing: a general approach to approximate and adaptive MLSE. *Proc. GLOBECOM'91*, Phoenix AZ.

[163] Rissanen J. (1986). Stochastic complexity and statistical inference. *Proc. 7th Conf. on Analysis and Optimization of Systems*, Antibes, France, pp. 393–407.

[164] Rissanen J. (1986). A predictive least-squares principle. *IMA J. Math. Contr. Inform.*, vol. 3, pp. 211–222.

[165] Robinson E.A. (1982). A historical perspective of spectrum estimation. *Proc. IEEE*, vol. 70, pp. 885–907.

[166] Saelid S. and Foss B. (1983). Adaptive controllers with a vector variable forgetting factor. *Proc. 22nd IEEE Conf. on Decision and Control*, San Antonio TX, pp. 1488–1494.

[167] Salgado M.E., Goodwin G.C. and Middleton R.H. (1988). Modified least squares algorithm incorporating exponential resetting and forgetting. *Int. J. Control*, vol. 47, pp. 477–491.

[168] Sayed A.H. and Kailath T. (1994). A state-space approach to adaptive RLS filtering. *IEEE Signal Process. Mag.*, vol. 11, pp. 18–60.

[169] Sayed A.H. and Kailath T. (1998). Recursive least-squares adaptive filters. In Madisetti V.K. and Williams D.B. (eds), *The Digital Signal Processing Handbook.* CRC Press, Boca Raton FL, pp. 21/1–21/37.

[170] Sethares W.A., Lawrence D.A., Johnson C.R. Jr and Bitmead R.R. (1986). Parameter drift in LMS filters. *IEEE Trans. Acoust. Speech Signal Process.*, vol. ASSP-34, pp. 868–879.

[171] Silverman R.A. (1957). Locally stationary random processes. *IRE Trans. Inform. Theory*, vol. IT-3, pp. 182–187.

[172] Slock D.T.M. and Kailath T. (1991). Numerically stable fast transversal filters for recursive least squares adaptive filtering. *IEEE Trans. Signal Process.*, vol. 39, pp. 92–114.

[173] Slock D.T.M. (1992). Backward consistency concept and round-off error propagation dynamics in recursive least-squares algorithms. *Opt. Eng.*, vol. 31, pp. 1153–1169.

[174] Smith J.O. III (1992). Physical modeling using digital waveguides. *Computer Music Journal*, special issue on physical modeling of musical instruments, Part I, vol. 16, pp. 74–91.

[175] Smith J.O. III (1995). Digital waveguide models for sound synthesis based on musical acoustics. *Proc. 15th International Congress on Acoustics*, Trondheim, Norway, vol. 3, pp. 489–492.

[176] Smith J.O. III (1996). Physical modeling synthesis update. *Computer Music Journal*, vol. 20, pp. 44–56.

[177] Snyder R.L. (1967). A partial spectrum approach to the analysis of quasi-stationary time series. *IEEE Trans. Inform. Theory*, vol. IT-13, pp. 579–587.

[178] Söderström T.. Ljung L. and Gustavsson I. (1978). A theoretical analysis of recursive identification methods. *Automatica*, vol. 14, pp. 231–244.

[179] Söderström T. and Stoica P. (1988). *System Identification*. Prentice Hall, Englewood Cliffs NJ.

[180] Solo V. (1994). Averaging analysis of the LMS algorithm. In *Control and Dynamic Systems: Advances in Theory and Applications*, Leondes C.T., (ed.), vol. 65, Part 2. Academic Press, Englewood Cliffs NJ, pp. 379–397.

[181] Spanias A. (1994). Speech coding: a tutorial review. *Proc. IEEE*, vol. 82, pp. 1541–1562.

[182] Subba Rao T. (1970). The fitting of non-stationary time-series models with time-dependent parameters. *J. R. Statist. Soc. B*, vol. 32, pp. 312–322.

[183] Therrien C.W. (1992). *Discrete Random Signals and Statistical Signal Processing*. Prentice Hall, Englewood Cliffs NJ.

[184] Tjøstheim D. (1976). Spectral generating operators for non-stationary processes. *Adv. Appl. Prob.*, vol. 8, pp. 831–846.

[185] Tront R.J., Cavers J.K. and Ito M.R. (1984). Performance of Kalman decision-feedback equalization in HF radio modems. *Proc. ICC*, pp. 50.7.1–50.7.5.

[186] Tsatsanis M.K. and Giannakis G.B. (1996). Modeling and equalization of rapidly fading channels. *Int. J. Adaptive Contr. Signal Process.*, vol. 10, pp. 159–176.

[187] Tsatsanis M.K., Giannakis G.B. and Zhou G. (1996). Estimation and equalization of fading channels with random coefficients. *Signal Process.*, vol. 53, pp. 211–229.

[188] Verhaegen M. and Van Dooren P. (1986). Numerical aspects of different Kalman filter implementations. *IEEE Trans. Automat. Contr.*, vol. AC-31, pp. 907–917.

[189] Verhaegen M. (1989). Round-off error propagation in four generally applicable, recursive least-squares estimation schemes. *Automatica*, vol. 25, pp. 437–444.

[190] Viterbi A.J. (1967). Error bounds for convolutional codes and asymptotically optimum decoding algorithm. *IEEE Trans. Inform. Theory.*, vol. IT-13, pp. 260–269.

[191] Weiss A. and Mitra D. (1979). Digital adaptive filters: conditions of convergence, rates of convergence, effects of noise and errors arising from the implementation. *IEEE Trans. Inform. Theory*, vol. IT-25, pp. 637–652.

[192] Widrow B., McCool J., Larimore M.G. and Johnson C.R. Jr (1975). Stationary and nonstationary learning characteristics of the LMS adaptive filter. *Proc. IEEE*, vol. 64, pp. 1151–1161.

[193] Widrow B. and Walach A. (1984). On the statistical efficiency of the LMS algorithm with nonstationary inputs. *IEEE Trans. Inform. Theory*, vol. IT-30, pp. 211–221.

[194] Willsky A.S. and Jones H.L. (1976). A generalized likelihood ratio approach to the detection and estimation of jumps in linear systems. *IEEE Trans. Automat. Contr.*, vol. AC-21, pp. 108–112.

[195] Willsky A.S. (1976). A survey of design methods for failure detection in dynamic systems. *Automatica*, vol. 12, pp. 601–611.

[196] Willsky A.S. (1979). *Digital signal processing and control and estimation theory: points of tangency, areas of intersection and parallel directions*. MIT Press, Cambridge MA.

[197] Xianya X. and Evans R.J. (1984). Discrete time stochastic adaptive control for time-varying systems. *IEEE Trans. Automat. Contr.*, vol. AC-29, pp. 638–640.

[198] Yaglom A.M. (1958). Correlation theory of processes with stationary random increments of order *n*. *Am. Math. Soc. Transl. Ser. 2*, vol. 8.

[199] Ydstie B.E. (1985). Adaptive control and estimation with forgetting factors. *Proc. 7th IFAC/IFORS Symp. Identification and System Parameter Estimation*, York, UK, pp. 1761–1766.

[200] Ydstie B.E. and Sargent R.W.H. (1986). Convergence and stability properties of an adaptive regulator with variable forgetting factor. *Automatica*, vol. 22, pp. 749–751.

[201] Young P.C. (1984). *Recursive Estimation and Time-Series Analysis*. Springer-Verlag, New York.

[202] Yule G.U. (1927). On the method of investigating periodicities in disturbed series with special references to Wolfer's sunspot numbers. *Phil. Trans. R. Soc. London*, vol. A226, pp. 267–298.

[203] Zetterberg L.H., Kristiansson L. and Mossberg K. (1978). Performance of a model for a local neuron population. *Biol. Cybernetics*, vol. 31, pp. 15–26.

[204] Zetterberg L.H. (1978). Recent advances in EEG data processing. *Contemp. Clin. Neurophysiol.* (EEG suppl. no. 34), pp. 19–36.

[205] Zhang Q. and Haykin S. (1983). Tracking characteristics of the Kalman filter in a nonstationary environment for adaptive filter applications. *Proc. Int. Conf. Acoust. Speech Signal Process.*, Boston MA, pp. 671–674.

[206] Ziegler R.A. and Cioffi J.M. (1992). Estimation of time-varying digital radio channels. *IEEE Trans. Vehicular Tech.*, vol. 41, pp. 134–151.

[207] Zonn W. (1975). The Titius–Bode law. *Problemy*, 1 (346), pp. 9–12 (in Polish).

Index